The Mediterranean Basin, California, Chile, the western Cape of South Africa and southern Australia share a mediterranean climate characterised by cool wet winters and hot dry summers. These five regions have differing patterns of human settlement but similarities in natural vegetation and some faunal assemblages. This unique documentation of the introduced floras and faunas in these five regions of mediterranean climate both increases our understanding of the ecology of biological invasions, and points the way to more effective management of the biota of these regions.

Biogeography of Mediterranean Invasions

Biogeography of Mediterranean Invasions

Edited by

R. H. GROVES
CSIRO Division of Plant Industry, Canberra, Australia

and

F. DI CASTRI
Centre L. Emberger, CNRS, Montpellier, France

CAMBRIDGE UNIVERSITY PRESS
Cambridge
New York Port Chester
Melbourne Sydney

CAMBRIDGE UNIVERSITY PRESS
Cambridge, New York, Melbourne, Madrid, Cape Town, Singapore, São Paulo

Cambridge University Press
The Edinburgh Building, Cambridge CB2 8RU, UK

Published in the United States of America by Cambridge University Press, New York

www.cambridge.org
Information on this title: www.cambridge.org/9780521360401

First published 1991
This digitally printed version 2008

A catalogue record for this publication is available from the British Library

ISBN 978-0-521-36040-1 hardback
ISBN 978-0-521-06390-6 paperback

Contents

Contributors

P. Arce, Laboratorio de Botanica, Facultad de Ciencias Biológicas, Pontificia Universidad Católica de Chile, Casilla 114-D, Santiago, Chile

H. Aschmann, Department of Earth Sciences, University of California, Riverside, California 92521, USA

R. C. Bigalke, Department of Nature Conservation, University of Stellenbosch, Stellenbosch 7600, South Africa

J. Blondel, Centre L. Emberger, CNRS, BP 5051, 34033 Montpellier Cedex, France

R. K. Brooke, Percy Fitzpatrick Institute of African Ornithology, University of Cape Town, Rondebosch 7700, South Africa

G. Cheylan, Musée d'Histoire naturelle d'Aix-en-Provence, 6 rue Espaniat, 13100 Aix-en-Provence, France

H. T. Clifford, Department of Botany, University of Queensland, St Lucia, Queensland 4067, Australia

B. J. Coman, Department of Conservation, Forests & Lands, PO Box 48, Frankston, Victoria 3199, Australia

H. J. Deacon, Department of Archaeology, University of Stellenbosch, Stellenbosch 7600, South Africa

M. Debussche, Centre L. Emberger, CNRS, BP 5051, 34033 Montpellier Cedex, France

F. di Castri, Centre L. Emberger, CNRS, BP 5051, 34033 Montpellier Cedex, France

E. R. Fuentes, Laboratorio de Ecologia, Facultad de Ciencias Biológicas, Pontificia Universidad Católica de Chile, Casilla 114-D, Santiago, Chile

P.-H. Gouyon, Laboratoire d'Evolution et Systématique Végétales, Université de Paris-Sud, Bât 362, 91405 Orsay-Cedex, France

R. H. Groves, CSIRO Division of Plant Industry, GPO Box 1600, Canberra, ACT 2601, Australia

J. L. Guillerm, Centre L. Emberger, CNRS, BP 5051, 34033 Montpellier Cedex, France

F. M. Jaksic, Laboratorio de Ecologia, Facultad de Ciencias Biológicas, Pontificia Universidad Católica de Chile, Casilla 114-D, Santiago, Chile

A. A. Kirk, Laboratoire de Zoogéographie, Université Paul Valery, BP 5043, 34032 Montpellier Cedex, France

P. M. Kloot, Department of Agriculture, GPO Box 1671, Adelaide, South Australia, 5001 Australia

E. Le Floc'h, Centre L. Emberger, CNRS, BP 5051, 34033 Montpellier Cedex, France

H. N. Le Houérou, Centre L. Emberger, CNRS, BP 5051, 34033 Montpellier Cedex, France

J. Lepart, Centre L. Emberger, CNRS, BP 5051, 34033 Montpellier Cedex, France

W. Z. Lidicker Jr, Museum of Vertebrate Zoology, University of California, Berkeley, California 94720, USA

J. L. Long, Agriculture Protection Board of Western Australia, Bougainvillea Ave., Forrestfield, Western Australia 6058, Australia

J.-P. Lumaret, Laboratoire de Zoogéographie, Université Paul Valery, BP 5043, 34032 Montpellier Cedex, France

P. R. Mawson, Agriculture Protection Board of Western Australia, Bougainvillea Ave., Forrestfield, Western Australia 6058, Australia

G. Montenegro, Laboratorio de Botanica, Facultad de Ciencias Biológicas, Pontificia Universidad Católica de Chile, Casilla 114-D, Santiago, Chile

K. Myers, CSIRO Division of Wildlife & Ecology, PO Box 84, Lyneham, ACT 2602, Australia

M. L. Navas, Laboratoire de Pathologie et Biologie Végétales, Ecole Nationale Supérieure d'Agronomie, 34060 Montpellier, France

Z. Naveh, Faculty of Agricultural Engineering, Technion City, Haifa 32000, Israel

I. Olivieri, INRA, Centre de Montpellier, Domaine de Melgueil, 34130 Mauguio, France

D. Pepler, Department of Nature Conservation, University of Stellenbosch, Stellenbosch 7600, South Africa

I. D. Peters, Departmento de Biologia General, Universidade Federal de Vicosa, 36570 Vicosa, Brazil

V. Poblette, Laboratorio de Botanica, Facultad de Ciencias Biológicas, Pontificia Universidad Católica de Chile, Casilla 114-D, Santiago, Chile

J.-M. Prosperi, INRA, Centre de Montpellier, Domaine de Melgueil, 34130 Mauguio, France

L. D. Pryor, Department of Forestry, Australian National University, GPO Box 4, Canberra, ACT 2601, Australia

T. D. Redhead, CSIRO Division of Wildlife & Ecology, PO Box 84, Lyneham, ACT 2602, Australia

M. Rejmanek, Department of Botany, University of California, Davis, California 95616, USA

J. Roy, Centre L. Emberger, CNRS, BP 5051, 34033 Montpellier Cedex, France

W. R. Siegfried, Percy Fitzpatrick Institute of African Ornithology, University of Cape Town, Rondebosch 7700, South Africa

G. R. Singleton, CSIRO Division of Wildlife & Ecology, PO Box 84, Lyneham, ACT 2602, Australia

L. Sonié, Centre L. Emberger, CNRS, BP 5051, 34033 Montpellier Cedex, France

R. L. Specht, Department of Botany, University of Queensland, St Lucia 4067, Queensland, Australia

S. Teillier, Laboratorio de Botanica, Facultad de Ciencias Biológicas, Pontificia Universidad Católica de Chile, Casilla 114-D, Santiago, Chile

C. D. Thomsen, Department of Agronomy & Range Science, University of California, Davis, California 95616, USA

L. Trabaud, Centre L. Emberger, CNRS, BP 5051, 34033 Montpellier Cedex, France

J.-L. Vernet, Laboratoire de Paleoecologie, Université des Sciences et Techniques du Languedoc, Place Eugene Bataillon, 34060 Montpellier Cedex, France

F. Vuilleumier, Department of Ornithology, American Museum of Natural History, Central Park West at 79th St., New York, New York 10024, USA

M. J. Wells, Botanical Research Institute, Private Bag X101, Pretoria 0001, South Africa

Preface

The idea for this volume originated in 1983 at the first meeting of the Scientific Advisory Committee for the program on the Ecology of Biological Invasions. At that time, the program had been only recently launched by the Scientific Committee On Problems of the Environment (SCOPE) as a major new initiative. Because at least half the members of the Scientific Advisory Committee were from regions of mediterranean climate and had interacted in the earlier SCOPE program on the Ecological Effects of Fire, the idea of a book on biogeographical aspects of invasions in mediterranean regions was agreed to as an attractive one. Now, almost six years later, that idea is a reality.

An association of mediterranean ecologists is also now a reality. The grouping is formalised as the International Society of Mediterranean Ecologists (ISOMED), the society being a member body of the International Union of Biological Sciences. If its previous publication record is any indication, mediterranean ecology as a discipline is destined to be an especially productive subdiscipline of botany, zoology and biogeography. We hope that publication of this volume adds to the already impressive publication lists on the subjects of mediterranean ecology and of biological invasions. Both topics have received considerable stimulus because of the interest and enthusiasm of our colleague Dr Harold A. Mooney. It is to him as Chairman of the Scientific Advisory Committee that we dedicate this book and in doing so we thank him for his wise counsel, his support and his continuing friendship over the years.

Many other colleagues and friends also deserve our thanks. To our fellow members of the Scientific Advisory Committee – James Drake, Fred Kruger, Marcel Rejmanek, José Sarukhan and Mark Williamson – we express our gratitude for their support. We thank all authors of the chapters that follow for their patience and forbearance. Martin Walters and, more recently, Alan Crowden at Cambridge University Press shared our earlier vision and helped make it a reality. So too did Peter Raven in the early days of the book. Throughout the life of the book Tricia Kaye helped in many ways, especially with advice on word processing. All these colleagues and many others have helped us produce this book, which because of its international authorship has required editorial skills greater than usual. But the challenge to us as editors is trivial compared to the challenge to all biological scientists to better understand the invasion process as it is worked out by many plants and animals in regions of mediterranean climate. If this book helps others to meet that challenge in their scientific research then it will have served its purpose as well as meeting our aims in creating a book from the original idea voiced by Hal Mooney in Paris in 1983.

R. H. Groves & F. di Castri
Canberra & Montpellier
January 1991

Part I Introduction

I

An ecological overview of the five regions of the world with a mediterranean climate

F. DI CASTRI

The five regions of the world with a mediterranean-type climate, i.e. parts of the Mediterranean Basin, California, central Chile, southern Africa and south-western and southern Australia, are situated between parallels 30° to 40° North and South. A mediterranean climate can be very broadly defined as a transitional regime between temperate and dry tropical climates, with a concentration of rainfall in winter and the occurrence of a summer drought of variable length (di Castri, 1981). It is a 'young' climate, firmly established only in the Pleistocene (Axelrod, 1973).

In these five regions of mediterranean climate, the vegetation is characterised by the dominance of woody shrubs with evergreen leaves that are broad and small, stiff and sticky (sclerophyllous). An overstorey of small trees may sometimes be present as well as an understorey of annuals and herbaceous perennials. This vegetation type is usually called 'maquis' in France, 'chaparral' in California, 'matorral' in Chile, 'fynbos' in South Africa and 'heath' or 'mallee' in Australia (di Castri, 1981).

Over the last 20 years there has been an unprecedented amount of research on the biota and ecosystems of the five regions of mediterranean climate. Mooney (1984) and di Castri *et al.* (1988*a*) outlined the main aspects of this particularly productive 'inter-mediterranean' research. Co-operation has been achieved because of a mixture of spontaneous collaborative research and several more structured scientific programs. The scientific attractiveness of the inter-mediterranean model is that research can be carried out on geographically disjunct areas of land (thereby implying that most biota are phylogenetically diverse). These areas share a similar type of climate and it is assumed that current selective forces and available resources are similar. To develop hypotheses for testing in the field about convergence and divergence of structural and functional patterns of ecosystems and biota, there is an implicit need to differentiate as far as possible that which is attributable to phylogeny from that which derives from current environmental driving forces (di Castri & Hadley, 1985).

Separation of the genetic from the environmental component is only a gross approximation in the field, compared with results (and the oversimplification) obtained from the use of controlled climate chambers. Furthermore, homoclimatic comparisons all too often do not take into account the great variety shown by mediterranean climates, even within the same climatic region. Climate may act only indirectly by shaping a certain type of vegetation or by facilitating the application by humans of specific land-use patterns. It is not surprising, therefore, that the homoclime approach is being challenged (Roy *et al.*, this volume) and that the overall inter-mediterranean convergence has also been questioned (Blondel *et al.*, 1984). Climate is only one of the factors to be considered and homoclimatic studies, when applied, should go much further in depth than the simple analysis of a few climatic diagrams. Above all, such studies should address a very specific question (Nazar *et al.*, 1966; di Castri, 1973).

Prior to 1973, a number of articles were published which compared the vegetation (Naveh, 1967; Specht, 1969; Mooney & Dunn, 1970) and soil fauna (di Castri, 1963) of several regions of mediterranean climate. The first volume that covered physical, ecological and evolutionary characteristics of all five mediterranean-climate regions was that of di Castri & Mooney (1973). Subsequent volumes (Mooney & Conrad, 1977; di Castri *et al.*, 1981; Kruger *et al.*, 1983; Dell *et al.*, 1986; di Castri *et al.*, 1988*b*) addressed more specific issues of mediterranean-climate regions, such as fire effects, soil nutrient deficiencies, resilience and water stress. Cody & Mooney (1978) and di Castri (1981) reviewed and discussed evolutionary and ecological aspects of the five regions; in what follows I refer mostly to the latter two papers to provide an overview of the five regions with a mediterranean climate.

During the implementation of the Scientific Committee on Problems of the Environment (SCOPE) project on the ecology of biological invasions, the interest of many participants often shifted towards mediterranean-climate ecosystems because of several factors: there was a wealth of data already available, the interchange of introduced species was greater than for other ecosystems, and because of the invasive potential of species originating in the Mediterranean Basin. Kruger *et al.* (1989) and Fox (1990) discussed these factors from the viewpoint of the ecology of biological invasions in mediterranean-climate regions, although the latter did not include the Mediterranean Basin.

In this introductory overview, in order to avoid a tedious description of the geographic and ecological characteristics of the five individual regions of mediterranean climate, I shall try to cluster them into 'similarity complexes' according to different criteria: climate and biogeography, geology and soil, and disturbance regimes, against a background of evolutionary and historic trends. Whenever possible I shall emphasise the environmental conditions that are likely to have facilitated a species becoming invasive, or a given ecosystem being invaded.

Climate and biogeography

The geographic locations of the five regions are given in Figure 1.1 together with their main climatic patterns, typified by 'idealised' climatic diagrams. The intra-regional variability of each climate is not shown, however, and particularly the many climatic types existing in the Mediterranean Basin.

The three regions in the Southern Hemisphere show the most temperate conditions. In addition, in South Africa and southern Australia, but not in Chile, the occurrence of summer (tropical) precipitation is not uncommon. Mediterranean

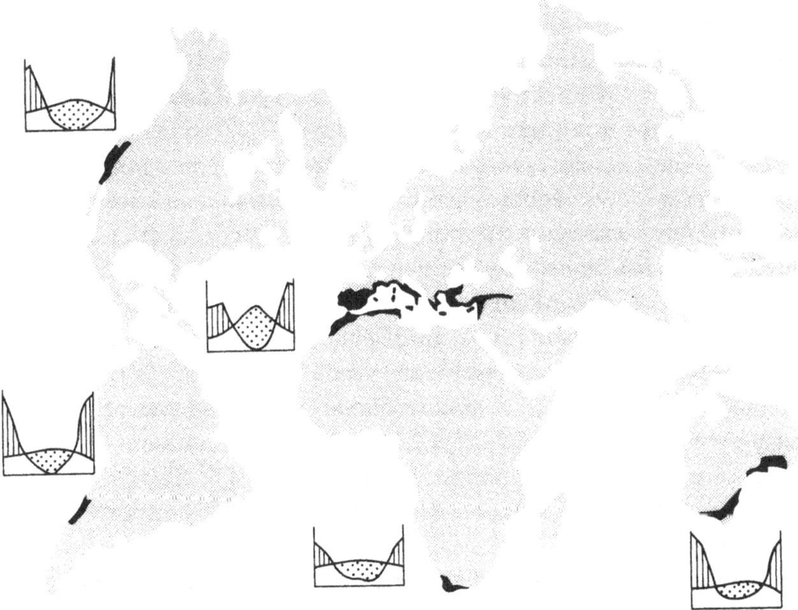

Figure 1.1. Geographic location of the five regions of the world with a mediterranean-type climate, together with schematic climatic diagrams (Walter *et al.*, 1975), typifying the main trends of each region. For the two Northern Hemisphere sites, the abscissa is months from January to December and for the three Southern Hemisphere sites, July to June. Left hand ordinate, mean monthly temperature (°C) and the right hand ordinate, mean monthly precipitation (mm). The dotted field represents the relatively droughty season and the vertical hatching the relatively humid season.

trends from the west can overlap with tropical intrusions from the east in both regions but most commonly in South Africa.

The two regions of the eastern Pacific, Chile and California, represent a typical west-coast mediterranean climate that is buffered by the frequency and intensity of marine fogs and by proximity to the Humboldt and Californian cold currents. Changes in mean temperature with latitude are almost negligible in Chile.

The climate of the Mediterranean Basin, particularly in the north and east, is the most continental-like, where killing frosts happen unpredictably and periodically. This feature may limit the success of some invasions, especially of those taxa from the more tropical-like South African and southern Australian regions. For instance, in the northern Mediterranean Basin, many apparently naturalised invasive plants from South Africa (mainly *Lampranthus* – syn. *Mesembryanthemum* – and other succulents) and from Australia (species of *Acacia* and *Eucalyptus*) were eradicated from large areas when killing frosts of −20 °C to −30 °C during the winter of 1984–85 were followed by unusually intense snowstorms in 1986–87.

Australia and South Africa resemble 'truncated' continents, in that they lack polarward continuity between the mediterranean ecosystems and wet temperate forests, cold taigas and tundras at higher latitudes. This feature may help to explain why some conifers (which are largely absent as a group in the South African flora) are so successful as invasive trees in Australia and South Africa. The case of *Pinus radiata* is noteworthy. Originally from coastal California, where *P. radiata* has a limited distribution, it has been introduced to Chile, Australia, South Africa and the Mediterranean Basin as well as to a number of other countries. In Chile *P. radiata* is naturalised but is not invasive; it replaces the indigenous very wet Valdivian forests and yields highly in such environments. In Australia and South Africa *P. radiata* is highly invasive and even penetrates into natural ecosystems in South Africa. In the Mediterranean Basin *P. radiata* is not invasive. A gradient in progressive invasibility of *P. radiata* would be therefore the Mediterranean Basin, Chile, southern Australia and South Africa, where it is most invasive.

The Mediterranean Basin is at the crossroads biogeographically, having been covered by repeated waves of invasion during successive latitudinal shifts of vegetation in glacial and interglacial periods (di Castri, 1989*a*). This aspect even makes it difficult to define precisely the meaning of the term 'Mediterranean species'. The term can mean: (i) a species that has evolved *in situ* under conditions of mediterranean climate; (ii) a species that has evolved *in situ* prior to the emergence of the mediterranean climate, but one that was pre-adapted to survive the constraints of such a climate, partly because of the extreme physical heterogeneity of the region; (iii) a species that evolved outside the region, but which colonised Mediterranean ecosystems subsequently as an 'escape response' from either extremely arid or extremely cold conditions; or (iv) an invasive species that arrived 'recently', such as *Opuntia ficus-indica* from southern North America which has become such a conspicuous component of Mediterranean landscapes that it is usually considered to be an indicator of mediterranean climate. All these meanings imply that the Mediterranean biota is a mosaic of species of different biogeographic origins, and that several species should be considered as naturalised 'old' invasive species. The most common geographic trend for invasions, both from outside the region and within the region itself, is from east to west.

An example that synthesises the main attributes of an 'ideal' invasive species is that of the grasses that evolved in the eastern Mediterranean under conditions of mediterranean climate concomitantly with the emergence of the first human activities and disturbances. In the widest sense the grasses have co-evolved with humans and some of the adaptations they show to human disturbance may now be genetically based. The Mediterranean grasses are pre-adapted to displace the native grasses in other regions that were never in close contact with human impacts. Mediterranean grasses have invaded central Europe where they have shifted their growing season from winter to summer (Kornas, 1983). A particularly well studied example concerns mainly ruderal or pioneering species from the Mediterranean Basin that have become dominant in Californian grasslands where they replace native annual and perennial grasses (Jackson, 1985; Jackson & Roy, 1986). Invasion by Mediterranean grasses is also characteristic of vast areas of Chile and Australia.

Following on from some of the previous points, the degree of evolutionary and biogeographic isolation has been greatest in Australia, then in South Africa and then Chile (where the arid Atacama desert to the north and the very high Cordillera to the east have acted as biogeographic barriers) followed by California and lastly the Mediterranean Basin. It is no accident that, when connected by way of new and human transportation systems since 1500 AD, these five regions have shown a similar sequence of susceptibility to invasion, with Australia and South Africa ranking highest. Islands within the Mediterranean Sea have also proved to be highly vulnerable to biological invasions in more ancient times (di Castri, 1989*a*).

Species diversity and the rate of endemism are exceptionally high in some parts of Western Australia and the Cape region of South Africa (with values approaching those of tropical rain forests), followed by Chile, California and the Mediterranean Basin. Even so, in Mediterranean Europe species diversity of plants and many animal groups is much higher than in any other European region. The evidence for high species diversity and of great vulnerability to invasion in South Africa and Australia does not seem to support one of the hypotheses of Fox & Fox (1986) – their species-richness hypothesis – in the way that species-rich ecosystems should be more resistant to invasion. Perhaps this hypothesis has to be validated at a smaller scale of approximation, such as for intra-regional and inter-ecosystem comparisons.

Finally, the main biogeographical aspects considered in all the above comparisons should take into account the facilities provided by humans to change the original areas of distribution of species. It is undeniable that the social and economic situation of Europe in the 1500s provided Mediterranean European species (some of which were 'old' invaders from the east) with many more opportunities to reach distant continents where their invasiveness was facilitated by new human-caused disturbance of otherwise pristine habitats (di Castri, 1989*a*). Such has been the history of many invasive species (di Castri, 1990; see also Crosby, 1986). Two main lines of transportation may be recognised as regards mediterranean-climate regions: the first, from the Iberian Peninsula to Chile and California until the time of the opening of the Panama Canal; and the second, from England to South Africa and Australia until the opening of the Suez Canal. It is unquestionable that the cultural characteristics of the Spanish colonisers on the one hand and the English and Dutch on the other have progressively increased the biotic differences between the Chile–California complex and the Australia–South Africa complex (di Castri, 1981; Fox, 1990).

Geology and soils

From a topographical viewpoint, South Africa has very diverse landscapes, with a series of ranges parallel to the south coast. Western Australia shares some of these characteristics, even as regards some lithological aspects, whilst southern Australia exhibits a flatter landscape than any other region of mediterranean climate in the world (together with the southern Levant in the Mediterranean Basin). Chile and California, with their longitudinal central valleys bordered by two mountain ranges, have the most similar terrains and climates of the world. Together, they constitute what may be termed a north–south mirror image. As regards the Mediterranean Basin, almost all geomorphological types can be found in such a large and heterogeneous region. Nevertheless, as shown in Figure 1.2, the most conspicuous differences are in soil conditions. Here again, two groups or complexes can be clearly identified. On the one hand, Chile and California are characterised by moderately fertile soils, especially Chile, where the abundance of grasses even in the understorey to matorral is evidence of moderate soil fertility. On the other hand, many soils of South Africa and Western or southern Australia have developed from nutrient-poor, very old parent materials or from Quaternary infertile siliceous sands. These acidic soils have been heavily weathered, highly leached and often podzolised. According to Specht (1979), the natural vegetation covering these soils should be considered as a true heathland and not as a mediterranean shrubland. Undoubtedly, soil status rather than climate is the main ecological factor controlling the distribution of mediterranean-climate vegetation in Australia and South Africa. The Mediterranean Basin, whilst largely dominated by soils on calcareous rocks, has also several areas with acidic soils and a typical cover of plants belonging to the family Ericaceae. It remains to be demonstrated if these soils are more susceptible to invasion than the neighbouring neutral or alkaline soils.

Disturbance and invasion

Up to this point, the two main factors that appear to play a major role in explaining the processes of biological invasions, at least at this coarse level of resolution, are the degree of biogeographical isolation in evolutionary terms and the natural or human-induced opportunities for long-distance dispersal. These factors differ according to timing and chance.

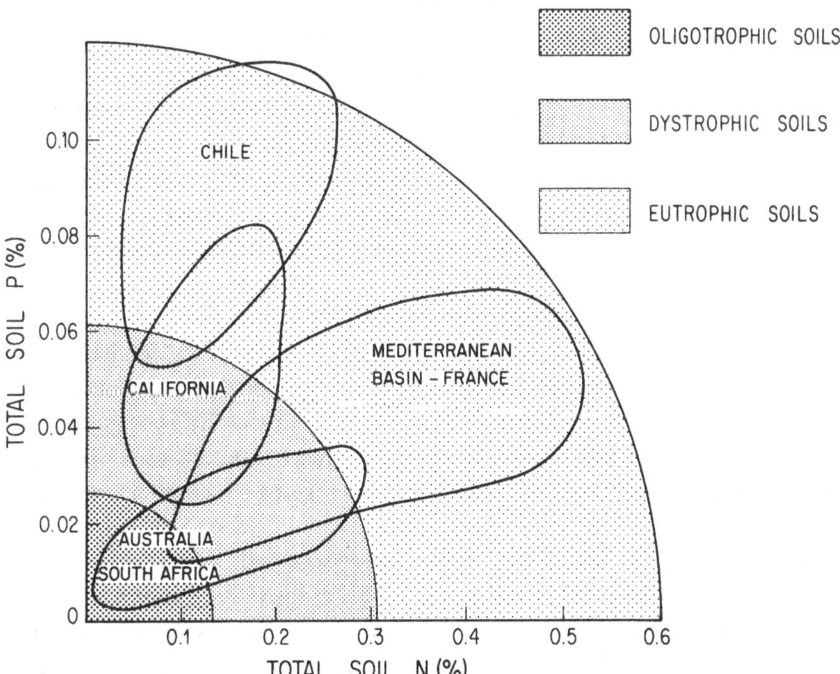

Figure 1.2. Levels of phosphorus and nitrogen in the soils of the five regions with a mediterranean-type climate (largely modified and redesigned from Rundel (1978) and di Castri (1981).

Timing and chance also determine the meaning and effects of the third driving force in biological invasions, viz. the endogenous (natural) and exogenous (human-caused) disturbances, terms that are interpreted here according to Fox & Fox (1986).

It has been correctly postulated that there is no invasion without some degree of previous or concomitant disturbance – the 'disturbance hypothesis' of Fox & Fox (1986). On the other hand, it is well known that ecosystems and regions that have been submitted in geological and historical times to long periods of continuous or intermittent disturbance are more resistant to invasion, and even that some of the component species show a high invasive potential (di Castri, 1990).

There is no contradiction between these two postulates, so long as the timing of disturbance is clearly defined: at one extreme, a sudden and unpredictable new disturbance; and at the other, a long-lasting regime of repetitive disturbance. In the latter case, disturbance is a selective factor that has been progressively incorporated as an adaptive response of species and, through a different kind of process, as a pattern controlling ecosystem dynamics (di Castri, 1989*b*).

The Mediterranean Basin has been submitted to continuous tectonic and climatic vicissitudes, especially during the last glaciations. It was one of the first regions of the world to face the impacts of relatively high-density human populations, who initiated the practices of primaeval agriculture and animal husbandry. Some of the 'new' human disturbances, such as forest clearing, grazing pressure and fires lit by humans, represented a kind of 'relay' as regards those natural disturbances already existing (di Castri, 1989*a*). A blend of new species assemblages occurred, frequently because of migrations within the region and from outside.

California and Chile have also been submitted to recent tectonic processes and to glaciations, although at a lower level compared to those of the Mediterranean Basin. They are the two mediterranean-climate regions that were colonised last by human populations, albeit with a very low level of impact. The occupation by Spaniards, which occurred much earlier and was more significant in Chile than in California, represented for each of these two regions a disruption by a new disturbance regime.

Finally, Australia and South Africa have developed their vegetation on ancient and stable basement complexes. Part of their native vegetation has not yet acquired a mediterranean-type phenology and maximum growth still occurs in summer (Specht, 1973). Invasive species from other mediterranean-climate regions may have, therefore, a competitive advantage in this respect since they better match the present-day climate. South Africa is, in absolute terms, the region that has been exposed first to the presence of primitive human populations, although human impact has been minimal, as has that of the Australian aborigines. The more recent colonisation by Europeans has, in a few centuries, modified the landscapes of these two regions at least as much as the thousands of years of human occupation in the Mediterranean Basin.

In conclusion, this analysis of disturbance regimes, following as it does those sections dealing with the biogeographical and evolutionary isolation and the human-induced opportunities for dispersal, confirms the gradient already discussed whereby the Mediterranean Basin shows the greater relative resistance to invasion (and the greater wealth of invasive species). The Mediterranean Basin stands in a pivotal position between the western California–Chile complex and the southern South Africa–Australia complex. The latter is in general the most vulnerable to invasion whilst the former occupies an intermediate position.

Clustering of the five regions

Clustering of the five regions into three complexes (the Mediterranean Basin, California–Chile and Australia–South Africa) was proposed earlier (di Castri, 1981), irrespective of any consideration of their relative proneness to biological invasions. It now seems to apply also to their relative invasibility.

The degrees of similarity between and among the five regions are illustrated in Figure 1.3. Similarities between Australia and South Africa depend mostly but not exclusively on phylogenetic commonalities of several taxa. For California and Chile the similarity depends on strong evolutionary convergence of taxa from a different origin and on similar patterns of functioning at the ecosystem level. For some groups, e.g. soil animals, Chile is more linked to Australia and, to a lesser degree, South Africa, because of old Gondwanan relationships. Similarities between California and the Mediterranean Basin respond both to phylogenetic affinities (North America and Eurasia separated only in the Tertiary) and to evolutionary convergence. Most of the similarities between the Mediterranean Basin and Chile are attributable to the effect of similar land-use patterns. This same clustering has also been adopted by Fox (1990) when discussing exchanges of Mediterranean weeds between mediterranean-climate regions.

Naveh & Whittaker (1980) proposed another subdivision into only two groups – the older 'Gondwanan' group of South Africa and southern Australia, and the more recent 'Pleistocene' group of California, Chile and the Mediterranean Basin. Clearly, there is a large measure of agreement between these two groupings. The Gondwanan connotation is somewhat misleading to the present discussion because the appearance of a mediterranean-type climate has occurred comparatively recently in South Africa and Australia. In addition, it is important to single out the 'crossroads' position of the Mediterranean Basin. The latter region is so large and so heterogeneous and there is such a multiplicity of ecosystem types, that it seems very difficult to combine it with other much more defined and restricted regions, such as California and Chile.

The main factors underlying the similarities and the differences between the five regions are synthesised in Figure 1.4. The figure illustrates how many differences there are when studying, at a finer level of resolution, the regions that have been classically considered the most similarly convergent disjunct pieces of land in the world. It also illustrates how important it is to avoid overgeneralisations on aspects such as convergent and divergent evolution, homoclimes and also, in the chapters that follow, biological invasions.

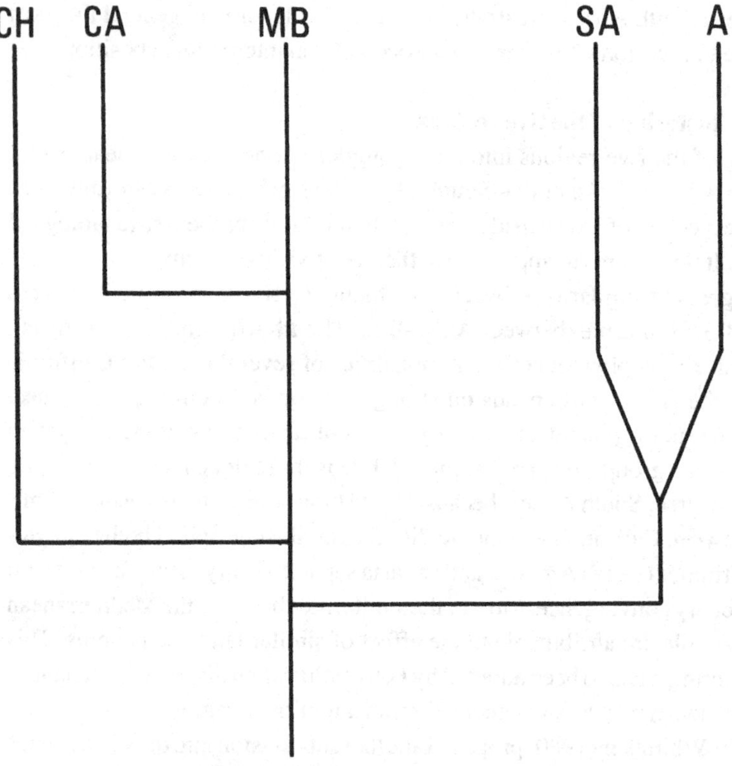

Figure 1.3. Degrees of affinity between the five regions of the world with a mediterranean-type climate. CH, Chile; CA, California; MB, Mediterranean Basin; SA, South Africa; A, Australia.

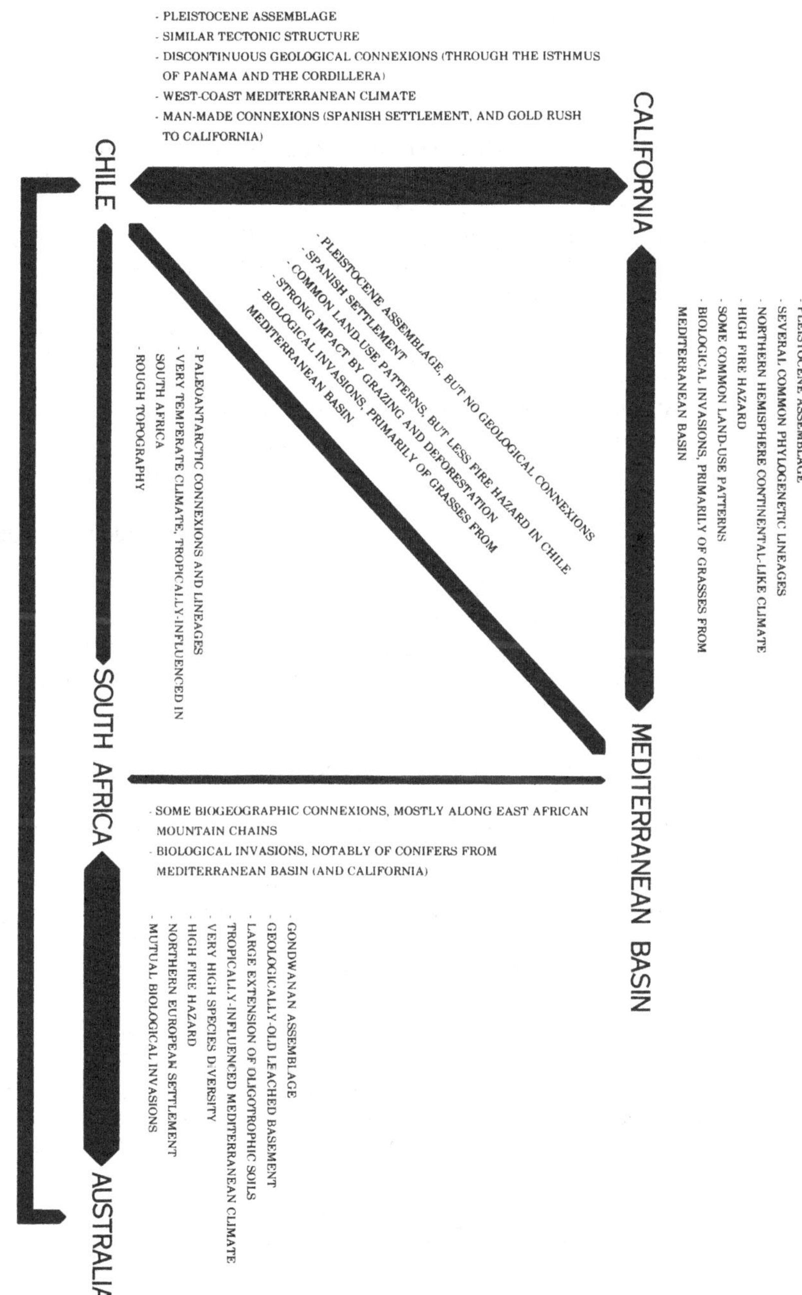

Figure 1.4. Similarities among the five regions of the world with a mediterranean-type climate, due mostly to geological and topographical features, evolutionary convergence patterns, phylogenetic commonalities and parallel human impacts. The degree of similaritiy is proportional to the thickness of connecting bars (after di Castri (1981; 1989*b*) largely modified and redesigned).

References

Axelrod, D. I. (1973). History of the mediterranean ecosystem in California. In *Mediterranean-Type Ecosystems. Origin and Structure*, ed. F. di Castri & H. A. Mooney, pp. 225–77. Berlin: Springer-Verlag.

Blondel, J., Vuilleumier, F., Marcus, L. F. & Terouanne, E. (1984). Is there ecomorphological convergence among Mediterranean bird communities of Chile, California, and France? *Evolutionary Biology*, **18**, 141–213.

Cody, M. L. & Mooney, H. A. (1978). Convergence versus nonconvergence in mediterranean-climate ecosystems. *Annual Review of Ecology & Systematics*, **9**, 265–321.

Crosby, A. W. (1986). *Ecological Imperialism. The Biological Expansion of Europe, 900–1900*. Cambridge: Cambridge University Press.

Dell, B., Hopkins, A. J. M. & Lamont, B. B. (ed.) (1986). *Resilience in Mediterranean-type Ecosystems*. Dordrecht: Junk.

di Castri, F. (1963). Estado biológico de los suelos naturales y cultivados de Chile Central. *Boletín de Producción Animal (Chile)*, **1**, 25–31.

di Castri, F. (1973). Climatographical comparisons between Chile and the western coast of North America. In *Mediterranean-Type Ecosystems. Origin and Structure*, ed. F. di Castri & H. A. Mooney, pp. 21–36. Berlin: Springer-Verlag.

di Castri, F. (1981). Mediterranean-type shrublands of the world. In *Mediterranean-Type Shrublands*, ed. F. di Castri, D. W. Goodall & R. L. Specht, pp. 1–52. Amsterdam: Elsevier.

di Castri, F. (1989*a*). History of biological invasions with special emphasis on the Old World. In *Biological Invasions. A Global Perspective*, ed. J. A. Drake, H. A. Mooney, F. di Castri, R. H. Groves, F. J. Kruger, M. Rejmanek & M. Williamson, pp. 1–30. Chichester: Wiley.

di Castri, F. (1989*b*). The evolution of terrestrial ecosystems. In *Ecological Assessment of Environmental Degradation, Pollution and Recovery*, ed. O. Ravera, pp. 1–30. Amsterdam: Elsevier.

di Castri, F. (1990). On invading species and invaded ecosystems: a play of historical chance and biological necessity. In *Biological Invasions in Europe and the Mediterranean Basin*, ed. F. di Castri, A. J. Hansen & M. Debussche, pp. 3–16. Dordrecht: Kluwer.

di Castri, F. & Hadley, M. (1985). Enhancing the credibility of ecology: can research be made more comparable and predictive? *GeoJournal*, **11**, 321–38.

di Castri, F. & Mooney, H. A. (ed.) (1973). *Mediterranean-Type Ecosystems. Origin and Structure*. Berlin: Springer-Verlag.

di Castri, F., Goodall, D. W. & Specht, R. L. (ed.) (1981). *Mediterranean-Type Shrublands. Ecosystems of the World*, vol. 11. Amsterdam: Elsevier.

di Castri, F., Floret, C., Rambal, S. & Roy, J. (1988*a*). Preface. In *Time Scales and Water Stress. Proceedings of the 5th International Conference on Mediterranean Ecosystems*, ed. F. di Castri, C. Floret, S. Rambal & J. Roy, pp. v–viii. Paris: International Union of Biological Sciences.

di Castri, F., Floret, C., Rambal, S. & Roy, J. (ed.) (1988*b*). *Time Scales and Water Stress. Proceedings of the 5th International Conference on Mediterranean Ecosystems*. Paris: International Union of Biological Sciences.

Fox, M. D. (1990). Mediterranean weeds: exchanges of invasive plants between the five mediterranean regions of the world. In *Biological Invasions in Europe and the Mediterranean Basin*, ed. F. di Castri, A. J. Hansen & M. Debussche, pp. 179–200. Dordrecht: Kluwer.

Fox, M. D. & Fox, B. J. (1986). The susceptibility of natural communities to invasion. In *Ecology of Biological Invasions: An Australian Perspective*, ed. R. H. Groves & J. J. Burdon, pp. 57–66. Canberra: Australian Academy of Science.

Jackson, L. E. (1985). Ecological origins of California's Mediterranean grasses. *Journal of Biogeography*, **12**, 349–61.

Jackson, L. E. & Roy, J. (1986). Growth patterns of mediterranean annual and perennial grasses under simulated rainfall regimes of southern France and California. *Acta Oecologica, Oecologia Plantarum*, **7**, 191–212.

Kornas, J. (1983). Man's impact upon the flora and vegetation in central Europe. In *Man's Impact on Vegetation*, ed. W. Holzmer, M. J. A. Werger & I. Ikusima, pp. 277–86. The Hague: Junk.

Kruger, F. J., Mitchell, D. T. & Jarvis, J. U. M. (ed.) (1983). *Mediterranean Type Ecosystems. The Role of Nutrients*. Berlin: Springer-Verlag.

Kruger, F. J., Breytenbach, G. J., Macdonald, I. A. W. & Richardson, D. M. (1989). The characteristics of invaded mediterranean-climate regions. In *Biological Invasions. A Global Perspective*, ed. J. A. Drake, H. A. Mooney, F. di Castri, R. H. Groves, F. J. Kruger, M. Rejmanek & M. Williamson, pp. 181–213. Chichester: Wiley.

Mooney, H. A. (1984). Collaborative research among ecologists in the mediterranean-climate regions of the world. *Intecol Bulletin*, **10**, 51–5.

Mooney, H. A. & Conrad, C. E. (ed.) (1977). *Proceedings of the Symposium on the Environmental Consequences of Fire and Fuel Management in Mediterranean Ecosystems*. Washington: USDA Forest Service.

Mooney, H. A. & Dunn, E. L. (1970). Convergent evolution of mediterranean-climate evergreen sclerophyll shrubs. *Evolution*, **24**, 292–303.

Naveh, Z. (1967). Mediterranean ecosystems and vegetation types in California and Israel. *Ecology*, **48**, 445–59.

Naveh, Z. & Whittaker, R. (1980). Structural and floristic diversity of shrublands and woodlands in N. Israel and other mediterranean areas. *Vegetatio*, **41**, 171–80.

Nazar, J., Hajek, E. R. & di Castri, F. (1966). Determinación para Chile de algunas analogías bioclimáticas mundiales. *Boletín de Producción Animal (Chile)*, **4**, 103–73.

Rundel, P. W. (1978). Ecological impact of fires on mineral and sediment pools and fluxes. In *Fire and Fuel Management in Mediterranean-climate Ecosystems: Research Priorities and Programmes*, ed. J. K. Agee, pp. 17–21, MAB Technical Note 11. Paris: UNESCO.

Specht, R. L. (1969). A comparison of the sclerophyllous vegetation characteristic of mediterranean-type climates in France, California and southern Australia, I & II. *Australian Journal of Botany*, **17**, 277–308.

Specht, R. L. (1973). Structure and functional response of ecosystems in the mediterranean climate of Australia. In *Mediterranean-Type Ecosystems. Origin and Structure*, ed. F. di Castri & H. A. Mooney, pp. 113–20. Berlin: Springer-Verlag.

Specht, R. L. (ed.) (1979). *Heathlands and Related Shrublands. A. Descriptive Studies. Ecosystems of the World*, vol. 9A. Amsterdam: Elsevier.
Walter, H. Harnickell, E. & Mueller-Dombois, D. (1975). *Climate – Diagram Maps* (English edn). Berlin: Springer-Verlag.

Part II Historical background

The climate of five regions of the world is characterised by a regular period of winter rainfall followed by a summer drought of varying length. Such a climate is termed 'mediterranean' because it is the typical climate of most countries of the Mediterranean Basin. This climatic feature is also characteristic of four other regions of the world – namely: extensive areas of California, Chile, South Africa and southern Australia.

A biological invasion of any region is basically a process rather than an event and thus, to quote Deacon in his chapter in this section, 'it is not restricted in time'. But for the reader to better understand the biogeographic aspects of invasions, the five chapters in this section provide some historical background to the biota of the five regions of mediterranean climate, especially as that background relates to human settlement of the regions.

2

The palaeohistory of the Mediterranean biota

Z. NAVEH & J.-L. VERNET

The historical background to invasions in the Mediterranean Basin is an integral part of the evolution of the region's natural and cultural landscapes. Vegetation changes were induced by a combination of climatic stress, natural and constant human disturbances and a resulting resilience to invasion. Recently, the problems of human impact on Mediterranean vegetation have been discussed in detail (see, e.g. Le Houérou, 1981; Pignatti, 1983; Pons & Quézel, 1985; Vernet & Thiébault, 1987; Naveh & Kutiel, 1990). In this chapter we shall describe, briefly, the natural and cultural processes which shaped the Mediterranean landscapes from the Pleistocene onwards. We shall then present some of the archaeobotanical evidence for evolutionary and historical changes in the vegetation.

The evolution of natural and semi-natural Mediterranean landscapes in the Pleistocene

The landscapes of the Mediterranean Basin are relatively young geological systems; they gained their present geomorphological forms by violent uplift in the late Tertiary and early Quaternary periods (di Castri & Mooney, 1973; di Castri *et al.*, 1981). Final shaping of the landscapes took place in the Pleistocene in a highly dynamic period of climatic fluctuations, and tectonic and volcanic activity. Site conditions, both physical and biological, diversified increasingly. From the Middle Pleistocene onwards, early Mediterranean peoples further diversified these landscapes on both a regional and local scale, thereby contributing to their present multi-dimensional heterogeneity (di Castri, 1981). Because of the continuing interaction between natural and anthropogenic processes, di Castri (1981, p. 18) has commented appropriately '. . . that man has partly co-evolved with these ecosystems and that co-evolutionary features are present in a number of ecological and cultural characteristics of these regions'.

Co-evolution of landscape and the biota is certainly true for the Mediterranean Basin where the final stages of geological and biological evolution coincided with the major phases of biological and cultural human evolution. This co-evolution has been described in more detail for Mount Carmel in Israel (Naveh, 1984) as the emergence of what has been called a total human ecosystem of early Mediterranean peoples. Humans evolved from a palaeolithic *Homo erectus* hunter and food-gatherer to the more advanced neanderthaloid and the epi-palaeolithic *Homo sapiens*, who collected food intensively, to, finally, the food-producing neolithic *Homo sapiens*.

Phases in co-evolution

Three phases in the evolution of Mediterranean landscapes can be distinguished. In commenting on the early phase, Pignatti (1983) emphasised the important role of human disturbance in the evolution of new habitats and hence, in the early phases of co-evolution, in also stimulating the evolution of the flora. During the very long period of 300000 to 500000 years of human presence as the late Acheulian and middle palaeolithic cultures in the Levant, these disturbances were confined most probably to widely scattered habitations; they have been mostly overlooked. Tectonics and erosion have obliterated most archaeological evidence (see Bar-Joseph, 1984), including ash deposits.

Over this co-evolutionary period, fire was one of the most important factors in natural disturbance. From recent archaeological findings in the Petralona limestone cave in northern Greece, the use of fire by Acheulian hunter-gatherers has been dated to about 1 million years in the lowest levels and to 500000 to 600000 years in the upper levels. Petralona can, therefore, be considered as the oldest existing proof of fire culture in the world (Ikeya & Poulianos, 1979). In France, the first evidence of prehistoric charcoal and fire is from an Acheulian site near the Mediterranean Sea occupied about 400000 years ago (Vernet, 1975).

In Israel, the first archaeological evidence of fire, and probably also of its use by *Homo erectus*, has been provided from an Acheulian assemblage in the Upper Jordan Rift Valley (Bar-Joseph, 1984). Perles (1977) claimed that the mastering of fire was perfected about 100000 years ago by Mousterian Neanderthaloids who produced lamps to light their caves and torches to carry fire. Using fire these humans could open up dense forest and brush thicket to facilitate hunting and food collecting. They increased edible food supplies for themselves and their beasts by encouraging the regeneration of trees and shrubs and by promoting the invasion of grasses and bulbous and tuberous plants (Naveh, 1974, 1984).

At about the same time, the Mediterranean vegetation was already well established. We can assume, therefore, that those woody and herbaceous genotypes which developed efficient means of regeneration, both vegetatively and sexually, to overcome the stresses imposed by natural and human-induced fires, had the best chances of survival (Naveh, 1975). For instance, all sclerophyllous species are obligatory root-sprouters, whereas all dwarf shrubs, as well as perennial herbaceous plants, are mostly facultative root-sprouters and have dual vegetative and sexual regenerative mechanisms. *Pinus halepensis*, the only indigenous conifer, relies only on post-fire seed germination, just like an annual species. Post-fire seed germination is enhanced in many other perennial and annual species that follow fire (Naveh, 1974, 1975).

In the major phase of co-evolution, pre-agricultural landscapes were modified by the culturally rich and advanced upper and epi-palaeolithic populations of *Homo sapiens* – a phase which probably reached its peak late in the last pluvial, 10000–15000 years ago. The mild climate, with its winter rainfall, favoured the development of a rich Mediterranean flora and fauna, and the spread of food-collecting, hunting and fishing populations of humans in the ecotones of mountains, foothills, coastal plains and river valleys. Carmel Natufians used carefully prepared, tiny and sharp microliths to optimise hunting of swift game (especially gazelle), flint sickles to cut wild grasses, and mortars and pestles as pounding tools to prepare staple foods from roasted cereals and acorns. They also collected fruits and bulbs. They constructed houses and developed a complex and rich communal, cultural and spiritual life (Bar-Joseph, 1984).

It can, therefore, be surmised that great sophistication was achieved also in the use of fire as a tool for vegetation management. Fire was used to convert dense, pristine forests, woodlands and shrublands into more open and richer semi-natural woody and herbaceous vegetation. Humans probably created heterogeneous fire-induced mosaics of forests, woodlands and shrublands, dominated by sclerophyll trees and shrubs with a rich understorey of light-demanding herbaceous plants able to spread rapidly. The latter became dominant in the marshy or drier or rocky sites; some served as the progenitors of domesticated cereals, pulses and vegetables, as well as a source of pasture plants and agricultural weeds.

The transition from intensive food collection to food production and domestication of plants and animals can be regarded as the culminating final phase of the co-evolution of early humans and plants in the Mediterranean Basin. It has been termed 'specialised domestication' by Rindos (1984). This transitional period was followed several thousand years later by 'agricultural domestication'. In the process the Natufian and other epi-palaeolithic cultures served most probably as an important link because of their intensive environmental manipulations, chiefly through the use of fire. In the transition from food collection by 'passive cultivation' to food production by 'active cultivation' and domestication, prescribed burning, and the favourable seedbeds created thereby (Kutiel & Naveh,

1987), acted as a major trigger and had a multiplier effect on the co-evolutionary process.

In southern France in the Rhône valley, the existence of slash-and-burn agricultural practices in the Mediterranean is supported by findings of charcoal, dated 7350 BC (Pons & Quézel, 1985). The presence of charcoal corresponds with a decrease in the percentage of deciduous oak pollen, and is accompanied by high percentages of pollen of the families Labiatae and Leguminosae, as well as that of Compositae, species of *Plantago* and others considered to be weeds of cereal cultivation in prehistoric times. Cereal pollen also appeared at the same time. Pons & Quezel (1985) interpreted earlier evidence for the temporary clearance of forest of *Quercus pubescens* and a simultaneous increase in *Plantago*, Ericaceae and light-demanding herbaceous species as evidence of pre-neolithic forest clearing for grazing animals.

Evolution and degradation of the agro-pastoral landscape and the meta-stable vegetation of the Mediterranean Basin

This period spans the early Holocene and most of historic time. It can be subdivided into two major phases, each of which we shall now discuss briefly.

The final formation of the agricultural–pastoral Mediterranean landscape

The narrowing of the broad spectrum of neolithic agriculture into cereal cropping and pastoral livestock husbandry in the early Holocene apparently caused the complete destruction of the pristine vegetation in the lowlands and along riverbeds and was followed by severe soil erosion. Such a process can still be witnessed in the exposure by erosion of ancient, calcified palaeosols on the Israeli coastal plain (Dan & Yaalon, 1971).

The final formation of the Mediterranean agro-pastoral upland landscape was initiated by the domestication of fruit trees in the Bronze Age about 5000 BP (Zohary & Spiegel-Roy, 1975). It was closely connected with the rise of proto-urban civilisations and their improved implements and artefacts, and was aided by the development of transportation and trade. This led to much more intensive and far-reaching ecological changes. The physical, biological and cultural features of the co-evolution of palaeolithic Mediterranean humans and their semi-natural landscapes was replaced by unilateral human dominance, henceforth governing the fate of Mediterranean landscapes for better or worse. This process was completed only after the invention of iron tools several thousand years later at about the first millennium BC. Iron tools enabled the uprooting of shrubs and trees on the slopes and the clearing and terracing of upland fields. Wherever these slopes were too steep or rocky, favourable microsites between rock outcrops were also planted with fruit trees, such as olives, figs, almonds, pomegranates and grapes. Such intensive activities may be regarded as one of the very few instances in which agriculture improved the initial factors of topography, soil parent material and moisture regime on a long-term basis (Naveh & Dan, 1973); but, alas, at the expense of the natural flora and fauna.

All the other non-arable uplands served as pastures for goats and other livestock – as well as wildlife – and in addition to this use as fodder, wild plants were collected also for specialised uses such as herbs, spices, medicinal plants and as sources for fuel and tool construction (including wooden ploughs). It was the unfortunate combination of tree and wood cutting, fire and grazing which led gradually to landscape desiccation, especially in the drier regions and on steeper and less fertile slopes. Invasion of the more xeric elements of herbaceous and woody colonisers from adjacent drier regions was probably encouraged thereby, a process which in turn provided a further source of weeds of cultivated land.

The agro-pastoral landscape and its decline during historic times

Since the decline of the Roman empire and subsequently of the Byzantine empire, cultivated uplands have undergone dramatic changes in their soil–vegetation systems. Throughout the long history of agro-pastoral utilisation, the semi-natural and pastoral ecotopes of open forests, shrublands, woodlands and grasslands, together with the agricultural ecotopes of terraces, patch- and hand-cultivated rock polycultures, have created a mosaic of landscapes. The transfer of fertility, by way of grazing animals, and of seeds, by way of grazing, wild herbivores and insects, created ideal conditions for introgression and spontaneous hybridisation of wild and cultivated plants and biotypes. Genotypes with a high level of adaptation to these human-modified habitats evolved (Zohary, 1969).

Many of these herbaceous and chiefly annual plants had the best chances to spread as weeds in Europe. Many centurics later some of the same plants spread to other continents with similar mediterranean-climates such as California, Chilc and southern Australia. These plants were pre-adapted to disturbance in both natural upland pastures and in cultivated fields. Some of these plants, such as the grasses and legumes, became valuable pasture plants (Naveh, 1967).

The agricultural and pastoral landscapes resemble in many aspects the shifting mosaic of landscape, described by Forman & Godron (1986) for systems exhibiting a pattern of long-term change (in this case, changes in land-use patterns during historic times) along with short-term, internal spatial conversion (in this case, climatic fluctuations and cultural rotational cycles). These environmental and cultural processes have created complex and highly dynamic patterns of degradation and regeneration. Furthermore, these patterns do not fit any of the deterministic models of succession (see also Lepart & Debussche, this volume).

Palaeobotanical evidence on evolutionary and historic changes of the Mediterranean vegetation

Following the broad overview of the history of Mediterranean landscapes and their anthropogenic modifications presented in the previous section, we shall now deal with the origin, evolution and transformation of the vegetation of these Mediterranean landscapes during geologic, prehistoric and historic times. This information has been derived chiefly from palaeobotanical and palynological studies and from the dating of charcoal deposited in prehistoric and historic times.

The origins

The early Miocene saw the alternation of phases with open plant associations predominant in times of semi-arid climate and forested phases dominant in times of subhumid climate (Roiron, 1979, 1984; Bocquet, 1980; Bessedik, 1985). In Israel, according to Horowitz (1979), Mediterranean vegetation began to develop in the late Miocene when the country was still connected with areas of Africa having a tropical climate. Palaearctic elements penetrated in Pliocene times, when the Jordan valley was open to the north. The flora throughout the Quaternary seems to have been of the Mediterranean type.

The Plio-Pleistocene period

Suc (1984) found a pollen zonation in the western Mediterranean corresponding to the Pliocene and to the beginning of the Pleistocene. Mediterranean conditions could be identified from about 3 million years BP (before present), when 'xerophytic' plant groups began to appear, e.g. *Phillyrea*, *Olea*, *Cistus*, *Pistacia* and *Quercus ilex*. The first definite evidence for 'mediterraneity' has been found prior to the first cold periods dated at 2.2 million years BP (Zagwin, 1960, 1974). This is the time when the planktonic foraminifer *Neogloboquadrina atlantica*, common in more northern latitudes, first appeared in the Mediterranean region. At this time trees were few, except for *Pinus*, although they became more common in a later, more humid period (the 'Tiglian' period, Baggioni *et al.*, 1981, Roiron, 1983). This forested phase came between two steppic phases, with several minor and regional variations in which temperatures and rainfall levels varied. For instance, in a recent palynological study of the late Pliocene and early Pleistocene palynozone in the Jordan–Dead Sea Rift Valley, Levin & Horowitz (1987) found higher percentages of arboreal pollen (mainly of *Picea orientalis*) in all pollen spectra. The climate of this region was in general temperate and influenced by the onset of glacial and interglacial conditions in Europe.

The Middle and late Quaternary

During the Middle Quaternary period, which began about 0.8 million years BP, climatic oscillations rapidly followed one another with intensely cold periods alternating with dry, hot periods (see e.g. Wijmstra & Smit, 1976). At the site studied by Wijmstra & Smit (1976), the interglacial phase was dominated by deciduous oaks (see also data for a Spanish site in Pons & Reille, 1986). These and other results reveal a pattern of vegetation dynamics controlled by temperature changes with an approximate amplitude of 8 °C (Vernet, 1986).

An extensive amount of palynological information has been obtained from the southern Levant from bores in sediments of the now dry Hula Lake Basin at the northern end of the Upper Jordan Rift Valley (Horowitz, 1979, 1987). The main characteristics of the Quaternary palynozones derived from this site are presented in Table 2.1.

From results obtained from a number of sites in both the western and eastern Mediterranean, a number of generalisations can be developed (Fortea Perez *et al.*, 1987). The beginning of the Holocene (preboreal, boreal) was characterised by conditions favourable to soil erosion on slopes. The vegetation at this time was predominantly herbaceous and the climate rather cold and dry. This vegetation was succeeded by more closed forests of pines which were replaced in their turn by oak forests during the Atlantic period that reached an optimum about 5000 years BP. During the sub-boreal period, climatic conditions were more typically mediterranean with alternate humid and dry phases and renewed erosion. Documented human influences began long before the Roman period, in the middle neolithic (see, for instance, a chronology of human impact in southern France in Vernet & Thiébault, 1987).

Palaeobotanical evidence for the evolution of domestic crops and weeds in the Mediterranean Basin

The origin and spread of agriculture in the eastern Mediterranean is well known (Zohary, 1986) and the domestication of plants equally well known (Zohary & Hopf, 1988). Much less is known of these aspects for the western Mediterranean, however, as we shall attempt to show in this section.

The few botanical remains of the Natufians of Mount Carmel (see earlier) do not demonstrate the use of domesticated cereals, but rather the collection of wild forms. Wild einkorn wheat (*Triticum dicoccoides*), wild barley (*Hordeum spontaneum*) and wild rye (*Secale secale*), as well as wild legumes, have all been identified from late Natufian layers at a site in northern Syria (Kislev, 1984).

Table 2.1. *Main characteristics of the Quaternary palynozones in Israel. The climate indicated may have been interrupted by one or more interstadials for each of the palynozones, and only the general trend is given*

Palynozone	Vegetation	Climate	Age
QX	Evergreen oak maquis	Dry Mediterranean interstadial	Holocene
QIX	Deciduous oak forest	Wet Mediterranean pluvial	Würm
QVIII	Garrigue	Dry Mediterranean interpluvial	Riss/Würm
QVII	Deciduous oak forest	Wet Mediterranean pluvial	Riss
QVI	Evergreen oak and pine stands	Dry Mediterranean interpluvial	Mindel/Riss
QV	Deciduous oak forest	Wet Mediterranean pluvial	Mindel
QIV	Oak and pine stands	Dry Mediterranean interpluvial	Günz/Mindel
QIII	Oak, pine and spruce forest	Wet Mediterranean pluvial to interstadial	Günz
QII	Oak and spruce stands	Dry Mediterranean interpluvial	Preglacial Pleistocene

Source: Horowitz, 1987.

The earliest evidence for domestication of barley in Israel comes from a recent excavation of a neolithic site in the Jordan Valley, dated 10250–9790 BP (O. Bar-Joseph & M. E. Kislev, personal communication). Here, charred wild barley was mixed with two-row barley (*Hordeum distichum*). Domesticated emmer wheat (*Triticum turgidum* subsp. *dicoccum*) has been identified from a site near Damascus dated 9790–9590 BP (van Zeist & Bakker-Heeres, 1979; see also Zohary, 1986). The most frequent pulses to appear as constant companions of these cereals are lentil (*Lens culinaris*) and pea (*Pisum sativum*) and less commonly, bitter vetch (*Vicia ervilia*) and chickpea (*Cicer arietinum*). Clear indications of their cultivation appear only about 8000 BP (Zohary, 1986). A recent discovery of broadbean seeds (*Vicia faba*) at a Neolithic site in northern Israel adds this legume to the list of earliest cultivated pulses for the eastern Mediterranean region (Kislev, 1985). According to van Zeist & Bakker-Heeres (1979) flax (*Linum usitatissimum*) was also cultivated in this region before 8000 BP.

Kislev (1984) described the emergence and spread of wheat cultivation from Israel to western Anatolia and into the eastern Mediterranean Basin and then further northwards. He found that the rate of spread during the first 3200 years of the expansion of wheat was about 1.2 km per year. Most likely not all species spread at the same rate into temperate Europe as that for emmer and einkorn wheats. Other crops were probably only secondary or occasional immigrants.

In the western Mediterranean region at a site in the southern part of the Massif Central of southern France *Ervum ervilia*, *Triticum aestivo compactum*, *Hordeum*, *Lathyrus cicera* and *Pisum* were found in a neolithic layer; in an epipalaeolithic layer *Ervum ervilia*, *Lathyrus cicera*, *Vicia*, *Lens*, *Pisum* and *Vitis sylvestris* were found. Amongst these plants, it is particularly surprising to find peas and lentils, since they are usually considered to be domesticated taxa of the eastern Mediterranean. This result either poses a problem in chronology or else it allows us to envisage from the beginning of about the seventh millennium BP an embryonic stage of agriculture in the western Mediterranean similar to that found several thousand years earlier in the eastern Mediterranean. Observations made at the same site (Vaquer *et al.*, 1979) showed that the mesolithic groups who came to the site were not really hunters and they willingly gathered the plants that they found. Their tools, belonging mostly to the 'armature' (framework) type, were probably reserved for the reaping of plants. It is difficult to know whether the 'gatherer' stage had not already evolved or if these people could not have, by a process of primitive mass selection, preserved some of the plants they needed at certain times of the year. A proto-agriculture could then develop by 'cropping' some unused seeds which may sprout the following year on either trampled soils near the site or – as described above – on soils after fires lit intentionally.

Van Zeist (1987), however, reminds us of the oriental origin of the cultivated legumes and asserts that their introduction to the western Mediterranean was in the neolithic. This is so, but nevertheless wild leguminous plants often have been found in the south-eastern part of France since mesolithic times. It may be that some farmers came to the western Mediterranean from the east by boat. Such an explanation is certainly credible for south-eastern Italy. In fact, there was already a nascent agriculture of cereals in this region of Italy during the eighth millennium BP. For instance, at Rendina, Follieri (in Guilaine *et al.*, 1987) found *Triticum aestivum*/*durum* and *Hordeum* in a deposit older than 7150 BP. *Triticum monococcum* and *T. dicoccum* have been identified from another site in levels of the same age. Legumes (*Vicia faba* and *Lens* cf. *culinaris*) first appeared in the seventh millennium above a layer dated 6480 BP at an Italian site. Thus the components for a rotation of crops were present in the western Mediterranean from about this time.

Fruit trees (olives and grapes) came later. It is only in the second half of the seventh millennium BP that the wheats (einkorn – *Triticum monococcum*, starch – *T. dicoccum*, soft – *T. aestivo-compactum*) and barley with naked seeds (*Hordeum vulgare* var. *nudum*) appeared in continental Mediterranean France (Marinval, in Guilaine *et al.*, 1987). On the Iberian Peninsula, the first cultivated plants were probably introduced by humans directly from the east and not necessarily by way of northern Italy and France (Hopf, in Guilaine *et al.*, 1987).

It is thus during the middle neolithic that a new economic order was established, especially as larger villages developed. It was also the time of the beginning of an agro-pastoral system with a stable agriculture together with a developed system of animal breeding (Cipolloni-Sampo, in Guilaine *et al.*, 1987).

Archaeological evidence for the existence of ancient field weeds in the eastern Mediterranean is based upon the findings of carbonised plant remains. As discussed in detail by van Zeist (1987), these species represent only a part of the weeds which supposedly grew together with cultivated plants in such fields. Nevertheless, van Zeist (1987) identified a considerable number of segetal plants which may have occurred already in pre-pottery neolithic sites, such as Tel Ramad near Damascus, dated 8200–7900 BP, together with emmer wheat, free-threshing wheat, hulled and naked barley, pea, lentil and linseed. Amongst possible weeds are *Adonis*, *Ammi maius*, *Androsace maxima*, *Avena barbata*, *A. sterilis*, *Bellevalia*, *Cephalaria syriaca*, (presently a particularly noxious weed), *Galium*, *Lithospermum arvense*, *Lolium rigidum*, *Medicago*, *Melilotus*, *Phalaris paradoxa*, *Silene*, *Trigonella*, *Vaccaria pyramidata* and many others.

Much later, at a Bronze Age site on the right bank of the Euphrates river, dated 4390–3890 BP, samples of charred grain deposits of threshed barley seeds also contained seeds of weeds such as *Vaccaria pyramidata* and species of *Aegilops*, *Bromus*, *Bupleurum*, *Coronilla*, *Gypsophila*, *Malva*, *Medicago* and others.

Various species of eastern Mediterranean origin, such as *Lithospermum arvense*, *Anagallis* and *Fumaria* apparently migrated to Europe only in later periods and the archaeological evidence from European neolithic fields indicates that the majority of the weeds at that time were still of European origin.

The Recent period

Since the Gallo-Roman period, profound changes in agrarian systems have resulted from human colonisation. E. Grau Almero (personal communication) showed the nature of some of these changes for a site near Valencia, a city founded by the Romans in 2128 BP. Oak forest gave way to *Pinus halepensis* and maquis; various horticultural species appeared and disappeared. Changes also occurred in other areas, e.g. in Languedoc in the Middle Ages (Planchais, 1982) and in Israel around the Lake of Galilee (Baruch, 1987).

In summary, the biota of the Mediterranean Basin has undergone many changes over millennia. Invasions and extinctions of fauna are described subsequently by Cheylan (this volume) and Blondel (this volume). Floral change has been profound, especially the effects of domestication of crop species and the deliberate use of fire. The total result of all these biotic changes has been the creation of a mosaic of human-modified landscapes all over the Mediterranean Basin. Some 'commensal' species spread extensively within the Mediterranean Basin and are still spreading. Some of these species became prime candidates to invade other regions of mediterranean climate when Europeans arrived and settled and tried to grow crops and graze animals (Groves, 1986). Currently, plants from some of these other regions are invading Mediterranean landscapes (see Guillerm, this volume).

Acknowledgements. We are grateful to Dr M. Weinstein-Evron, Haifa University, for her helpful comments on a draft of Z. Naveh's part of this chapter. Z. Naveh was supported by the Funds for Promotion of Research of the Technion, Israel Institute of Technology. J.-L. Vernet's contribution to this chapter was translated from French to English by Ms A. Dao, CNRS, Montpellier, and reviewed by Dr L. Miller, Ithaca, N.Y., both of whom deserve our thanks.

References

Baggioni, M., Suc, J.-P. & Vernet, J.-L. (1981). Le Plio-Pléistocène de Camerota (Italie méridionale), géomorphologie et paléoflores. *Géobios*, **14**, 229–37.

Bar-Joseph, O. (1984). Near East. *Neue Forschungen zur Altsteineit. Forschungen zur Allgemeinen und Vergleichenden Archeologie*, **4**, 232–98.

Baruch, V. (1987). The Late Holocene vegetational history of Lake Kinnereth (Sea of Galilee), Israel. *Paleorient*, **12**, 37–48.

Bessedik, M. (1985). Reconstitution des environnements miocènes, des regions nord-ouest méditerranéennes à partir de la palynologie. Thèse, Université des Sciences et Techniques du Languedoc, Montpellier.

Bocquet, G. (1980). Crise de salinité messinienne et floristique méditerranéenne. *Naturalia Monspeliensia (hors serie)*, **36**, 21–31.

Dan, J. & Yaalon, O. H. (1971). On the origin and nature of the paleopedological formations in the central coastal fringe areas of Israel. In *Paleopedology: Origin, Nature and Dating of Paleosols*, ed. O. H. Yaalon, pp. 245–60. Jerusalem: Israel Universities Press.

di Castri, F. (1981). Mediterranean-type shrublands of the world. In *Ecosystems of the World*, vol. 11, *Mediterranean-type Shrublands*, ed. F. di Castri, D. W. Goodall & R. L. Specht, pp. 1–52. Amsterdam: Elsevier.

di Castri, F., Goodall, D. W. & Specht, R. L. (ed.) (1981). *Ecosystems of the World*, vol. 11, *Mediterranean-type Shrublands*. Amsterdam: Elsevier.

di Castri, F. & Mooney, H. A. (ed.) (1973). *Mediterranean-type Ecosystems. Origin and Structure*. Berlin: Springer-Verlag.

Forman, R. T. T. & Godron, M. (1986). *Landscape Ecology*. New York: Wiley.

Fortea Perez, J., Marti Oliver, B., Fumanal Garcia, M. P., Dupré Ollivier, M. & Perez Ripoll, M. (1987). Epipaleolitico y neotlitisacion en la zona oriental de la peninsula iberica. In *Actes Premières Communautés Paysannes en Méditerranée Occidentale*, ed. J. Guilane, J. Courtin, J. L. Roudil & J.-L. Vernet, pp. 581–91. Paris: CNRS.

Groves, R. H. (1986). Invasion of mediterranean ecosystems by weeds. In *Resilience in Mediterranean-type Ecosystems*, ed. B. Dell, A. J. M. Hopkins & B. B. Lamont, pp. 129–46. Dordrecht: Junk.

Guilaine, J., Barbaza, M., Gasco, J., Geddes, D., Jalut, G. & Vernet, J.-L. (1987). L'abri du Roc de Dourgne, écologie des cultures du Mésolithique et du Néolithique ancien dans une vallée montagnarde des Pyrénées de l'Est. In *Actes Premières Communautés Paysannes en Méditerranée Occidentale*, ed. J. Guilaine, J. Gourtin, J. L. Roudil & J.-L. Vernet, pp. 545–54. Paris: CNRS.

Horowitz, A. (1979). *The Quaternary of Israel*. New York: Academic Press.

Horowitz, A. (1987). Subsurface palynostratigraphy and paleoclimates of the Quaternary Jordan Rift Valley Fill, Israel. *Israel Journal of Earth Sciences*, **36**, 31–44.

Ikeya, A. & Poulianos, A. N. (1979). ESR age of the trace fire at Petralona. *Anthropos*, **6**, 44–7.

Kislev, M. E. (1984). Emergence of wheat agriculture. *Paleorient*, **10**, 61–70.

Kislev, M. E. (1985). Early Neolithic horsebean from Yiftach'el, Israel. *Science*, **278**, 319–20.

Kutiel, P. & Naveh, Z. (1987). The effect of fire on soil nutrients of *Pinus halepensis* forests in Israel. *Plant & Soil*, **104**, 269–74.

Le Houérou, H. N. (1981). Impacts of man and his animals on Mediterranean vegetation. In *Ecosystems of the World*, vol. 11, *Mediterranean-type Shrublands*, ed. F. di Castri, D. W. Goodall & R. L. Specht, pp. 497–522. Amsterdam: Elsevier.

Levin, N. & Horowitz, A. (1987). Palynostratigraphy of the Early Pleistocene QI palynozone in the Jordan–Dead Sea Rift, Israel. *Israel Journal of Earth Sciences*, **36**, 45–58.

Naveh, Z. (1967). Mediterranean ecosystems and vegetation types in California and Israel. *Ecology*, **48**, 445–59.

Naveh, Z. (1974). The ecology of fire in Israel. *Annual Tall Timbers Fire Ecology Conference, Tallahassee, Florida*, **13**, 131–70.

Naveh, Z. (1975). The evolutionary significance of fire in the Mediterranean region. *Vegetatio*, **9**, 199–206.

Naveh, Z. (1984). The vegetation of the Carmel and Nahal Sefunim and the evolution of the cultural landscape. In *The Sefunim Prehistoric Sites, Mount Carmel, Israel*, ed. A. Ronen, pp. 23–63. Oxford: BAR International Series 230.

Naveh, Z. & Dan, J. (1973). The human degradation of Mediterranean landscapes in Israel. In *Mediterranean-type Ecosystems. Origin and Structure*, ed. F. di Castri & H. A. Mooney, pp. 370–90. Berlin: Springer-Verlag.

Naveh, Z. & Kutiel, P. (1990). Changes in the Mediterranean vegetation in Israel in response to human habitation and land uses. In *The Earth in Transition.*

Patterns and Processes of Biotic Impoverishment, ed. G. M. Woodwell, in press. New York: Cambridge University Press.

Perles, C. (1977). *Préhistoire du Feu*. Paris: Masson.

Pignatti, S. (1983). Human impact on the vegetation of the Mediterranean. In *Man's Impact on Vegetation*, ed. W. Holzner, M. J. A. Werger & I. Ikusima, pp. 151–62. The Hague: Junk.

Planchais, N. (1982). Palynologie lagunaire de l'étang de Mauguio, paléoenvironnement végétal et évolution anthropique. *Pollen & Spores*, **24**, 93–118.

Pons, A. & Quezel, P. (1985). The history of the flora and vegetation and past and present human disturbance in the Mediterranean. In *Conservation of Mediterranean Plants*, ed. C. Gomez-Campo, pp. 25–43. The Hague: Junk.

Pons, A. & Reille, M. (1986). Nouvelles recherches pollenanalytiques à Padul (Granada), la fin du dernier glaciaire et l'Holocène. In *Proceedings of a Symposium on Climatic Fluctuations during the Quaternary in the Western Mediterranean Regions*, ed. F. Lopez-Vera, pp. 405–20. Madrid: Universidad Autónoma de Madrid.

Rindos, D. (1984). *The Origin of Agriculture. An Evolutionary Perspective*. New York: Academic Press.

Roiron, P. (1979). Recherches sur les flores plio-quaternaires méditerranéennes, la macroflore pliocène de Pichegu près de Saint-Gilles (Gard). Thèse, Université de Montpellier.

Roiron, P. (1983). Nouvelle étude de la macroflore plio-pléistocène de Crespia, Catalogne, Espagne. *Geobios*, **16**, 687–715.

Roiron, P. (1984). Les macroflores messiniennes de Méditerranée nord-occidentale et la crise de salinité. *Paléobiology*, **14**, 415–22.

Suc, J.-P. (1984). Origin and evolution of the Mediterranean vegetation and climate in Europe. *Nature*, **307**, 429–32.

van Zeist, W. (1987). Some reflections on prehistoric field weeds. In *Palaeoecology of Africa and the Surrounding Islands*, vol. 18, ed. J. A. Coetzee, pp. 405–27. Rotterdam: Balkema.

van Zeist, W. & Bakker-Heeres, J. A. H. (1979). Some economic and ecological aspects of the plant husbandry of Tell Aswad. *Paleorient*, **5**, 161–9.

Vaquer, J., Barbaza, M. & Vigneron, E. (1979). La grotte du Rec des Tremouls, près l'Abeurador, Félines-Minervois (Hérault), premiers résultats. In *Congrès Préhistorique de France, 21ème session, Montauban Cahors, Sept 1979*, vol. II, 298–301.

Vernet, J.-L. (1975). Les charbons de bois des niveaux mindeliens de Terra Amata (Nice, Alpes-maritimes). *Comptes Rendus de l'Académie Sciences, Paris*, **208D**, 1535–7.

Vernet, J.-L. (1986). Changements de vegetations, climats et action de l'homme au Quaternaire en Méditerranée occidentals. In *Proceedings of a Symposium on Climatic Fluctuations during the Quaternary in the Western Mediterranean Regions*, ed. F. Lopez-Vera, pp. 535–47. Madrid: Universidad Autónoma de Madrid.

Vernet, J.-L. & Thiébault, S. (1987). An approach to northwestern Mediterranean recent prehistoric vegetation and ecologic implications. *Journal of Biogeography*, **14**, 117–27.

Wijmstra, T. A. & Smit, A. (1976). Palynology of the middle part (30–78 m) of the 120 m deep section in northern Greece (Macedonia). *Acta Botanica Neerlandica*, **25**, 297–312.

Zagwin, W. H. (1960). Aspects of the Pliocene and early Pleistocene vegetation and climate in the Netherlands. *Mededelingen Geologische Stichting*, **(C)3**, 1–78.

Zagwin, W. H. (1974). The Pliocene–Pleistocene boundary in western and southern Europe. *Boreas*, **3**, 75–97.

Zohary, D. (1969). The progenitors of wheat and barley in relation to domestication and agricultural dispersal in the Old World. In *The Domestication and Exploitation of Plants and Animals*, ed. P. J. Uco & G. W. Dimbleby, pp. 35–46. Chicago: Aldine.

Zohary, D. (1986). The origin and early spread of agriculture in the Old World. In *The Origin and Domestication of Cultivated Plants*, ed. C. Barigozzi, pp. 3–20. Amsterdam: Elsevier.

Zohary, D. & Hopf, M. (1988). *Domestication of Plants in the Old World.* Oxford: Oxford University Press.

Zohary, D. & Spiegel-Roy, P. (1975). Beginning of fruit growing in the Old World. *Science*, **187**, 319–27.

3

Human impact on the biota of mediterranean-climate regions of Chile and California

H. ASCHMANN

In the widest sense, the areas of mediterranean climate in Chile and California have similar histories in that the Europeans who came to dominate the native peoples of both regions were from Spain, another area with a mediterranean climate, and they were familiar with plants already adapted to that climate. The native peoples of the two areas, however, were different. Those of Chile cultivated crops, most of which were derived from the Andean highlands or the Peruvian coastal desert. Being more numerous and engaged in farming they had long disturbed the native vegetative cover. Further, it is likely that the comparatively recent Inca conquest had introduced large herbivores, e.g. the llama and alpaca, and probably plants such as *Nicotiana glauca* and *Schinus molle* (Bahre, 1979).

Significant entry of Europeans into Chile was accomplished more than two centuries earlier than in California and it was by territorial conquest. European entry into California was by the Franciscan missions and involved the establishment of distant points of control and the gradual domination of the neighbouring Indian groups. Through both thc Spanish and the Mexican periods, until 1846, extensive parts of California's region of mediterranean climate were not controlled or even visited by Europeans. Thus the introduction of plants into a large part of California came much later than in Chile and can sometimes be documented.

The colonial period

In Chile the conquistadores found a farming system that could be exploited. To this they added wheat and grapes, because of taste preferences as well as adaptability, and all European livestock. In California the missionaries and their escorts had to initiate farming and the raising of stock until the Indian peoples could be trained in these new skills. Both Mexican and European crops were introduced, but wheat did much better than maize until irrigation could be developed. As in Chile, the European livestock did very well in California.

Taking over the Inca conquests as far south as the Río Maule was relatively easy, despite a few rebellions that were crushed. Between that river and the Río Bío-Bío the native peoples showed more resistance, and in the non-mediterranean forested areas further south, the Indians were able to maintain their independence until 1880. North of the Río Maule just about all of the Indians were put in 'encomienda' almost immediately, and the Spaniards and their 'yanacona' allies brought from Peru could requisition food from them. The handling of rapidly increasing livestock herds was a favoured and responsible (though somewhat dangerous) job and seems to have devolved on the 'mestizo' population as soon as one developed. The real interest on the part of the conquerors, however, was in precious metals, and the working of not very rich gold placers began immediately, using Indians from encomiendas and others drafted from villages (McBride, 1936).

Disease, to some extent maltreatment in mines, and the fact that the Indians were not particularly numerous meant that by the end of the sixteenth century labour was scarce with respect to land. No more food was needed than in aboriginal times, but dry-farming of wheat on a shifting basis probably made for more clearing of flat lands. Irrigated farming in the valleys from Santiago northwards was maintained but with reduced extent and intensity. The introduction of European domestic animals, especially cattle, and their rapid increase in numbers, exposed lands covered with annuals to a new level of grazing pressure and provided a reason to burn to convert shrub-covered areas to herbaceous growth. Again, this was most effective on the relatively level land of the coastal terraces and in the Central Valley (Aschmann & Bahre, 1977).

For the next 200 years there was a slow but steady increase in the human population combined with an increase in the concentration of land in large holdings or 'estancias', the non-irrigated ones being used largely for grazing. Smallholders, often with communal land tenure, cultivated for subsistence, by dry-farming in the coastal ranges, often by 'curben' or shifting cultivation, and by grazing goats and sheep in the steeper areas. In the eighteenth century a modest market for hides, tallow and dried beef had been developed in Peru, and even some wheat was shipped there. The large landholders met this market which permitted some imports and maintained living standards, which though modest were far superior to those of the smallholders or the landless. The latter were increasingly tied to the large properties as 'inquilinos' in the labour force. Except for some of the northern communes, growing enough food was not a problem in the colonial period (McBride, 1936).

It is not known when European grasses and herbs invaded and then replaced the native annuals on the gently sloping surfaces, but with livestock raising spreading rapidly throughout mediterranean Chile, it could have happened by the end of the sixteenth century. By the eighteenth century, with a growing population, extension of at least temporary cultivation, and deliberate burning to increase pasture, upper alluvial fans and even some steep hill slopes in areas north of Santiago lost their broad-leaved sclerophyll shrubs to herbaceous plants, many of which were accidental introductions from Europe.

Entry of the Franciscan missionaries and their military escorts into California began in 1769. Trekking overland from the area of the former Jesuit missions in Baja California, they brought livestock, seeds and weeds. Missions were established at points, first at San Diego and Carmel, with presidios at the former and at Monterey near the latter. Filling in the system and extending it to north of San Francisco Bay continued until 1823 with 21 missions ultimately established. All the missions were close to the coast, although some were behind a range of coastal hills. Since the Indians knew no agriculture, their conversion to Christianity and training as farmers proceeded together. Irrigated cultivation of a wide variety of field crops and fruit trees in the south and dry-farming further north were successful, but the Indians were dying of disease so fast that only limited areas were needed or could be cultivated. After 1790, raids into the interior to capture fugitive neophytes were combined with recruitment or capture of gentile Indians to maintain the labour force.

Livestock generally flourished, especially horses and cattle, but the mission neophytes could not be trusted to herd far from the missions. A neophyte herdsman had to be back at the mission, or at least to a permanent 'asistencia' settlement, and account for his livestock daily, thereby restricting grazing to a radius of about 8 kilometres. Soldiers in the presidios and at the missions sometimes grazed their mounts more extensively, but it was the civilian inhabitants of pueblos, beginning at Los Angeles in 1783 and supplemented by retired soldiers, who expanded cattle raising throughout the whole coastal belt then under missionary influence. Extensive land grants or grazing rights had been given under Spanish rule, but the rate of granting land for grazing accelerated rapidly after Mexican independence in 1821 and through the Mexican period to 1846 (Dana, 1840; Donley *et al.*, 1979).

Explorers' descriptions from the late eighteenth century indicate that the distribution of grassland and chaparral differed little from the present, being governed by slope and rock type and maintained by Indian burning practices that were continued by the ranchers. But European grasses and herbs steadily displaced native annuals as areas in the coastal valleys and the less steep hills were subjected to grazing pressure (Aschmann, 1959).

Nineteenth century developments

During the first half of the nineteenth century Chile became independent, and the country was opened to foreign investment. British investors focused primarily on foreign commerce and on silver and copper mining in the Norte Chico, an area at the dry end of the spectrum of mediterranean climates. Demands for fuel around mining and smelting centres devastated the local woody plants, nearly eradicating the gallery forests along stream courses and destroying the woody matorral on the interfluves. In the drier north and on the overgrazed communes it has not come back, leaving a degraded vegetation on the steeper slopes (Bahre, 1979). From the vicinity of Santiago southward there was an increase in wheat farming with some wheat being exported to Peru.

The economic awakening following Chile's independence, although mining offered the initial impetus, provoked substantial investment in agriculture by some, but not all, of the large landholders from the Aconcagua Valley to the Central Valley well south of Santiago. Canals for irrigation were constructed and extensive lands were converted to vineyards and other fruit trees. The extensive marshy lake of San Vincente Tagua-Tagua was converted by drainage into excellent lacustrine soil that has been farmed intensively ever since. Rangeland was ploughed and planted to wheat. A substantial number of large 'fundos' were ready to expand rapidly the area cultivated and production increased when the Californian and, shortly afterward, the Australian gold rushes expanded explosively the demand for Chilean wheat (McBride, 1936).

In California, annexation by the United States was followed immediately by the discovery of gold in 1848. Two hundred thousand immigrants were attracted in a decade, mostly to the central part of the state. They came from the eastern United States and most of the rest of the world. In the Mother Lode and other mining districts there was heavy logging and increased burning, but except for actual tailings piles and valley bottoms affected by placers, and later by hydraulicking, the forest and woodland vegetation of the Sierra Nevada has recovered to something close to its original state. The demand for meat meant that grazing pressure was applied to grasslands of the Central Valley and other valleys that had been beyond mission and Mexican influence. The replacement of native bunch grasses and annuals by European annuals was extremely rapid. The market for wheat extended cultivation through the Sacramento Valley and the valleys and gentler hills on either side of San Francisco Bay (Robbins *et al.*, 1951).

Local production could not meet the Californian demand and Chile was the country best positioned to fill the market. The Victorian gold rush in Australia in 1856 offered Chile an additional market. Both large landholders and smallholders ploughed and planted wheat wherever slopes permitted and in parts of the coastal ranges where they did not. Grazing was relegated to steep slopes even though the demand for meat was rising. The matorral of much of the Norte Chico was burned to improve pasture, and, in part because much of the land there was communally held by poor people, a common sequence was clearing, planting of wheat, pasturing the abandoned land followed by permanent degradation of woody vegetation. Except for the Araucanian-held areas south of the Río Bío-Bío (forested areas with a non-mediterranean climate), cultural land use as either cultivation or grazing, had reached its physiographic limits by 1870. Any further increase in agricultural productivity to provide for a growing population would have to come from intensification of production, especially by irrigation (McBride, 1936).

Whilst gold mining in California fell off rapidly after 1860 few of the immigrants who came to mine went home. Some moved to other mining districts but many turned to agriculture. Arable land became scarce throughout the state, and hillsides were ploughed or put under orchards and vineyards in many localities.

Two mutually reinforcing developments affected land use in California during the last third of the nineteenth century. The completion of a transcontinental railroad in 1869, and its extension to southern California in the next decade, opened the United States market to Californian agricultural products, initially wheat but increasingly to perishable specialities such as raisins, prunes and, later, citrus fruits. The potential profitability of such activity justified heavy investment in irrigation facilities. Until the twentieth century irrigation works were limited to drainage basins, but California's topography readily permitted intensive cultivation of the eastern two thirds of the San Joaquin Valley and the coastal lowlands of southern California. Many of these areas were too dry for dry-farming and were only poor pasturelands. The area around Riverside is a good example, it changing from a xerophytic scrub to intensive citrus culture at this time. In the coastal valleys in the central part of the state, vineyards and deciduous fruits and nuts occupied moderate slopes and needed little irrigation. Truck crops irrigated by groundwater, in part for local and in part for national markets, occupied the valley bottoms (Donley *et al.*, 1979).

Thus, only a few decades after Chile's agriculture and grazing had expanded to utilise all suitable land in areas of mediterranean climate, the same situation was arrived at in California. The rangelands (in both countries) were not readily irrigated, and interstices in the cultivated areas were taken over by weedy annuals, almost all of them of European origin. In Chile two wild perennial plants, one native and one introduced, have extended their ranges. The 'espino' (*Acacia caven*) seems to have expanded southward in relatively flat but unirrigable lands in the lee of the coast ranges both in dry-farmed and grazed areas. As a shrub of open woodland it appears to be tolerated as a source of fuel and emergency browse. Even in the best parts of Chile's Central Valley the introduced blackberry (*Rubus* spp. – see also Montenegro *et al.*, this volume) has preempted an amazing amount of good farmland. With or without human assistance blackberries are established along fencelines and road and rail rights-of-way. Probably encouraged as an effective fence in a socially insecure countryside, the impassable linear thickets have spread to widths of up to ten metres. Whilst there are continuing attacks on individual thickets large landholders have not been willing to go to the considerable and unpleasant effort and expense needed to eradicate the weed. Native and introduced blackberries exist in California as well, but they show their land-consuming propensities only locally and they have generally been controlled adequately.

In the last part of the nineteenth century canals continued to be constructed on the sides of river valleys from the Aconcagua northward in Chile. The effort was to start farther upstream and with a higher canal irrigate progressively higher terraces. The more northerly of these valleys are too dry to be considered as having a mediterranean climate and they required irrigation for any cultivated crop to be able to grow. The initial impetus was to grow feed to maintain and fatten cattle driven over the Andes from Argentina, cattle which ultimately would supply nitrate mines in the desolate north. The limited flow of these northern rivers was over-committed, and since the earlier water users had better rights, the canals often are empty and the newly cleared higher terraces are rarely planted or, if planted, do not receive sufficient water to sustain a crop (McBride, 1936; Bahre, 1979).

World wheat prices fell after 1870, and wheat growing remained profitable only in the more level lands of the Central Valley. These lands were generally in large estates whose owners had options of converting to improved pastures, vineyards, other fruit crops, or irrigated truck crops near the larger cities. The smallholders in the Coast Ranges from the Aconcagua south to the Río Bío-Bío and beyond and the communal holders of the southern Norte Chico could only try to maintain their incomes by continuing to grow wheat, something impossible in all but the wettest years. Areas on moderate slopes that were cleared of matorral are subject to strong erosion and are over-grazed when they cannot be cultivated. Annuals, including introduced species, xeric shrubs and cacti comprise the degraded vegetation (Winnie, 1965). Holders of hill lands and coastal terraces, even north of the Aconcagua, who were not immediately dependent on them for income, had another option. Plantations of Monterey pine (*Pinus radiata*) from California and *Eucalyptus* spp. from Australia appeared on both moderate and steep slopes with linear boundaries that mark property lines. Free of their native predators, these trees do better in Chile than in their homelands. Whilst by world standards neither tree has very good wood, in Chile where softwoods for construction are scarce there is great demand for timber of *P. radiata*; moreover, for owners who can wait 20 to 40 years for a harvest they are profitable. Thus these plantations of introduced trees may have permanently displaced native matorral or woodland with a profitable crop which also protects the soil resource (Fuentes & Hajek, 1979).

Modern developments

In California the early twentieth century marked the beginning of the construction of great dams and aqueducts to transfer water between drainage basins, often several hundreds of kilometres. Curiously, although they made possible a great expansion of agricultural production, these great engineering works had very little effect on the wild vegetation in areas of mediterranean climate. Most of the water was destined for desert areas, such as the Imperial and Coachella Valleys and later the west side and south end of the San Joaquin Valley, or for major urban areas such as San Francisco and Los Angeles. Areas of mediterranean climate were already dry-farmed and many of them changed to more intensively cultivated crops with irrigation. Dams came later in Chile and initially were placed in the dry Norte Chico where they could stabilise irrigated cultivation in their own valleys, but they could not collect enough water in dry years to permit significant expansion of irrigated land. Recent dam construction in Chile has focused on developing hydroelectric power rather than irrigation. In both countries, of course, vegetation was eliminated from reservoir basins (Donley *et al.*, 1979).

In 1800 the population of Chile was perhaps ten times that of California and it continued to grow steadily by natural increase. In California waves of immigration, beginning with the Gold Rush and continuing to the present, greatly accelerated population growth. Early in the 1920s the population of California came to equal that of Chile at a little less than 4 million people. Since then California's population has increased seven-fold and Chile's about three-fold. In both Chile and California the urban population has exceeded the rural for many decades, although urbanisation has been much more intense in California (Corporación de Fomento, 1967; Donley *et al.*, 1979).

'Urban sprawl' has been especially characteristic of California since before World War II and accounts for almost all of the state's population growth. It is most developed around Los Angeles and San Francisco Bay and, more recently, around San Diego, but it also occurs around middle-sized centres such as Fresno and Sacramento. The land taken over by houses, shopping centres, streets and parking lots was primarily agricultural, but it included both dry-farmed areas and irrigated, intensively cultivated orchards almost indiscriminately. Citrus groves are on their way to being eliminated from the entire Los Angeles lowland.

Santiago is also spreading out as it grows, but the best irrigated farmland offers more resistance to urbanisation.

California's urban sprawl has long included a feature not represented significantly in Chile: exclusive and expensive suburbs have sought building sites on steep hillsides that in southern California were typically covered by chaparral. Around San Francisco Bay both woodland- and chaparral-covered slopes are utilised. Typically, building lots are large and many owners have chosen to leave the original vegetation on much of their land. This practice creates a severe fire hazard and catastrophic conflagrations have occurred. Political pressure to improve fire protection has reduced fire frequency, but fuel accumulation makes the delayed next fire even more disastrous (Minnich, 1987). The original vegetation returns and people rebuild houses. Sections of the Santa Monica Mountains in and west of Los Angeles have gone through as many as three building-burning-rebuilding cycles in this century.

In northern San Diego County a recent entry into extremely steep chaparral-covered slopes has been orchards of frost-sensitive avocados and lemons. Enormous investments in clearing, planting, developing pipe irrigation to each tree using expensive water, and even cable systems to get to each tree for harvesting have been required. These plantings cannot be economic and may be attributed to a mixture of income tax shelters and conspicuous consumption in an affluent society. But they have made avocados abundant and cheap.

In both California and Chile slopes covered by woodland, chaparral or matorral have resisted invasion by introduced perennials. The native shrubs could be displaced by Monterey pine plantations or houses, but even where annuals have come in after repeated burning, the native shrubs tend to return, although often in degraded form. Introduced perennials, such as *Nicotiana glauca*, *Ricinus communis* (castor bean) and *Cytisus scoparius* (Scotch broom), have established themselves in places with severely disturbed soil such as on road cuttings, actively gullied sites and steep areas cleared for orchard planting. Because they occur along transportation routes, these plants seem to be more prevalent than they really are. Two trees that were introduced to both California and Chile, either for economic reasons or as ornamentals, have become naturalised enough to reproduce – viz. *Schinus* spp. (pepper tree) and *Eucalyptus* spp. Their reproduction, however, requires specialised circumstances and the trees never seem to travel far from their point of introduction (Howard & Minnich, 1989; see also Pryor, this volume).

In summary, both in California and Chile, fairly level lands that could be pastured or ploughed had their grasses and herbs replaced by introduced annuals almost as soon as Europeans entered an area. The native bushy or woody vegetation in both countries has been far more resistant to invasions. Only if the native vegetation were replaced by another land use did it disappear. There is a difference, however, in the treatment of the vegetation on steep slopes in the two countries. In California, the chaparral is left alone or lightly pastured by cattle. Burning is its only disturbance, one from which it is adapted to recover, although perhaps with subtle alterations in species composition. In Chile the matorral or woodland is used, both generally for firewood and charcoal and selectively for medicinals, food, saponin (from *Quillaja saponaria*), tannin or other uses. The quillay and the Chilean palm (*Jubaea chilensis*) which are destroyed in exploitation have become extinct locally (Bahre, 1979). South of Santiago the matorral or woodland maintains its integrity, although thinned and with some alterations to species composition. To the north, especially in the Norte Chico, the combination of woodcutting and over-grazing by goats has eliminated both perennial and annual species, leaving the uplands of some communes virtual deserts.

References

Aschmann, H. (1959). The evolution of a wild landscape and its persistence in southern California. *Annals of the Association of American Geographers*, **49**, 34–56.

Aschmann, H. & Bahre, C. (1977). Man's impact on the wild landscape. In *Convergent Evolution in Chile and California: Mediterranean Climate Ecosystems*, ed. H. A. Mooney, pp. 73–84. Stroudsburg, Pennsylvania: Dowden, Hutchinson & Ross.

Bahre, C. J. (1979). Destruction of the natural vegetation of north-central Chile. *University of California Publications in Geography*, **23**, 1–116.

Corporación de Fomento de la Producción (Chile) (1967). *Geografiá Económica de Chile*. Santiago: Corporación de Fomento de la Producción (Chile).

Dana, R. H. (1840). *Two Years before the Mast*. New York: Harper & Bros.

Donley, M. W., Allan, S., Caro, P. & Patton, C. P. (1979). *Atlas of California*. Culver City, California: Pacific Book Center.

Fuentes, E. R. & Hajek, E. R. (1979). Patterns of landscape modification in relation to agricultural practice in central Chile. *Ecological Conservation*, **6**, 265–71.

Howard, L. F. & Minnich, R. A. (1989). The introduction and naturalization of *Schinus molle* (Pepper Tree) in Riverside, California. *Landscape and Urban Planning*, **18**, 77–89.

McBride, G. M. (1936). *Chile: Land and Society*. New York: American Geographical Society.

Minnich, R. A. (1987). Fire behavior in southern California chaparral before fire control: the Mount Wilson burns of the turn of the century. *Annals of the Association of American Geographers*, **77**, 599–618.

Robbins, W. W., Bellue, M. K. & Ball, W. S. (1951). *Weeds of California*. Sacramento: State of California Printing Division.

Winnie, W. W. Jr (1965). Communal land tenure in Chile. *Annals of the Association of American Geographers*, **55**, 67–86.

[illegible] for more [illegible] (C [illegible] Mediterranean [illegible] is a [illegible] the [illegible] species [illegible] the [illegible] (Catha, 1979) [illegible] Salvia [illegible] in the North, especially [illegible] and annual grass [illegible]

References

[illegible] Ann. Assoc. Amer. Geogr. [illegible]

[illegible] California [illegible]

[illegible] New York: Harper & Row.

[illegible] Atlas of California [illegible] Book Center.

[illegible] California Landscape [illegible]

[illegible] New York: American Geographical Society.

[illegible] Annals of the Association of American Geographers [illegible]

[illegible] State of California [illegible]

[illegible] American Geographers [illegible]

4

Central Chile: how do introduced plants and animals fit into the landscape?

E. R. FUENTES

Within the Chilean region with a mediterranean climate *sensu stricto* (Aschmann, 1973; di Castri, 1973) there are two main types of landscapes: the low flatlands and the lower mountainous areas (Fuentes, 1988). The flatland areas are mainly found along the coast, in major river valleys, and constitute the Intermediate Depression (Figure 4.1). The second type of landscape, the low mountainous, occurs as the coastal ranges and the lower part of the Andes. Above these two landscapes are the desolate high Andean landscapes in which winter temperature is low and precipitation high. These high Andean landscapes are very important for land use in the areas with mediterranean climate because they serve as snow accumulators during the winter and supply water during the non-winter seasons for agriculture and the urban centres of the lowland areas (Weischet, 1970). It is this high altitude water reservoir, all along central Chile, that allowed the lowland flatter areas to be developed and transformed into various types of cultivated fields.

The transformation of the low flatlands has been radical, to the extent that the original vegetation is frequently unknown and still a topic of some debate (see Oberdorfer, 1960; Mann, 1968; Fuentes *et al.*, 1989, 1990). These lowlands are undoubtedly the most 'domesticated' or 'artificial' areas, where introduced plants and animals are most frequent and native species scarcest. These are also the areas where the overwhelming bulk of the Chilean human population lives and has lived since pre-Columbian times (García Vidal, 1982).

The intermediate, mid-altitude mountainous landscapes, have been less influenced by humans (Borde & Gongorá, 1956; Baraona *et al.*, 1961). These latter landscapes are nowadays covered with sclerophyllous shrubs and trees in the southern part (Mooney, 1977) and with drought-deciduous shrubs in the more northern sector (Mooney & Dunn, 1970). Here, in this intermediate landscape, human use is frequently more extensive than on the lowland areas, the original vegetation is more frequent (Fuentes *et al.*, 1984) and introduced plants and animals are less common. In addition, population is scarce and people live nowadays mostly at a subsistence economy level in which small scale dry-farming, goat-grazing, wood-cutting and charcoal-making predominate (Fuentes & Hajek, 1979).

After the Spanish conquest (c. 1550), there was a period, however, in which these low mountainous landscapes were used for extensive dry-farming (Gastó & Contreras, 1979) and to obtain various animal and plant products (Cunnill, 1971). Because of the mountainous relief and the severe winter storms, these land uses have led to heavy soil erosion (Fuentes & Hajek, 1979; Gastó & Contreras, 1979). Consequently, in the North Chico (UNCOD, 1977; Bahre, 1979) and along the coastal ranges south of Valparaiso severe desertification has occurred (Fuentes & Hajek, 1979). There are some recent re-afforestation efforts in these regions, using mainly introduced species of *Atriplex* in the coastal zones of the North Chico and *Pinus radiata* south of Valparaiso (García-Vidal, 1982).

In summary, in central Chile three types of altitudinal landscapes can be distinguished, and they differ in population density, the intensity of their use by humans, and in the relative frequency of invasive species. For all three variables there is a decrease with increasing altitude.

A gradient in invasibility

At a finer scale of resolution some additional patterns can be found. In the second part of this chapter I shall refer mostly to the intermediate altitude landscapes because, as mentioned previously, the flatter areas are nowadays either completely transformed into fields with irrigated agriculture or, on a much smaller scale, are covered by what seems to be a heavily transformed savanna type of vegetation known locally as 'espinal' (Rundel, 1981). (The opposite extreme, the high altitude landscape, seems to have very few introduced animals and plants and is outside the main area with a mediterranean-type climate.)

As with the coarser level of landscape resolution described above, on a finer scale consideration of the relation between human settlements and population density on the one hand, and the distribution of invasive animals and plants on the other, is useful. Basically, invasive species are of two kinds, which are differentiated by how far from human settlements they can be found (Table 4.1). Domesticated species live around settlements or in situations where constant human activity allows their presence. (The entire low flatland landscapes seem to fit into this category.) The other extreme is represented by species that are naturalised and as such have become part of the overall landscape. This might not be a strict dichotomy because there are species that seem to occupy intermediate

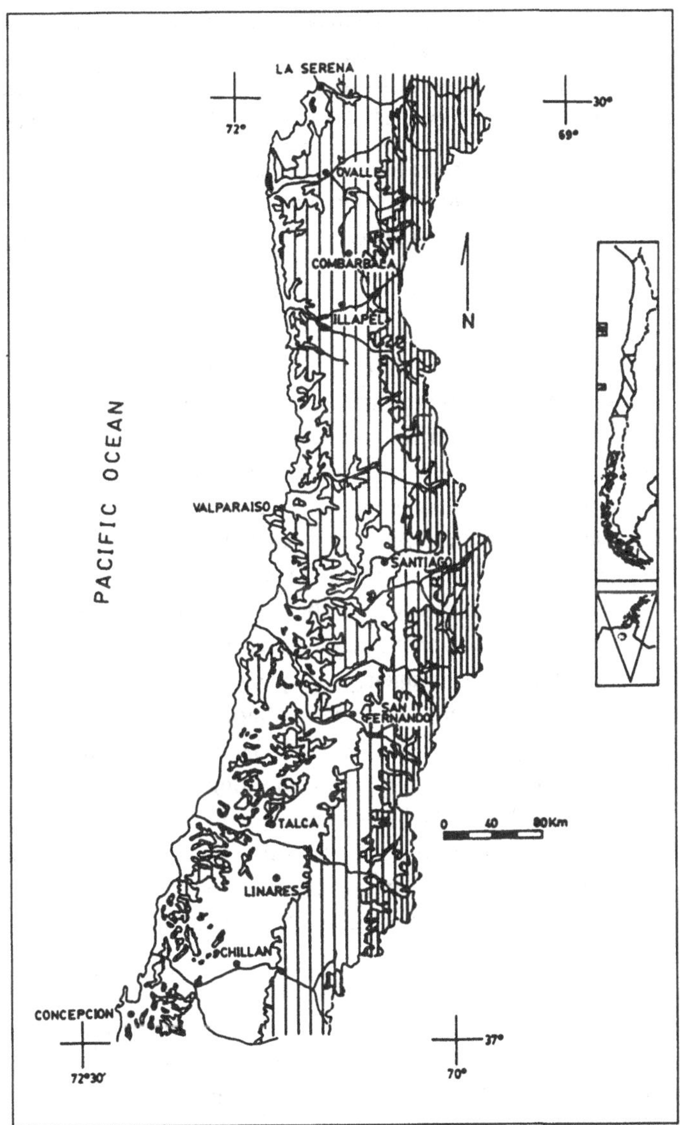

Figure 4.1. Three types of landscapes in Chile. The map shows the spatial distribution of the high Andean (dark-hatched), low mountainous (light-hatched) and flatland (unhatched) landscapes within the geographical bounds of the Chilean area with a true mediterranean-type climate. Larger cities and rivers are also shown. Whereas to the south of Santiago the coastal ranges, Intermediate Depression and Andean ranges can be clearly separated, to the north of Santiago, in the region known as Norte Chico, the two ranges merge into each other (see the text for further explanation and discussion).

Table 4.1. *Examples of four types of introduced species. Generation times are relative to either plants or animals in the same column (see text for discussion)*

Generation time	Domesticated species	Naturalised species
Short	*Eschscholtzia californica*	*Bromus hordeaceus*
	Iridomyrmex humilis	*Drosophila suboscura*
Longer	*Pinus radiata*	*Rubus ulmifolius*(?)
	Capra hircus	*Oryctolagus cuniculus*

positions (e.g. *Rattus rattus*, see Simonetti, 1983) in what could in fact prove to be a gradient of situations. Nevertheless, the dichotomy is heuristically useful, not only because it allows for the classification of invasive species but it also follows what must have been the historical sequence in most cases, and thus raises the question as to what prevents the expansion of introduced but not invasive species or the full release of the so-called intermediate species.

A ranking of the frequency with which invasive species fall into categories seems to indicate that invaders with a short generation time are considerably more frequent than ones with a long generation time (see also Montenegro *et al.*, this volume). The relative frequency of species in the domesticated versus naturalised categories favours the former, although a full analysis cannot be made because the situation with insects is still largely unknown. There is, however, little doubt that naturalised species with a long generation time are fewest, both among plants and animals. In fact, vertebrates known to have become truly naturalised include only *Oryctolagus cuniculus*, *Lepus capensis* and *Callipepla californica*, and among woody plants there are no species that can be said unequivocally to have become naturalised. Perhaps the closest would be *Rubus ulmifolius*.

The ecology of these invasions is still obscure and, with few exceptions, we do not know what enables an introduced organism to successfully naturalise in central Chile. Some species such as *Iridomyrmex humilis*, the Argentine ant, seem to be restricted to the vicinity of human settlements (Figure 4.2), but the reasons for this pattern are unknown. We have statistical evidence (Fuentes, Eissman & Ipinza, unpublished data) that none of the native ant species can by themselves exclude the Argentine ant, but given that there is a negative correlation between its distribution and the sum of abundances of all native ant species (Figure 4.2), it might well be that diffuse competition can partly explain its restriction to the vicinity of human settlements.

Competition with native species does not always restrict the expansion of introduced species, however, as Brncic & Budnik (1987) showed for

Drosophila suboscura. This species seems not only to use resources that native congeners do not utilise, but in addition seems capable of at least 'pushing' itself into a community of native *Drosophila* species. What is different in these two cases? Why is *I. humilis* a restricted, domesticated species, whereas *D. suboscura* has been capable of invading the landscape?

Similar questions can be asked regarding the restriction of *Pinus radiata* and *Eucalyptus* spp. to domesticated situations. Personal observations have shown that near to established plantations and when the land is well cleared (e.g. road cuttings), seedlings of both taxa can be found, but they have not yet expanded into the 'wild' landscape (see also Pryor, this volume). Why has *P. radiata* naturalised in both South Africa and Australia (Groves, 1986), but not in Chile?

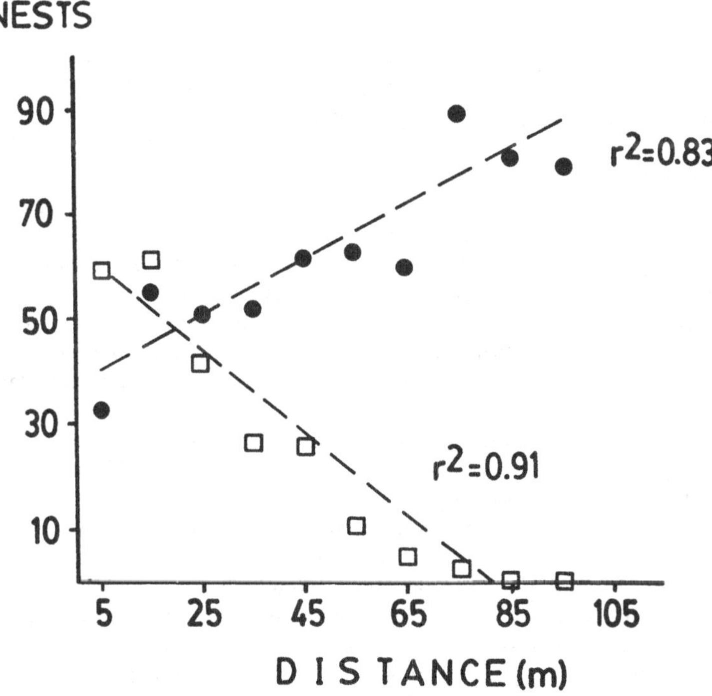

Figure 4.2. Abundance of ants in the vicinity of human settlements. The number of nests of native (dark dots) and number of openings of nests of the Argentine ant (*Iridomyrmex humilis*) (light squares) are shown as a function of distance to isolated houses located in a matorral landscape near Santiago. Results are for 10 × 10 metre quadrats at each distance. Each curve is the sum of nine independent transects. Native species are *Brachymyrmex giardii*, *Camponotus morosus*, *Conomyrme hipocritus*, *Solenopsis gayi* and *Tapinoma antarticum*.

Not all cases of 'release' from the domesticated situation are related to competition, however. Jaksic and Fuentes (this volume) document that at least in the case of the European rabbit (*Oryctolagus cuniculus*), it is low predation pressure in Chile that can largely account for its wide distribution in the landscape. Such an explanation could also be valid for the only other leporid in central Chile, *Lepus capensis*, but this is not likely to be true for the California quail (*Callipepla californica*) that is also ubiquitous and rather secretive. Future research should seek to further clarify the patterns of invasions and the underlying mechanisms.

Acknowledgements. F. M. Jaksic made useful comments that improved an earlier version of the manuscript. Grants FONDECYT 748/1987 and DIUC 93/87 generously supported the work of the author.

References

Aschmann, H. (1973). Distribution and peculiarity of Mediterranean ecosystems. In *Mediterranean-type Ecosystems: Origin and Structure*, ed. F. di Castri & H. A. Mooney, pp. 11–19. Berlin: Springer-Verlag.

Bahre, C. (1979). Destruction of the natural vegetation of north-central Chile. *University of California Publications in Geography*, **23**, 1–116.

Baraona, R., Aranda, X. & Santana, R. (1961). *Valle de Putaendo, Estudio de Estructura Agraria*. Santiago: Editorial Universitaria, Universidad de Chile.

Borde, J. & Gongorá, M. (1956). *Evolutión de la Propiedad Rural en el Valle del Puanque*. Santiago: Editorial Universitaria, Universidad de Chile.

Brncic, D. & Budnik, M. (1987). Some interactions of the colonizing species *Drosophila suboscura* with local *Drosophila* fauna in Chile. *Genética Iberica*, **39**, 249–67.

Cunnill, P. (1971). Factores en la destruccion del paisaje chileno: recoleccion, caza y tala coloniales. *Informaciones Geográficas*, **20**, 235–64.

di Castri, F. (1973). Climatological comparison between Chile and the western coast of North America. In *Mediterranean-type Ecosystems: Origin and Structure*, ed. F. di Castri & H. A. Mooney, pp. 21–36. Berlin: Springer-Verlag.

Fuentes, E. R. (1988). Landscape change in mediterranean-type habitats of Chile: patterns and processes. In *Changing Landscapes: An Ecological Perspective*, ed. I. S. Zonneveld & R. T. T. Forman, pp. 165–90. New York: Springer-Verlag.

Fuentes, E. R., Aviles, R. & Segura, A. (1989). Landscape change under indirect effects of human use: the savanna of central Chile. *Landscape Ecology*. **2**, 73–80.

Fuentes, E. R., Aviles, R. & Segura, A. (1990). The natural vegetation oe a man-transformed landscape: the savanna of central Chile. *Interciencia*, **15**, 293–5.

Fuentes, E. F. & Hajek, E. R. (1979). Patterns of landscape modification in relation to agricultural practice in central Chile. *Environmental Conservation*, **6**, 265–71.

Fuentes, E. R., Otaiza, R. D., Alliende, M. C., Hoffmann, A. J. & Poiani, A. (1984). Shrub clumps of the Chilean matorral vegetation: structure and possible maintenance mechanisms. *Oecologia*, **62**, 405–11.

García-Vidal, H. (1982). *Chile. Esencia y Evolución*. Santiago: Instituto de Estudios Regionales, Universidad de Chile.

Gastó, J. & Contreras, D. (1979). Un caso de desertificatión en el norte de Chile. El ecosistema y su fitocenosis. *Facultad de Agronomía, Universidad de Chile, Santiago, Boletín Técnico 42.*

Groves, R. H. (1986). Invasion of mediterranean ecosystems by weeds. In *Resilience in Mediterranean-type Ecosystems*, ed. B. Dell, A. J. M. Hopkins & B. B. Lamont, pp. 129–45. The Hague: Junk.

Mann, G. (1968). Die ökosysteme Südamerikas. In *Biogeography and Ecology in South America*, ed. E. J. Fittkau, J. Illies, H. Klinge, G. H. Schwabe & H. Sioli, pp. 171–229. The Hague: Junk.

Mooney, H. A. (ed.) (1977). *A Study of Convergent Evolution in Chile and California. Mediterranean-type Ecosystems.* Stroudsburg, Pennsylvania: Dowden, Hutchinson & Ross.

Mooney, H. A. & Dunn, E. (1970). Photosynthetic systems of mediterranean climate shrubs and trees of California and Chile. *American Naturalist*, **104**, 447–53.

Oberdorfer, E. (1960). Pflanzensoziologische Studien in Chile – Ein Vergleich mit Europa. *Flora et Vegetatio Mundi*, **2**, 1–208.

Rundel, P. W. (1981). The matorral zone of central Chile. In *Ecosystems of the World*, vol. 11, *Mediterranean-type Shrublands*, ed. F. di Castri, D. W. Goodall & R. L. Specht, pp. 175–201. Amsterdam: Elsevier.

Simonetti, J. (1983). Occurrence of the black rat (*Rattus rattus*) in central Chile. *Mammalia*, **47**, 131–2.

United Nations Conference on Desertification (UNCOD) (1977). *Case-study on Desertification. Region of Combarbalá (Chile).* Nairobi: United Nations Environment Program.

Weischet, W. (1970). *Chile, Siene Länderkundliche Indivualität und Struktur. Wissenschaftliche Länderkunden*, Band 2/3. Darmstadt: Wissenschaftliche Buchgessellschaft.

5

Historical background of invasions in the mediterranean region of southern Africa

H. J. DEACON

The area of mediterranean-type climate *sensu stricto* at the south-western tip of the African continent is restricted and is centred on Cape Town (Aschmann, 1973). Northwards along the western coast and inland from Cape Town, precipitation decreases sharply over the ranges of the Cape fold mountains. Eastwards towards Port Elizabeth rainfall tends to be bi-seasonal with equinoxial maxima and thus the climate is not strictly mediterranean, although there may be a deficit in summer precipitation.

The physiographic character of the region has a pronounced effect on precipitation, with both the width of the coastal platform and the strike of the bordering ranges of the mountains being of particular importance. It is noteworthy that the syntaxis of the north–south and east–west trending ranges is the Cape Town–Caledon mediterranean-type climate area which has the highest terrain diversity and is the most prominent centre of species richness (Deacon, 1983*a*).

The Capensis floral region as defined by Taylor (1978) corresponds to the Cape fold mountains and the attendant coastal platform. Typical montane and coastal fynbos vegetation is associated with quartz arenite substrates that are poor in bases. Afromontane forest inliers occur in restricted favourable localities on equally nutrient-poor substrates whilst renosterveld shrublands and thicket are found on more base-rich substrates. Phytogeographers have been impressed by the complexity of the vegetation associations in the Capensis region, in their distributions, diversity and the degree of endemism found. The character of Capensis is not simply the product of climates or the occurrence of extensive areas of nutrient-poor substrates, but an interplay between habitat factors including these and historical factors. An outline of the history of the Capensis region or the fynbos landscape is a necessary background to understanding the propensities for invasion.

Invasion is a process, not an event, and thus it is not restricted in time. Viewed from a palaeoecological perspective, the biota of any area should show change over time because of adjustments in the geographic range, evolution and extinction of component taxa. In the normal course of events, it is the taxa with the greatest tolerance of habitat factors, the generalists in the sense of Vrba (1980), that are the potential invaders. In a region of high habitat diversity it can be postulated that selection over time would favour the evolution of specialists from generalist ancestors. This is essentially what has happened in the Capensis region. The fynbos vegetation, for example, is a late specialised vegetation in relative terms and is the product of the specialisation of invaders into an older forest flora. The natural process of range increase, naturalisation and specialisation is slow but it has been promoted considerably by the greater diversity of habitats that evolved in the later Cainozoic. Again, this process has been accelerated in historic times by the jump dispersal of taxa through human intervention. Whether generalists or specialists in their native habitats, taxa from similar geographic contexts function as dedicated generalists when the constraints of co-evolution are lifted in jump dispersal. The historical factor is very obvious when it is noted that the successful plant neo-invasions of both South Africa and Australia are usually southern hemisphere taxa, a fact which underscores the importance of pre-adaptation of potential invaders.

Cainozoic history

The break-up of the Gondwana landmass was initiated in the Jurassic and completed in the early Cretaceous. The biota of southern Africa reflects the early separation of the subcontinent and the biota as a whole is distinct from that of formerly contiguous segments of Gondwana (Keast & Erle, 1972). The exceptions are in older invertebrate and vertebrate amphibian and reptile groups whose distribution can be explained by simple vicariance. Break-up was initiated before the main radiation of the angiosperms and considerably earlier than the radiation of the mammals (Deacon, 1983*b*). It is thus more parsimonious to assume that dispersal, that is invasion processes, rather than vicariance is involved in explaining the origins of the Capensis flora.

It is a gross oversimplification to consider the Capensis flora as a discrete entity with its own evolutionary history. It is rather the product of dispersal events involving taxa of tropical origin, *in situ* evolution and ultimately the local rates of extinctions in a mesic maritime island situation. Essentially its attributes are an accident of history and its designation as a discrete floral entity is an artificial product of criteria of classification. The Capensis flora, as pointed out by Nordenstam (1969), is different in degree (of specialisation) rather than in kind from the pan-tropical flora of Africa and the differences have been overemphasised. The fossil evidence essentially supports the view that the flora of the southern tip of the continent is the product of ongoing dispersal and replacement processes.

At the time of separation of the southern landmasses and the outlining of the southern African continental margin through shearing and rifting, early Cretaceous depositories show the flora to be dominated by gymnospermous bisaccate pollen (Scott, 1976). The invasion of this bisaccate province by angiosperms (Brenner, 1976) was initiated in the mid-Cretaceous (*c.* 100 m.y.) but was significant from the later Cretaceous (*c.* 80 m.y.) onwards with rapid modernisation of the palaeoflora close to the Cretaceous–Cainozoic boundary (*c.* 65 m.y.) (Coetzee *et al.*, 1983). The primary fossil evidence for the evolution of the palaeofloras in this time range comes from palynomorphs preserved in the epiclastic sediments of kimberlitic diatremes; among the youngest of these, the occurrence on the farm 'Banke', in Namaqualand, may date to the earliest Cainozoic and represent a *Podocarpus–Araucaria* gymnospermous forest largely replaced by possible Monimiaceous-dominated vegetation with a *Protea–Erica–Restio* understorey. In the maritime situation of the Capensis region, the fossil record is only from the early Miocene (*c.* 20 m.y.) onwards and this suggests the relatively late or Pliocene (5.2–2.0 m.y.) replacement of essentially a subtropical alliance of forest taxa by dominant sclerophyll vegetation (Coetzee *et al.*, 1983).

Axelrod & Raven (1978) provided a provocative model based on uniformitarian arguments for the origins of the Cape sclerophyllous vegetation. In some details this model is unsatisfactory, in particular in the assumption that the sclerophyllous vegetation was swept into the fynbos landscape in the last 10000 years or so (Taylor, 1980). A more acceptable scenario is that the ancestral sclerophyllous taxa had a long history in southern Africa and within and bordering the present Capensis region these taxa replaced a forest-dominated vegetation in the later Cainozoic as climates became cooler and drier. Within this region of high terrain and edaphic diversity, speciation has been promoted and the maintenance of species richness has been facilitated in mesic habitats. The fynbos vegetation is thus seen as a late Cainozoic and largely substrate-specialised vegetation that has suffered increasing invasion from subtropical generalist taxa as erosion has reduced the extent of duricrust (silcrete) cover on base-rich substrates in the intermontane valleys and coastal forelands.

Modern ecosystems are the product of historical dispersal accidents, substrate modification and long-term climatic change. As summer-dry climates came into being through the more stable position of the South Atlantic high pressure system in the Plio-Pleistocene, the basic parameters of modern ecosystems were established. During the Pleistocene, however, about the last 2 m.y., there have been marked fluctuations in precipitation and temperature. This has required a resilience to the forcing effects of periodic climatic change that has yet to be evaluated fully in respect of plant or animal communities in the Capensis region. What is clear, however, is that fully modern biogeographic distributions were achieved surprisingly recently in about the last 5000 years in the southern Cape, the area of bi-seasonal rainfall. There are indications that the area of strictly summer-dry climate may have contracted in the later Holocene, thereby allowing some invasion of subtropical pioneer species such as *Themeda triandra* and *Acacia karoo* (see Deacon & Lancaster, 1988, for discussion).

Human settlement

Africa has been the centre of hominid evolution and has the longest history of human settlement. Dispersion from the equatorial and tropical latitudes may have occurred in the Lower Pleistocene with human populations present in the Maghreb and Near East upwards of a million years ago. It is difficult to put a fixed time depth on the penetration of human populations into the mediterranean-type climate region of the Cape, but they were certainly present during the Middle Pleistocene as abundant Acheulian occurrences attest (Deacon, 1983*c*). Acheulian settlement in the fynbos region is almost exclusively in the intermontane valleys and coastal forelands. This contrasts with the late Pleistocene and Holocene (the last 125000 years), middle Stone Age and later Stone Age occupations which are widespread and may indicate a different perception of potential food resources in their habitat. In the fynbos region geophytes are prominent as is the underground productivity. Anatomically and behaviourally modern middle and later Stone Age populations with the ability to make fire at will were able to manage their underground food resources using fire. These indigenous populations were concerned with maintaining productivity through controlling the post-fire succession of the natural biota rather than the introduction of 'new' species.

Relative to the Mediterranean Basin region which was a centre for domestication of particular species of plants and animals, the Cape mediterranean-type region was a non-centre. There is no evidence that any indigenous species were domesticated. This does not gainsay a close relationship between people and particular animal and plant species of economic importance at the Cape and it would seem that none of these species was pre-adapted in this context for domestication. Whereas some cereals were potentially ready-made domesticates, geophytes which represented slower renewing resources did not lend themselves to other than the basic level of fire management. Thus whilst full domestication of plants and animals may have been in progress on the southern and eastern fringes of the Mediterranean from 12000 years ago, it is only when northern African domesticates are dispersed the length of the continent in the last 2000 years that evidence for food production based on introduced domesticates is found in southern Africa (Deacon, 1986). In the Cape mediterranean-type climate region, the evidence is solely for the introduction of small stock, primarily sheep, and later cattle after perhaps AD 300. None of the introduced African cultigens, such as sorghum, was suitable for cultivation in the winter-rainfall region and did not penetrate the region. The herders of sheep and cattle at the Cape were the Khoikhoi or Hottentots. It is evident that they had substantial herds at the time of European contact from 1488. By the 1700s internecine strife and introduced diseases had destroyed the social cohesion of the Khoikhoi who were thereafter functionally landless serfs.

The European colonisation of the Cape was a horticultural event because a new and different southern flora became accessible to northern hemisphere gardeners. Although Cape plants were cultivated in European gardens, some earning common household recognition, few have become invasive (but see Le Floc'h, this volume, for some exceptions such as *Oxalis pes-caprae*).

The Cape served as a staging post for the colonisation of Australia during the nineteenth century and through the high level of Australian participation in the Anglo-Boer war there was considerable contact into the present century. It is largely through such contacts in the last century that there was an interchange of plants, but no animals, between these two southern continents. The apparently vacant niche for trees in the fynbos vegetation has meant that woody Australian species have proved efficient colonisers. The converse is true in Australia: it is the more herbaceous plants introduced from the Cape that have become invasive in Australia.

Conclusions

The Cape flora is unique in its degree of endemism and species richness. These features can be explained by the long separation of Africa from the other southern continents and the particular later Cenozoic situation of the Cape mountains and attendant coast as a biogeographic island characterised by mesic maritime conditions where nutrient-poor substrates and terrain diversity favoured species richness under summer-dry climates. The initial invasion in this biome was the proliferation of forest understorey fynbos species in the later Cainozoic. Climatic forcing caused changes in community composition but initial Stone Age human management through fire control was essentially benign. The introduction of livestock herding in the last two thousand years led to more intensive utilisation of the Cape coastal lowlands and intermontane valleys. The grazing and fire regimes adopted may have facilitated the invasion of the Cape region by introductions during the period of European colonisation. In particular, it is the woody species of the Proteaceae and *Acacia* that have been particularly successful as invaders.

Acknowledgements. Palaeoecological research in the mediterranean-type region of the Cape has been funded by the CSIR, Pretoria through the Fynbos Biome Project. Archaeological research on the peopling of this region has been supported by the Human Sciences Research Council, Pretoria, the University of Stellenbosch and the L.S.B. Leakey Foundation, Pasadena.

References

Aschmann, H. (1973). Distribution and peculiarity of mediterranean ecosystems. In *Mediterranean-Type Ecosystems: Origin and Structure*, ed. F. di Castri & H. A. Mooney, pp. 11–19. Berlin: Springer-Verlag.

Axelrod, D. I. & Raven, P. H. (1978). Late Cretaceous and Tertiary vegetation history of Africa. In *Biogeography and Ecology of Southern Africa*, ed. M. J. A. Werger, pp. 77–130. The Hague: Junk.

Brenner, N. (1976). Middle Cretaceous floral provinces and early migrations of angiosperms. In *Origin and Early Evolution of Angiosperms*, ed. C. B. Beck, pp. 23–47. New York: Columbia University Press.

Coetzee, J. A., Scholtz, A. & Deacon, H. J. (1983). Palynological studies and the vegetation history of the fynbos. In *Fynbos Palaeoecology: A Preliminary Synthesis*, ed. H. J. Deacon, Q. B. Hendey & J. J. N. Lambrechts, pp. 156–73. Pretoria: South African National Scientific Progress Report No. 75, CSIR.

Deacon, H. J. (1983*a*). An introduction to the fynbos region, timescales and palaeoenvironments. In *Fynbos Palaeoecology: A Preliminary Synthesis*, ed. H. J. Deacon, Q. B. Hendey & J. J. N. Lambrechts, pp. 1–20. Pretoria: South African National Scientific Progress Report No. 75, CSIR.

Deacon, H. J. (1983*b*). The comparative evolution of mediterranean-type ecosystems: a southern perspective. In *Mediterranean-type Ecosystems: The Role of Nutrients*, ed. F. J. Kruger, D. T. Mitchell & J. U. M. Jarvis, pp. 3–40. Berlin: Springer-Verlag.

Deacon, H. J. (1983*c*). The peopling of the fynbos region. In *Fynbos Palaeoecology: A Preliminary Synthesis*, ed. H. J. Deacon, Q. B. Hendey & J. J. N. Lambrechts, pp. 183–204. Pretoria: South African National Scientific Progress Report No. 75, CSIR.

Deacon, J. (1986). Human settlement in South Africa and archaeological evidence for alien plants and animals. In *The Ecology and Management of Biological Invasions in Southern Africa*, ed. I. A. W. Macdonald, F. J. Kruger & A. A. Ferrar, pp. 3–19. Cape Town: Oxford University Press.

Deacon, J. & Lancaster, N. (1988). *Late Quaternary Palaeoenvironments of Southern Africa*. Cape Town: Oxford University Press.

Keast, A. & Erle, F. C. (ed.) (1972). *Evolution, Mammals and Southern Continents*. Albany: State University of New York Press.

Nordenstam, B. (1969). Phytogeography of the genus *Euryops* (Compositae). *Opera Botanica*, **23**, 1–77.

Scott, L. (1976). Palynology of Lower Cretaceous deposits from the Algoa Basin (Republic of South Africa). *Pollen et Spores*, **18**, 563–609.

Taylor, H. C. (1978). Capensis. In *Biogeography and Ecology of Southern Africa*, ed. M. J. A. Werger, pp. 171–229. The Hague: Junk.

Taylor, H. C. (1980). Phytogeography of fynbos. *Bothalia*, **13**, 231–5.

Vrba, E. S. (1980). Evolution, species and fossils: how does life evolve? *South African Journal of Science*, **76**, 61–84.

6

A short history of biological invasions of Australia

R. H. GROVES

Biogeographers have been pre-occupied with the concept of invasions since they began writing about the Australian biota. This theme has perhaps been emphasised so strongly because Australia has been an island continent since it broke from the Gondwanan landmass fifty or more million years ago. The invasion theory has been especially prevalent for Australian plants. In this chapter I shall outline briefly the history of this dominant theme in Australian phytogeography; Barlow (1981) wrote more comprehensively on the same subject and I shall rely heavily on his material in what follows.

Hooker (1860) recognised four elements in the Australian flora:

1. An Australian or autochthonous element which consisted of plants endemic to the Australian continent
2. An Antarctic element consisting of plants having close taxonomic affinities with the flora of other southern land masses
3. An Indo-Malayan element consisting of plants with tropical affinities
4. A cosmopolitan element consisting mainly of herbaceous plants found in the floras of many other regions.

The concept of invasion and colonisation of the Australian landmass by different floras, perhaps at different times, was inherent in Hooker's subdivisions. His basic ideas prevailed for about 100 years (see e.g. Burbidge, 1960). There were few dissenters from Hooker's concepts until the results of plate tectonic studies became more widely recognised in the 1960s. After all, the main weakness of the invasion theory was the lack of any geographical evidence for the required land bridges, especially that between Australia and South-east Asia.

Based on a knowledge of plate tectonics, it is unlikely that collision between Gondwanan continental plates would evoke an abrupt biological response. Rather, the response would be more general. Barlow (1981) has claimed credibly

Table 6.1. *The different groupings in the Australian flora as proposed by Hooker (1860) and Nelson (1981)*

<table>
<tr><th>Element</th><th>Sub-element</th><th>Hookerian element</th></tr>
<tr><td rowspan="3">Intrusive (recent)</td><td>Cosmopolitan</td><td>(Cosmopolitan)</td></tr>
<tr><td>Tropical</td><td>Indo-Malayan</td></tr>
<tr><td>Neoaustral</td><td rowspan="2">Antarctic</td></tr>
<tr><td rowspan="2">Gondwanic (ancient)</td><td>Relict</td></tr>
<tr><td>Autochthonous</td><td>Australian or Autochthonous</td></tr>
</table>

Source: Nelson, 1981 (Table 4).

that exchange between the floras on the approaching Gondwanan plates would commence with colonisation by the more dispersible taxa while the landmasses were still well apart. Competition and adaptation would thus occur progressively even before contact, from which follows the corollary that a simple 'invasion' of one biota into the area of another would be unlikely anyway.

Nelson (1981) regrouped the elements in the present Australian flora into two distinct groups (Table 6.1) which better reflect tectonic history, and especially the apparent absence of land connections between about 50 m.y. BP and the establishment of closer connections between Australia and South-east Asia by way of New Guinea after about 15 m.y. BP. Nelson recognised an ancient Gondwanan element and a recent intrusive element, the latter having evolved from taxa that have entered Australia since the late Miocene.

Before mentioning more recent changes in the Australian biota, particularly those associated with human colonisation of the region, I wish now to consider briefly the evidence for the origin of a mediterranean-climate flora in southern Australia, i.e. those autochthonous taxa which today we regard as typically Australian and characterised by scleromorphy and a high level of endemism. Johnson & Briggs (1981) reviewed available knowledge on the biogeography of three 'old' plant families of the Southern Hemisphere – viz. Myrtaceae, Proteaceae and Restionaceae. For these three predominantly scleromorphic families, they claimed that the principal environmental factor associated with scleromorphy was adaptation to a deficiency of nutrients, rather than an adaptation to a mediterranean-type climate. During the course of the Tertiary, these scleromorphs differentiated repeatedly in response to a range of physiographic, pyric, biotic and climatic factors but the superimposition of a winter wet–

summer dry climate came much later. In fact, Johnson & Briggs quote Axelrod as suggesting that no area of mediterranean climate dates from before the Quaternary. Over aeons, extensive tracts of scleromorphic woodlands and shrublands developed in what are now the mediterranean-climate regions of southern Australia. These vegetation types, and especially the shrublands, physiognomically resembled the shrublands of Mediterranean Europe. And it was from Mediterranean Europe, either directly or indirectly, that many of the plants and animals were to be brought to Australia which in 200 years have had such drastic ecological effects on the biota of southern Australia.

Although non-Europeans brought a few plants to the tropical shores of northern Australia from regions such as Macassar (Macknight, 1976), there is no evidence for the deliberate importation of plants to the south prior to 1788 when Europeans began to settle eastern and southern Australia. Aborigines or Asian seafarers brought the dingo (*Canis familiaris dingo*) to northern Australia about 3500 years ago, and perhaps more than once, from whence it has spread to the south (Corbett, 1985). For 200 years many plant and animal species are known to have entered southern Australia. Data for plants becoming naturalised in four Australian states show the rate of increase of from four to six species a year to be almost constant over the last 100 years (Specht, 1981; Kloot, this volume). When the efforts at deliberate plant introduction of the early settlers and the persistent efforts of the acclimatisation societies in Australian states such as Victoria (Gillbank, 1986) are considered, it is surprising that the rate of naturalisation has been so low (Groves, 1986*a*). This rate of increase is apparently low partly because of federal quarantine legislation enacted first in 1908 and the efforts of State legislatures for over 100 years to enact vermin and noxious plant control legislation, the latter dating from Victoria's 'An Act to make provision for the eradication of certain thistle plants and the Bathurst Burr' passed on 19 March 1856 (Parsons, 1973).

Over the 200 years of European settlement of southern Australia many plants and animals have arrived. Some have established but the majority has failed to do so. Of those that have established, a small proportion has become invasive. Groves (1986*b*) estimated that for every 100 plants introduced, perhaps 10 have established and become naturalised, and of these 10, perhaps 1 or 2 have become noxious, in that they interfere with human activities in some way. Whether a similar situation applies to animals is unknown (but see Table 21.1 for evidence that the proportion of mammals successfully introduced to southern Australia, compared with the failed introductions, may be higher than for plants). Kloot (this volume), Redhead *et al.* (this volume) and Long & Mawson (this volume) subsequently relate some of the historical details associated with the invasive biota of southern Australia, as does the earlier volume of Groves & Burdon

(1986). Thomson *et al.* (1988) reviewed specifically faunal introductions to Australia, including those of fish and invertebrates.

In conclusion I repeat the summary of Barlow (1981) for plants – viz. 'The Australian flora, as we see it today, thus tells the story of a hundred million years of history of Australia as a southern land mass. The alien plants which have become naturalised so widely since European settlement are legitimately included in the Australian flora. Their impact on the flora, however great, has nevertheless occurred almost instantaneously in terms of the long history of colonisation of the continent by plants' (Barlow, 1981, p. 66). An equivalent time-scale applies to those animals now invasive in southern Australia.

References

Barlow, B. A. (1981). The Australian flora: its origin and evolution. In *Flora of Australia*, vol. 1, ed. A. S. George, pp. 25–75. Canberra: Australian Government Publishing Service.

Burbidge, N. T. (1960). The phytogeography of the Australian region. *Australian Journal of Botany*, **8**, 75–212.

Corbett, L. K. (1985). Morphological comparisons of Australian and Thai dingoes: a reappraisal of dingo status, distribution and ancestry. *Proceedings of the Ecological Society of Australia*, **13**, 277–91.

Gillbank, L. (1986). The origins of the Acclimatisation Society of Victoria: practical science in the wake of the gold rush. *Historical Records of Australian Science*, **6**, 359–74.

Groves, R. H. (1986*a*). Plant invasions of Australia: an overview. In *Ecology of Biological Invasions: An Australian Perspective*, ed. R. H. Groves & J. J. Burdon, pp. 137–49. Canberra: Australian Academy of Science.

Groves, R. H. (1986*b*). Invasion of mediterranean ecosystems by weeds. In *Resilience in Mediterranean-type Ecosystems*, ed. B. Dell, A. J. M. Hopkins & B. B. Lamont, pp. 129–45. The Hague: Junk.

Groves, R. H. & Burdon, J. J. (ed.) (1986). *Ecology of Biological Invasions: An Australian Perspective*. Canberra: Australian Academy of Science.

Hooker, J. D. (1860). Introductory essay. In *The Botany of the Antarctic Voyage of H. M. Discovery Ships 'Erebus' and 'Terror', in the Years 1839–1843*, vol. III, *Flora Tasmaniae*. London: Lovell Reeve.

Johnson, L. A. S. & Briggs, B. G. (1981). Three old southern families – Myrtaceae, Proteaceae and Restionaceae. In *Ecological Biogeography of Australia*, vol. 1, ed. A. Keast, pp. 427–69. The Hague: Junk.

Macknight, C. C. (1976). *The Voyage to Marege*. Melbourne: Melbourne University Press.

Nelson, E. C. (1981). Phytogeography of southern Australia. In *Ecological Biogeography of Australia*, vol. 1, ed. A. Keast, pp. 735–59. The Hague: Junk.

Parsons, W. T. (1973). *Noxious Weeds of Victoria*. Melbourne: Inkata Press.

Specht, R. L. (1981). Major vegetation formations in Australia. In *Ecological Biogeography of Australia*, vol. 1, ed. A. Keast, pp. 165–297. The Hague: Junk.

Thomson, J. M., Long, J. L. & Horton, D. R. (1988). Human exploitation of, and introductions to, the Australian fauna. In *Fauna of Australia*, vol. 1A, ed. G. R. Dyne & D. W. Walton, pp. 227–49. Canberra: Australian Government Publishing Service.

Part III Biogeography of taxa

Part IIIa Higher plants

Some plants of the Mediterranean Basin have been introduced, either deliberately or accidentally, to the four other regions of the world with mediterranean climate, where many of them have flourished and some have become weeds. Other plants originating in one of these other regions have been introduced to the Mediterranean Basin where they have become established. Thus there have been exchanges in the floras of all mediterranean regions. The following five chapters document the magnitude of these exchanges in higher plants. The exchanges have tended to be concentrated in a few large plant families, of which the grasses, composites and legumes predominate – mainly, no doubt, because of the size and usefulness of these three families for agriculture and horticulture.

A further five chapters describe ecological aspects of some well-documented plant invasions. Mediterranean-climate regions are characterised not only by a certain climatic regime and by the long association with human disturbances described in the previous section, but also by nutrient-deficient soils and by frequent fires. This section concludes, appropriately, with a chapter that challenges the emphasis on a particular climate as the main determinant of invasiveness in mediterranean regions. Other ecological factors interact with a mediterranean climate to influence which plants have become permanent components of the flora of all mediterranean regions. Chance too is a factor in the evolution of all five floras. Many more exchanges are likely in the future, even though the rate of increase in naturalisation may seem to be reducing in some regions of mediterranean climate such as California.

7

Invasive plants of the Mediterranean Basin

E. LE FLOC'H

The natural vegetation of the Mediterranean Basin is disappearing rapidly. From cultivated areas, it disappeared long ago; in the rest of the Basin it is now threatened because of increasing human demand for space, food, energy and urban settlement. Native vegetation is also disappearing because of incidents related to the misuse of land or because of natural disasters such as floods and forest fires.

The various Mediterranean environments, whether cultivated or not, have been affected to varying extents, so that some situations may be more accurately described as being 'new' environments. The human desire to expand cultivation of specific plants (mainly introduced) and to increase their yields, results in disturbance and a severe change in the utilised environments because of the introduction of new species, their cultivation in glasshouses and the use of irrigation and/or herbicides. In such environments, which are exploited presently or have been used formerly, some plants can either establish or increase their density to the extent that they can be termed invasive. Plants which become invasive in particular environments within the region from which they originate are termed 'apophytes'. Plants invading after their deliberate introduction (for afforestation or for cropping), or after being introduced inadvertently (as contaminated seeds, ship ballast or in wool), are termed 'anthropophytes', after Quézel *et al.* (1990). Crop invaders, which eventually cause serious economic problems in crop management systems or by decreasing yields, I shall call 'weeds'; this class of invasive plant is often the most aggressive.

Date of introduction and invasion

This chapter will survey those plants which have invaded the Mediterranean Basin recently and not those which invaded in palaeo-ecological times (Pons *et al.*, 1990; see also Naveh & Vernet, this volume). Information on native plants

Table 7.1. *Plants which have become naturalised in the Mediterranean Basin but which are not necessarily invasive*

Species	Region and date of introduction	Reference
Agave americana	Spain, 1492	Sauvaigo (1899)
Oenothera biennis	Italy, 16th century	Godron (1854)
Helianthus annuus	Spain, 16th century	Godron (1854)
Opuntia ficus-barbarica	Europe, 1731	Sauvaigo (1899)
Ailanthus altissima	France, about 1751	Sauvaigo (1899)
Phytolacca dioica	Europe, 1768	Sauvaigo (1899)
Buddleja globosa	France, 1774	Sauvaigo (1899)
Paspalum paspalodes	Cultivated at Bordeaux (France) in 1802	Godron (1853)
Acacia dealbata	Europe, 1924	Sauvaigo (1899)
A. karoo	Europe, 1827	Sauvaigo (1899)
Nicotiana glauca	Europe, 1827	Sauvaigo (1899)
Melia azedarach	Europe, before 1843	Sauvaigo (1899)
Eucalyptus globulus	As seeds in France, 1854; sown, 1860; introduced to Algeria, 1861	Sauvaigo (1899)
Acacia farnesiana	Europe, 1856	Sauvaigo (1899)
Atriplex semibaccata	Tunisia, 1895	Trabut (1904)
Blakiella inflata	Algeria, 1895	Trabut (1904)
Atriplex muelleri	Israel, 1974	Dafni & Heller (1982)
Heterotheca subaxillaris	Israel, 1975	Dafni & Heller (1982)

is often limited or absent; more information is available for cultivated plants or, more generally, for those introduced deliberately (see Sauvaigo, 1899; Trabut, 1904; Picot, 1928; Chevalier, 1932, 1939; van Zeist, 1980; Danin, 1982; Le Floc'h *et al.*, 1990). To estimate the date of introduction of the oldest invaders, various points need to be considered: namely, their origin, the results of archaeological excavations (Kosinova, 1974), the crop type in which the species can be found, as well as chronicles of botanical gardens, etc.

Some plants, deliberately introduced as food, for horticulture or as crops, have become naturalised but have seldom become invasive (Table 7.1). The dates of plant introductions are seldom known for the Mediterranean Basin; we can, however, quote Dafni & Heller (1980, 1982) who give the dates of first mention for those plants introduced to Israel during the twentieth century. Using the methods outlined above, details are given in Table 7.2 of the inadvertent introduction of some other plants.

Relatively little information is available on the rate at which plants spread in

Table 7.2. *Region and date of introduction, or first mention, of some plants now naturalised in the Mediterranean Basin*

Species	Region and date of introduction	Reference
Asperula arvensis	Introduced to western Mediterranean with grain	Montegut (1984)
Adonis spp.	Introduced to western Mediterranean with grain	Montegut (1984)
Chenopodium ambrosioides	Introduced in about the 17th century	Bauhin (in Thellung, 1910)
Amaranthus retroflexus	Introduced in about the 16th century	Thellung (1910)
Conyza canadensis	Appeared in Europe in 1655	Bryner (in Thellung, 1910)
Crepis sancta	Appeared in Europe in 1763	Gonan (in Thellung, 1910)
Galinsoga parviflora	Appeared in Europe in 1785	Dizerbo &Nehou (1952)
Oxalis pes-caprae	Introduced to Sicily in 1796	Pignatti (1982)
Coronopus didymus	Introduced to France about 1800	Lamic (in Thellung, 1910)
Ludwigia uruguayensis	Introduced to France about 1840	Gillot (1900)
Solanum elaegnifolium	Introduced to Montpellier about 1855	Thellung (1910)
Senecio inaequidens	Appeared in France in 1935	Jovet & Bosserdet (1962)
Sesuvium verrucosum	Introduced to Bahrain in 1965	Verdcourt (1985)
Aster subulatus	Introduced to Israel in 1978	Dafni & Heller (1980)

the Mediterranean Basin. Rate of spread is usually related to the intensity of road use, to the migratory movements of people and their herds, to bird migration and to the distance between locations where the plant can establish. For the Mediterranean Basin, *Solanum elaegnifolium* is among the best documented of species; it was reported as early as 1855 in the Botanical Garden at Montpellier, and subsequently in Greece. The plant is now also found in Spain, Italy and Yugoslavia (Guillerm *et al.*, 1990). The spread of this species around Tadla (Morocco) is well documented in studies of Tanji *et al.* (1985) and of Bouhache & Tanji (1985). The spread of the plant to these several locations in the Mediterranean Basin did not occur necessarily from one single point but it probably represents the effects of several successive introductions.

Heliotropium curassavicum was first found by Desfontaines (1798–1800) at one site near Sousse (Tunisia) in 1774 and was still there in 1921 (Burollet & Boitel, 1921). The plant was present at five sites in 1955 (Labbe, 1955) and the species is now common on all saline soils around most oases (Le Floc'h *et al.*, 1990). Some information is also available on *Oxalis pes-caprae*, which was introduced into Sicily in 1796 and was reported subsequently in Malta. The plant was quite probably introduced on several occasions as it has been known for some time to be easily propagated (Chabrolin, 1934).

Although the formulation of conclusions from so few well-documented cases is difficult, it nevertheless seems that a plant may stay unnoticed for many years at a site to which it was introduced and then suddenly spread later. Plants may also be introduced concurrently to several locations within such a vast region as the Mediterranean Basin.

Ecological status and invasion

The ecological status of a plant is defined here as the types of environment (cultivated, ruderal, grazing land, etc.) which that plant will colonise most frequently. The invasive plants which grow in traditionally managed crops are generally the most widespread. This is especially true for weeds associated with grain crops (e.g. *Agrostemma githago* and *Centaurea cyanus*). Other invasive plants may also be very common in cultivated fields, e.g. *Aster squamatus*, *Conyza bonariensis* and *C. canadensis* (with the latter less widespread in the eastern part of the Mediterranean Basin; L. Boulos, personal communication). *Oxalis pes-caprae* may be distributed as a contaminant of nursery stock for planting in orchards and thus may become weedy. Invasive plants with more specific ecological needs are obviously more localised. One example of this latter is *Euphorbia nutans* which is weakly invasive in most localities except in the cotton fields of the valley of Guadalquivir (Spain), where it is very aggressive. This species is resistant to the particular herbicides used commonly in cotton cropping. In addition, *Euphorbia nutans* has a growth cycle similar to that of the cotton plant (Hernandez-Bermejo *et al.*, 1984).

Natural moisture in some locations, and also irrigation, favour the spread of plants such as *Paspalum* spp., *Polygonum persicaria* and *Equisetum telmateia* (Guillerm & Maillet, 1982). The cultivation of rice also allowed for a particular group of invaders to develop, whereas the specific area of each of them is normally discontinuous; instances are *Leersia oryzoides* in the Camargue (Sustina, 1980), *Elatine triandra* which has been observed recently in the rice fields of Sardinia (Arrigoni, 1983) and *Echinochloa oryzoides* in both the Camargue and Algeria since 1950 (Dubuis & Faurel, 1957). Some parasitic plants can also be invasive, with the host being either another invasive plant (as in the case of *Cuscuta campestris* which lives on *Amaranthus retroflexus*, *A. blitoides*, *A. lividus* or *Chenopodium album*; Caussanel *et al.*, 1984) or a crop species (as in the cases of *Cuscuta* spp. on *Medicago sativa* and *Thesium humile* on various grain crops).

Plants invading aquatic environments are similar to weeds in irrigated situations and can be either submerged in channels or emergent on channel banks. Khattab & El-Gharably (1984) mention as aquatic weeds of Egypt the submerged invaders *Eichhornia crassipes* and *Lemna gibba*. As weeds living on the banks they mention the following species: *Echinochloa stagineum*, *Typha domingensis* and *Phragmites australis*. *Eichhornia crassipes* has become invasive in Portugal since 1976 (Guerreiro, in Duarte *et al.*, 1984).

Ruderal environments (roadsides, ruins, etc.) are often nitrate-enriched, a situation which allows for the establishment of allogenous plants such as *Ailanthus altissima*, *Nicotiana glauca* and *Xanthium spinosum*. Local or naturalised species can also encounter an appropriate 'new' environment for successful establishment, as for instance with *Foeniculum vulgare*, *Dittrichia viscosa*, *Peganum harmala* and *Withania sommifera* (Le Floc'h *et al.*, 1990). Since such ruderal environments are expanding, these plants can more easily become invasive.

The steppes of northern Africa and the eastern Mediterranean have long been subject to high grazing pressure, leading to the endangered status of some plants and to the spread of others which are less palatable for livestock. The latter group are either spiny (e.g. *Astragalus armatus*, *Atractylis serratuloides*, *Calicotome villosa*), poisonous (e.g. *Calotropis procera*, *Hypericum triquetrifolium*, *Solanum nigrum*, *Thapsis garganica*) or unpalatable (e.g. *Cleome ambylocarpa*. *Diplotaxis harra*, *Hammada scoparia*, *Hertia cheirifolia*) (Le Floc'h *et al.*, 1990). Such steppe environments cover large areas. The northern African steppes have undergone intense floristic changes during prehistoric times, when plants originating from the eastern Mediterranean became naturalised in this region. But, remarkably enough, the present flora does not include many species introduced recently.

Like the pastoral zones, the forests of the Mediterranean Basin are confined to the less productive areas, such as rock outcrops, steep slopes and land at high elevation. For a long period these environments have been under sustained pressure from activities such as forest clearing, fires and/or overgrazing. Plants able to invade these environments are mostly native or species naturalised long ago. With forest fires, annual or biennial species, which establish first, are soon replaced by more competitive, indigenous woody plants. If burning is repetitive, those species which are favoured by fire or tolerate it can establish and impose their own particular physiognomies on the resulting vegetation, as for instance with *Quercus coccifera*, *Q. ilex*, *Erica arborea*, *Asphodelus ramosus* or species of *Cistus*. Reafforestation may introduce new species, or reintroduce trees which had disappeared long ago. The introduced or reintroduced trees can sometimes establish effectively to the extent that they become naturalised, or even expand and replace other species without becoming really invasive. Quézel *et al.* (1990)

include in this category *Pinus brutia* in southern Anatolia, *P. nigra*, *Cedrus atlantica* and various species of *Abies*. *Eucalyptus* × *trabutii* (described by Tutin *et al.* (1968) as a hybrid of *Eucalyptus botryoides* and *E. camaldulensis*) is a taxon from Algeria to which both parents had been introduced from Australia around 1860. *Eucalyptus camaldulensis* is, according to Greuter *et al.* (1985), a naturalised tree in the eastern Aegean Islands and in Anatolia, Turkey. Several *Acacia* species were introduced in the nineteenth century as ornamental plants, for afforestation or for sand dune stabilisation, and have subsequently become naturalised (Greuter *et al.*, 1985) in many parts of the western Mediterranean Basin. The species of *Acacia* include *A. cyanophylla* (Libya), *A. cyclops* (Portugal), *A. dealbata* (France and Portugal), *A. farnesiana* (Morocco), *A. ligulata* (Cyprus), *A. longifolia* (France), *A. melanoxylon* (Algeria, France and Portugal) and *A. retinoides* (France). Parde (1927) considered most *Acacia* species as being naturalised in Portugal, including *A. melanoxylon*, *A. decurrens*, *A. molissima* (= *A. mearnsii*) and *A. pycnantha*.

Biology and invasion

Plants with a short growing period are obviously favoured in repeatedly disturbed environments such as cultivated areas. The rate of success of invasion of annual and biennial weeds, already related to the short duration of their life cycle, is in addition a consequence of the timing of the various growth phases and of crop management. When weeds are destroyed by ploughing or harvesting before their seeds have been released, their invasion is hampered. This group of species can be classified according to the main types of cultivation, the duration of rotations, etc. (Guillerm, this volume). The major subdivisions are:

1. Winter annual crops (grains) invaded by species such as *Avena sterilis*, *Cirsium arvense*, *Lolium multiflorum* and *Papaver rhoeas*. These species are widely distributed around the Mediterranean Basin. They are predominantly annual weeds (Le Maignan, 1981).
2. Summer annual crops (maize, sunflower, cotton), which are more recent and may be invaded by *Amaranthus albus*, *Chenopodium album*, etc.
3. Perennial crops such as vines with different seasonal floras (Guillerm & Maillet, 1982), oranges or other orchard crops characteristically invaded by species such as *Convolvulus arvensis* and *Anagallis arvensis*.

The use of herbicides has sometimes resulted in a selection of species or infraspecific taxa which are resistant to or tolerant of a particular herbicide and whose infestation is the less restricted as other potentially competing weeds are controlled. Much research is presently being devoted to these invaders which are favoured by application of chemicals, e.g. *Euphorbia nutans* (Hernandez-Bermejo *et al.*, 1984) and *Sinapis arvensis* (Jauzein, 1980).

Hemicryptophytic weeds, such as *Paspalum paspalodes*, are commonest in summer crops (Le Maignan, 1981) and in perennial crops. Geophytes such as *Cynodon dactylon*, *Cyperus esculentus*, *C. rotundus* and *Oxalis pes-caprae*, however few, also occur mainly in summer and perennial crops. Ploughing often promotes the dominance of plants with bulbs, stolons or rhizomes, which can be very damaging. Chamaephytes and phanerophytes are rarely weeds in these cropping systems.

This grouping, valid for crop weeds, is roughly the same for other types of land use, although therophytes are not always so prominent among the various biological categories. For example, the percentage of hemicryptophytes is increasing in the ruderal environments of North Africa and becoming significant (Le Floc'h *et al.*, 1990). Phanerophytes and nanophanerophytes mainly invade cool and wet ruderal locations (along rivers, for instance); examples are *Amorpha fruticosa* in the Rhône delta as well as *Ailanthus altissima*, *Fraxinus ornus* and *F. americana*.

The steppes of northern Africa and the eastern Mediterranean have few invaders, although chamaephytes and hemicryptophytes are relatively numerous; phanerophytes remain rare.

Succulent plants (such as *Agave americana*, *Aloe* spp., *Opuntia* spp. and *Carpobrotus edulis*) represent a special case of introduced invaders.

Phytogeography and invasion

This aspect of plant invasions has been very much disputed and commented on, at least in the case of the most ancient naturalised plants and species which are now cosmopolitan. The problem becomes less for recently introduced plants which have escaped from fields.

The fact that, for early agriculture within the Mediterranean Basin, plants moved from east to west is well known. Other species originating from different parts of the same Mediterranean isoclimatic zone were mostly introduced during the nineteenth century. There were many introductions of cultivated plants, especially those from Australia and New Zealand, e.g. species of *Eucalyptus*, *Acacia*, *Atriplex*, *Casuarina*, *Myoporum* and *Blakiella inflata*. These deliberate introductions were accompanied by species introduced inadvertently, such as *Solanum laciniatum*. Some of the present invaders came from the Cape Province of South Africa, e.g. *Carpobrotus edulis*, *Oxalis* spp. and *Cotula coronopifolia*. Few species, however, came from the mediterranean-climate region of western America.

The Mediterranean Basin flora also includes many temperate plants (Quézel *et al.*, 1990). These temperate species came either from Europe (*Ailanthus altissima, Buddleja davidii, Artemisia verlotiorum*) or, more often, from North America (species of *Ambrosia, Chenopodium, Conyza, Cuscuta, Oenothera* and *Aster squamatus, Amorpha fruticosa, Argemone mexicana, Datura stramonium, Galinsoga parviflora* and *Phytolacca americana*). A few species also came from the temperate part of South America, viz. *Amaranthus deflexus, Conyza bonariensis, Solanum* spp. and *Salpichroa origanifolia*.

Invaders from tropical climates into the Mediterranean Basin mainly originate from tropical Africa; among them is a large number of grasses which grow with summer annual crops, i.e. *Chloris gayana* and species of *Paspalum, Panicum, Pennisetum* and *Sorghum*. Plants from the central American tropical zone are fewer and include *Amaranthus hybridus, A. viridis, Asclepias curassavica, Datura metel, Lantana camara* and *Nicotiana glauca*.

Discussion and conclusions

The rate of success of invasion is usually more easily measured for introduced plants than for native ones. In a given situation, the number of species competing for space is relatively large; if the environment is intact, the already established species are likely to prevent the establishment of new plants. When the environment has been disturbed, competition operates between all species which are represented by their propagules. If the inputs are repeated regularly, naturalisation or even infestation may occur during favourable years. The best known example is the *Flora Juvenalis* in Montpellier (France). The digging of the Lez canal in about 1686 enabled the wool stores to operate at Port Juvenal from 1700, whereby imported wool was washed in hot water and left to dry out on nearby meadows. The diverse origin of the wool (which contained seeds), as well as the procedure used for cleaning, boosted the chances of establishment of a rich introduced flora. The latter numbered 13 species in 1813 (De Candolle, in Rioux & Quézel, 1950), 386 species in 1853 (Godron, 1853) and as many as 458 species in 1859 (Cosson, 1860). The storing and washing of wool were discontinued after 1880. The introduced flora subsequently regressed, so that, between 1905 and 1910, only some ten species remained (Thellung, 1908–10). Rioux & Quézel (1950) could find only six species remaining in 1950 and the plants were confined to one specific ruderal community. One reason for this impoverishment of the flora could be that germination of the newly introduced seeds may have been favoured by processing of the wool in hot water, whereas the seeds which were produced subsequently in the nearby meadows were subject to different conditions for germination. Among the species which established at Port Juvenal and seem to have naturalised were steppe plants which had no potential to expand. One species, however, had a very different fortune, namely *Ludwigia uruguayensis*. The plant was first reported by Godron in 1854; in 1857, Touchy mentioned that *L. uruguayensis*, a North American plant, was spreading along the Lez river so that it had become a nuisance to navigation, but that it did not produce seeds in spite of its plentiful flowering in this location. *Ludwigia uruguayensis* was later described by Flahault (1899) and Gillot (1900) as naturalised and then mentioned by Molinier & Tallon (1974) as occurring in southern France; its encroachment is continuing, especially in the Camargue.

As for *Ludwigia uruguayensis*, many plants which are now termed invasive had a rather extended latent phase from the time of their introduction to the time when they became invasive. This is borne out by some further examples. For instance, *Euphorbia nutans*, from North America, was reported in the Savoie (part of the French Alps) by Jovet (1949) and is indicated by Tutin *et al.* (1968) as occurring on disturbed terrain, locally naturalised in south-central and southern Europe, and occasional elsewhere. Nevertheless, and probably after 1978 (Hernandez-Bermejo *et al.*, 1984), this species (whose tolerance to herbicides has already been mentioned) has become aggressively invasive in the cotton fields of the Guadalquivir Valley (Spain), where cotton is irrigated. *Diplotaxis erucoides* was once considered rare in the Languedoc province of France (Loret & Barrandon, 1886).The plant is now abundant in the vineyards of the same region, its spread being related to herbicidal action on its competitors and to the short duration of its growing cycle so that the seeds are scattered even when ploughing is frequent. Around Mazamet (France), *Senecio inaequidens* has been reported repeatedly since 1936 (Galavielle & Blanchet, 1939; Guillerm *et al.*, 1990). Its spread towards the Mediterranean was only obvious, however, after 1970; its present dissemination is very fast, especially along roads and river valleys. In newly occupied sites, *S. inaequidens* is apparently a ruderal or a fallow species which thereafter invades areas of vineyards or any other crop with a comparable growing cycle. These three examples do not provide a solution to the problem of the so-called 'latent phase' of invasion, but they do suggest that some introduced species can, with changing agricultural practices (e.g. use of herbicides, irrigation practice, etc.), find conditions which favour their expansion. Selection pressure can also promote the spread of a given ecotype of a species as, for instance, with herbicide-resistant ecotypes. Ruderal environments can serve as a refuge from which species can repeatedly disperse their propagules, which may then be able to establish successfully in fields or other neighbouring sites. If ruderal sites having such potential invaders also border roads then the chances for subsequent encroachment in other environments are thereby favoured.

The extent of the potential area of a plant species is ultimately a function of the extent of the environments suitable for growth of that plant. Consequently, weeds in winter annual crops are the group of invasive plants likely to have the broadest distribution within the Mediterranean Basin. C. S. A. Stinson (personal communication) has established a list of the 14 major weeds of Europe. Similar lists have been produced for some parts of the Mediterranean Basin (see references in Guillerm *et al.*, 1990, and Le Floc'h *et al.*, 1990; and specialised lists in Bischof, 1978; Chaudhary & Zawawi, 1983; Boulos & El-Hadidi, 1984). However the task is not yet completed for all invaders and for all regions of the Mediterranean Basin. Many species are widely distributed already, among which are *Amaranthus* spp., *Anagallis arvensis*, *Aster squamatus*, *Avena sterilis*, *Capsella bursa-pastoris*, *Chenopodium murale*, *Convolvulus arvensis*, *Conyza* spp., *Cyperus esculentus*, *Echinochloa crus-galli*, *Euphorbia* spp., *Oxalis* spp., *Paspalum paspalodes*, *Raphanus raphanistrum*, *Sinapis arvensis*, *Solanum nigrum* and *Sorghum halepense*. Other species may have a wide geographical spread, but mainly in the southern parts (e.g. *Peganum harmala*, *Diplotaxis harra*) or northern parts (e.g. *Polygonum aviculare*) of the Mediterranean Basin. Some other species such as *Euphorbia nutans* are still only locally invasive, as mentioned earlier.

The present distribution of plants should not exclude the fact that in the medium term it is possible for a local invader to become aggressive and very invasive on a much larger scale. This introduces the concept of plants which may be future invaders for the entire area or regions of the Mediterranean Basin. Examples of this group include *Chloris virgata*, *Euphorbia hypericifolia*, *E. nutans*, *Galinsoga parviflora*, *Salpichroa origanifolia*, *Senecio inaequidens* and *Solanum elaegnifolium*.

Humans are very obviously a major biogeographical factor, since they influence and strongly accelerate the expansion of particular 'commensal' plants. There have been many deliberate introductions to the Mediterranean Basin. For instance, Poletaeff (1953) estimated that about 1000 species were introduced every year into Tunisia, mainly as seeds. Unplanned introductions are likely to be even more numerous, considering for instance the special case of *Flora Juvenalis* already presented which showed that up to 458 species had successfully established in one locality. Successful establishment of plants, although noticeable, does not occur very often, which means that plants with appropriate biological characteristics for such an infestation are relatively few. True invaders seem to spread progressively so that they sometimes become one among many cosmopolitan plants. As a consequence, plants from the Mediterranean Basin which have become important weeds in California, Chile, Australia or South Africa (Groves, 1986) are also mostly invasive in their native area.

References

Arrigoni, P. V. (1983). *Elatine triandra* Schkuhr. In *Med-Checklist Notulae 7*, ed. W. Greuter & T. Raus. *Willdenowia*, **13**, 97.

Bischof, F. (1978). *Common Weeds from Iran, Turkey, The Near East and North Africa*. Rossdorf (G.D.R.): G. T. Z. Eschborn Publishers.

Bouhache, M. & Tanji, A. (1985). Evaluation du stock de semences de la morelle jaune (*Solanum elaeagnifolium* Cav.) dans le sol du Tadla (Maroc). *Weed Research*, **25**, 11–14.

Boulos, L. & El-Hadidi, M. N. (1984). *The Weed Flora of Egypt*. Cairo: The American University in Cairo Press.

Burollet, M. & Boitel, C. (1921). Présence de l'*Heliotropium curassavicum* sur un point de la côte orientale tunisienne. *Bulletin de la Société d'Histoire Naturelle de l'Afrique du Nord*, **12** (8), 178–9.

Caussanel, J. P., Kheddam, M. & Gahlem, M. (1984). Contribution à la lutte chimique contre la cuscute dans les cultures d'oignon de l'Algérois. *Proceedings 3rd International Symposium of E.W.R.S. on Weed Problems in the Mediterranean Area, Oeiras, Portugal, 2*, pp. 551–8.

Chabrolin, Ch. (1934). Les graines d'*Oxalis cernua* Thung. en Tunisie. *Bulletin de la Société d'Histoire Naturelle de l'Afrique du Nord*, **7**, 273–5.

Chaudhary, S. A. & Zawawi, M. A. (1983). *A Manual of Weeds of Central and Eastern Saudi Arabia*. Riyadh: National Herbarium, Ministry of Agriculture & Water.

Chevalier, A. (1932). Les productions végétales du Sahara et de ses confins Nord et Sud. Passé, présent et avenir. *Revue Botanique Appliquée et d'Agronomie Coloniale*, **12**, 669–919.

Chevalier, A. (1939). Les origines et l'évolution de l'agriculture méditerranéenne. *Revue Internationale de Botanique Appliquée et d'Agriculture Tropicale*, **27**, 470–83.

Cosson, E. (1860). Appendix Florulae Juvenalis ou liste des plantes étrangères récemment observées au Port Juvénal près de Montpellier. *Bulletin de la Société Botanique de France*, **6**, 605–15.

Cosson, E. (1864). Appendix Florulae Juvenalis Altera ou deuxième liste de plantes étrangères récemment observées per M. Touchy au Port Juvénal prés de Montpellier. *Bulletin de la Société Botanique de France*, **11**, 159–64.

Dafni, A. & Heller, D. (1980). The threat posed by alien weeds in Israel. *Weed Research*, **20**, 277–83.

Dafni, A. & Heller, D. (1982). Adventive flora of Israel. Phytogeographical, ecological and agricultural aspects. *Plant Systematics & Evolution*, **149**, 1–18.

Danin, A. (1982). *Atriplex muelleri* Bentham. In *Med-Checklist Notulae 5*, ed. W. Greuter & T. Raus. *Willdenowia*, **12**, 38.

Desfontaines, R. L. (1798–1800). *Flora Atlantica*, 3 vols. Paris: Blanchon.

Dizerbo, A. H. & Nehou, J. (1952). Apparition de *Galinsoga parviflora* Cav. et *Galinsoga aristulata* Bicknell (Composées) dans le Massif Armoricain. *Bulletin de la Société Scientifique de Bretagne*, **27**, 85–92.

Duarte, C., Agusti, S. & Moreira, I. (1984). Water hyacinth (*Eichhornia crassipes* (Mart.) Solms) and water milfoil (*Myriophyllum aquaticum* (Vell.) Verdc.) in Portugal. *Proceedings 3rd International Symposium of E.W.R.S. on Weed Problems in the Mediterranean Area, Oeiras, Portugal 3*, pp. 667–74.

Dubuis, A. & Faurel, L. (1957). Notes de floristique nord-africaine. I. *Bulletin de la Société d'Histoire Naturelle de l'Afrique du Nord*, **48**, 471–93.

Flahault, Ch. (1899). Naturalisation et plantes naturalisées. *Bulletin de la Société Botanique de France*, **46**, xli–cviii.

Galavielle, L. & Blanchet, G. (1939). Une nouvelle plante adventice pour la flore française '*Senecio linifolius*'. *Bulletin de l'Academie des Sciences et Lettres – Montpellier*, **69**, 30–1.

Gillot, X. (1900). Etude des flores adventices. Advencité et naturalisation. *Actes Congrès International Botanique*, pp. 370–6.

Godron, D. A. (1853). *Flora Juvenalis seu enumeratis et descriptis plantarum e seminibus exoticis interlaus altatis, enatarum in campestribus Portus Juvenalis, propre Monspelium*, 1ère édn. *Mémoires de l'Académie des Sciences et Lettres, Montpellier*, 1–48.

Godron, D. A. (1854). *Flora Juvenalis, ou Enumération des Plantes Etrangères qui Croissent Naturellement au Port Juvénal, près de Montpellier. Précédée de Considérations sur les Migrations des Végétaux*, 2ème édn. Nancy: Grimblot & Veuve Raybois.

Greuter, W., Burdet, H. M. & Long, G. (ed.) (1985). *Med-Checklist*. Geneve: Conservation et Jardin botanique de Geneve.

Groves, R. H. (1986). Invasion of mediterranean ecosystems by weeds. In *Resilience in Mediterranean-type Ecosystems*, ed. B. Dell, A. J. M. Hopkins & B. B. Lamont, pp. 129–45. The Hague: Junk.

Guillerm, J. L. & Maillet, J. (1982). Weeds of western Mediterranean countries of Europe. In *Biology and Ecology of Weeds*, ed. W. Holzner & M. Numata, pp. 227–43. The Hague: Junk.

Guillerm, J. L., Le Floc'h, E., Maillet, J. & Boulet, C. (1990). The invading weeds within the Western Mediterranean Basin. In *Biological Invasions in Europe and the Mediterranean Basin*, ed. F. di Castri, A. J. Hansen & M. Debussche, pp. 61–84. Dordrecht: Kluwer.

Hernandez-Bermejo, J. E., Saavedra, M., Hildalgo, B., Motoro, J. M. & Garcia-Torres, L. (1984). Weed flora in the irrigated crops of the Guadalquivir River valley. *Euphorbia nutans* Lag.: a new weed species. *Proceedings of 3rd International Symposium of E.W.R.S. on Weed Problems in the Mediterranean Area, Oeiras, Portugal, 3*, pp. 621–7.

Jauzein, P. (1980). Hétérgénéité des semences de *Sinapis arvensis* L. *Comptes Rendus 6ème Colloque internationale sur Ecologie, Biologie et Systematique des Mauvaises Herbes, 2*, pp. 327–35.

Jovet, P. (1949). A propos d'*Euphorbia nutans* Lag. et *maculata* L. trouvés en Savoie. *Le Monde des Plantes*, **256**, 16.

Jovet, P. & Bosserdet, P. (1962). *Senecio harveianus* MacOrvan relevé chronologique des observations en France. *Bulletin de la Centre Etudes Recherches Scientifiques de Biarritz*, **7**, 417–20.

Khattab, A. H. & El-Gharably, Z. A. (1984). The problem of aquatic weeds in Egypt and methods of management. *Proceedings of 3rd International Symposium of E.W.R.S. on Weed Problems in the Mediterranean Area, Oeiras, Portugal, 1*, pp. 335–44.

Kosinova, J. (1974). Studies on the weed flora of cultivated lands in Egypt. 4. Mediterranean and tropical elements. *Candollea*, **29**, 281–95.

Labbe, L. (1955). Contributions à la connaissance de la flore phanérogamique de la Tunisie. 5. Espèces subspontanées et naturalisées. *Bulletin de la Société des Sciences Naturelles de Tunisie*, **8**, 97–117.

Le Floc'h, E. Le Houérou, H. N. & Mathez, J. (1990). History and patterns of plant invasion in Northern Africa. In *Biological Invasions in Europe and the Mediterranean Basin*, ed. F. di Castri, A. J. Hansen, & M. Debussche, pp. 105–33. Dordrecht: Kluwer.

Le Maignan, I. (1981). Contribution a l'étude des groupements de mauvaises herbes des cultures en France. Aspects synsystématiques et biologiques. These 3ème cycle, Université de Paris-Sud, Orsay.

Loret, H. & Barrandon. A. (1886). *Flore de Montpellier ou Analyse Descriptive des Plantes Vasculaires de l'Hérault*, 2ème edn, revue et corrigée par H. Loret. Paris: Masson.

Maillet, J. (1981). Evolution de la flore adventice dans le Montpellierais sous la pression des techniques culturales. Thèse Dr. Ing., Université Sciences et Techniques du Languedoc, Montpellier.

Molinier, R. & Tallon, G. (1974). Documents pour un inventaire des plantes vasculaires de la Camargue. *Bulletin du Muse d'Histoire Naturelle de Marseille*, **34**, 7–165.

Montegut, J. (1984). Causalité de la répartition des mauvaises herbes, espèces indicatrices du biotope cultivé. *La Recherche Agronomique Suisse*, **23**, 15–46.

Parde, L. C. G. (1927). *L'Introduction des Essences Exotiques dans les Forêts de l'Europe Occidentale*. Paris: Presses Université de France.

Picot, G. (1928). La culture et l'utilisation des *Eucalpytus* dans l'Afrique du Nord (d'après le Dr L. Trabut). *Revue Botanique Appliquée et Agronomie Coloniale*, **8**, 715.

Pignatti, A. (1982). *Flora d'Italia*, 3 vols. Bologna: Edagricole.

Poletaeff, N. (1953). L'Introduction des plantes nouvelles par le Service Botanique et Agronomique de Tunisie. *Revue de la Société Horticulture de Tunisie*, **50** (n.s.), 18.

Pons, A., de Beaulieu, J. L., Couteaux, M. & Reille, M. (1990). Vegetal invasions in Mediterranean Europe and North Africa from the paleoecological point of view. In *Biological Invasions in Europe and the Mediterranean Basin*, ed. F. di Castri, A. J. Hansen & M. Debussche, pp. 169–76. Dordrecht: Kluwer.

Quézel, P., Barbero, M., Bonin, G. & Loisel, R. (1990). Plant species recent invasions in the circummediterranean region. In *Biological Invasions in Europe and the Mediterranean Basin*, ed. F. di Castri, A. J. Hansen & M. Debussche, pp. 51–60. Dordrecht: Kluwer.

Rioux, J. & Quézel, P. (1950). La '*Flora Juvenalis*' en 1950. *Le Monde des Plantes*, **272**, 73–4.

Sauvaigo, E. (1899). *Flora Mediterranea Exotica. Enumération des Plantes Cultivées dans les Jardins de la Provence et de la Ligurie*. Nice: Imprimerie J. Ventre et Cie.

Sustina, M. (1980). Répartition actuelle et écologie des espèces spontanées des rizières de la région camarguaise et cas particulier de *Leersia oryzoides* (L.) Sw. Thèse 3ème cycle, Université Sciences et Techniques du Languedoc, Montpellier.

Tanji, A., Boulet, C. & Hammoumi, M. (1984). Contribution à l'étude de la biologie de *Solanum elaeagnifolium* Cav. (Solanacées) adventice des cultures dans le périmètre irrigué du Tadla (Maroc). *Weed Research*, **24**, 401–9.

Tanji, A., Boulet, C. & Hammoumi, M. (1985). Etat actual de l'infestation par *Solanum elaeagnifolium* Cav. pour les différentes cultures du périmètre du Tadla (Maroc). *Weed Research*, **25**, 1–9.

Thellung, A. (1908–1910). La flore adventice de Montpellier. *Mémoire de la Société Nationale des Sciences Naturelles et Mathématiques de Cherbourg*, **37**, 4, 7, 57–728.

Touchy, Dr (1857). Sur quelques plantes étrangères à la flore de Montpellier trouvées aux environs de cette ville. *Bulletin de la Société Botanique de France (Session extraordinaire, Montpellier)*, **4**, 626–7.

Trabut, L. (1904). Naturalisation de deux *Atriplex* australiens dans le Nord de l'Afrique (*A. halimoides* Lindl., *A. semibaccata* R.Br.). *Bulletin de la Société Botanique de France*, **51**, 105–6.

Trabut, L. (1922). Naturalisation d'un *Eucalyptus* en Algérie. *Eucalyptus algeriensis* Trab. *Bulletin de la Société Horticulture de Tunisie*, **20**, 42–3.

Tutin, T. G., Heywood, V. H., Burges, N. A., Moore, D. M., Valentine, D. H., Walters, S. M. & Webb, D. A. (1968). *Flora Europaea*, Vol. 2. Cambridge: Cambridge University Press.

van Zeist, W. (1980). Aperçu sur la diffusion des végétaux cultivés dans la région méditerranéenne. Colloque Fondation L. Emberger, Montpellier, 9–10 Avril 1980 sur 'la Mise en Place, l'Evolution et la Caractérisation de la Flore et de la Végétation Circumméditerranéenne', *Naturalia Monspeliensis*, **9**, 1–9.

Verdcourt, B. (1985). An introduced *Sesuvium* (Aizoaceae) in Arabia. *Kew Bulletin*, **40**, 208.

8

Invasive vascular plants of California

M. REJMANEK, C. D. THOMSEN & I. D. PETERS

The flora of California is one of the most distinctive in the world. Of approximately 5050 native vascular plant species not less than 1500 species are endemic (Stebbins & Major, 1965; Raven & Axelrod, 1978). The continental State of the USA with the next highest number of endemic plant species is Florida with 385 (Gentry, 1986). As is the case for other areas with a mediterranean-type climate, California also has a high number of naturalised introduced species. This number is estimated, depending on the concept of naturalisation, to be between 650 and 750 species (Munz & Keck, 1959; Munz, 1968). (If categories such as 'garden escapes' and 'occasional escapes' are included, the total number of introduced vascular plant species exceeds 1000.) Interestingly, the continental State with the second highest number of naturalised vascular plant species is again Florida with about 380 (Long & Lakela, 1976; Wunderlin, 1983; Clewell, 1985). Hawaii with 883 endemic and 600 naturalised species (Smith, 1985; Gentry, 1986) is intermediate between California and Florida.

Compared to other regions of the world with a mediterranean climate, the number of naturalised species in California is about 400 more than in the Fynbos Biome of southern Africa (Wells, this volume) and about 200 less than that of South Australia (Kloot, this volume). It is not clear whether these numbers reflect anything more than a difference in area – viz. fynbos biome (70000 km^2), California (411031 km^2), South Australia (984000 km^2).

Most of the considerations in this chapter are related to the whole of California even if only about half of the state has a true mediterranean climate as defined by Aschmann (1984). This situation arises mainly because of practical considerations. All available floras and checklists are delineated either by State or County boundaries, or by some important geomorphological or geological criteria. Even the 'California floristic province' as defined either by Raven & Axelrod (1978) or by Cronquist (1982; see also Takhtajan, 1986) does not correspond completely to Aschmann's true mediterranean climate. Three other floristic provinces partly extend to California – i.e. Vancouverian, Sonoran and Great Basin. California, as referred to in this chapter, is therefore neither climatically nor phytogeographically homogeneous. Another limitation for any detailed study on plant invasions and invasibility of plant communities is the almost absolute lack of floristic classifications of Californian vegetation, i.e. those based on complete lists of species. Only classifications based on physiognomy or dominant species are available for the whole state (Knapp, 1981, 1982; Barbour & Major, 1988; Major & Rejmanek, 1988/9). The most widespread types of potential natural vegetation in California are Mojave and Sonoran desert creosote bush (65 420 km^2), mixed conifer forest (55 230 km^2), California steppe (52 890 km^2), California oak woodlands (38 230 km^2) and chaparral of all kinds (34 000 km^2) (Barbour & Major, 1988). The actual areas of these original vegetation types have been reduced to differing extents during the last 250 years by various anthropogenic influences. Practically nothing is left of the California steppe once dominated by native caespitose perennial grasses (Baker, 1978; Burcham, 1982; Bartolome *et al.*, 1986; Mooney *et al.*, 1986).

History of introductions

The initial entry of introduced plants into California coincides with the arrival of the Spanish colonists and missionaries at San Diego in 1769 (Frenkel, 1970). By 1823 a chain of 21 missions and accompanying ranchos, pueblos and presidios was established that stretched 500 miles north from San Diego to Sonoma (Burcham, 1982). The Spanish colonial era was followed by a series of other settlement periods during the 1800s: the Mexican period (1825–1848), the Gold Rush period (1849–1860), and the period of agricultural diversification (1865–1880) when fertile valley lands formerly devoted to grazing were diverted to crops, especially grain production (see Aschmann, this volume).

As settlement proceeded and agriculture intensified, the landscape underwent an unparalleled transformation: drainage patterns were modified, large tracts of land were cultivated and planted, forest and woodland areas were reduced, and millions of hectares of grazing land were heavily utilised for an ever-expanding livestock industry. With the importation of supplies for settlements, agricultural implements and domestic animals, came a wealth of plant introductions. Some of these were intentional imports for culinary, medicinal, ornamental, and land-mark purposes; others were accidental imports as impurities in agricultural seed, ballast and packing materials (Baker, 1962; Burcham, 1982). Throughout the State, where native vegetation was destroyed, pre-adapted species with a long history as Eurasian weeds (see Guillerm, this volume) quickly colonised and spread, thereby permanently altering the composition and physiognomy of many of California's native plant communities. We do not know if they moved in and took over where native vegetation was not destroyed. The extensive changes that occurred in California's herbaceous flora following the invasion of introduced weedy plants has been described by Mack (1989) as one of 'the great historical convulsions of the earth's biota: massive changes in the species composition of once vast communities through the transoceanic transport of alien organisms and their subsequent incursion into new ranges' (p. 156).

The rate of entry and period of establishment for many introduced species have been investigated by a number of botanists, agriculturalists and plant geographers. Frenkel (1970) and Raven (1988) summarised the approximate chronologies of immigrant plant species as follows: at least 16 species were established during the Spanish colonisation; 63 species were added during the Mexican period and an additional 55 species became established during the Gold Rush and 'American' pioneer periods. By 1860, at least 134 introduced plants had gained a permanent foothold in the State, and the number of recorded invasive plants has risen steadily since that time. Development of railroads during the latter part of the nineteenth century, population expansion and the proliferation of roads and automobiles in the twentieth century provided even greater opportunity for introduction and dispersal of invasive species. Parish (1920) listed 281 immigrant taxa for southern California alone: 76 were naturalised and generally distributed, 55 were naturalised and distributed locally, 55 were naturalised but nowhere abundant, and 95 were regarded as fugitives and waifs. Jepson (1925), including only those immigrants that successfully competed with natives, calculated that 292 introduced plants had become naturalised. Robbins *et al.* (1951) reported the total number of weedy species to be 497. Munz & Keck (1957) listed 797 species of introduced plants and Munz (1968) added another 178 to give a total of 975.

Raven, in agreement with Howell (1972), suggested that this latter number is inflated and that the actual number of invasive taxa now part of California's flora should be restricted to those plants that are fully naturalised and are capable of persisting without human interference. Thus, Raven does not include narrow garden escapes, taxa that are only sparingly naturalised or agricultural weeds that are maintained through such activities as summer irrigation, even though they may be widespread and abundant. With this conservative interpretation, Raven calculated the number of naturalised immigrant species to be 654. Therefore, depending on the criteria one follows, the rate of increase of successful invasive plants until 1968 may be either exponential (Figure 8.1) or almost linear.

Cumulative numbers of introduced vascular plant species (both naturalised and occasional escapes) in California during the period 1769–1957 were estimated by Frenkel (1970). His estimates correspond to the first six points in Figure

8.1, the sixth point being derived from Munz & Keck (1957). The last closed point in Figure 8.1 is based on the supplement to Munz & Keck (Munz, 1968; see also Howell, 1972). The estimate for 1988 ('Jepson', 1988) is based on draft manuscripts for the new edition of Jepson's *Manual of the Flowering Plants of California* as available in December 1988. We counted 43 new introductions within 1489 species belonging to 62 partially or completely updated treatments of plant families. By extrapolating this proportion to an expected number of species in the manual (about 6400 species), we surmise some 185 new introduced species recorded between 1968 and 1988 which gives an estimate of the total cumulative number of introduced species equal to 1160. Attempts to fit these data by a logistic (S-shape) function give dramatically different results depending on whether the last estimate ('Jepson', 1988) is included or not. Without this point, the increase of introduced species is practically exponential. Inclusion of the point results in a levelling off of the introduced species curve (Figure 8.1). Our review of records at the California Department of Food & Agriculture (CDFA) supports this last conclusion, and even if the most recent point in Figure 8.1 is not as reliable as the previous ones, it nevertheless suggests that the rate of arrival of introduced species has slowed down during the last 50 years. (There may be a delay of several decades between arrival of a species, its collection and publication.) However, whilst the rate of new introductions has slowed through improved seed-cleaning practices, border inspections and weed detection programs, many previously established weeds continue to expand their geographical range and local influence in California.

The indicated decline of the arrival rate is noteworthy. So far, most of the reports from other countries show either exponential or linear increase (Myers & Henry, 1979; Specht, 1981; Groves, 1986; Edgar & Shand, 1987; Esler & Astridge, 1987; Garnock-Jones, 1987; Webb, 1987). The only indication of a declining rate of arrival of invasive vascular plant species was found in grassland and savanna biomes of southern Africa (Henderson & Wells, 1986).

Taxonomic survey and geographical origins

Table 8.1 provides a taxonomic and biogeographic summary of introduced vascular plants in California. To distinguish between species classed as 'naturalised' and 'occasionally escaped' (also 'once reported', etc.) use has been made, with few exceptions, of Munz & Keck (1957) and Munz (1968) in classifying non-native species into these two categories.

There are 18 families with more than 10 introduced species in California. Gramineae and Asteraceae are represented by much higher numbers of introduced species than any other family. Leguminosae and Cruciferae form the second group with slightly more than 50 introduced species per family. The rest of the families have between 10 and 30 introduced species per family. Using only naturalised species, a similar pattern emerges.

Because the first three families in Table 8.1 are among the first four largest families of vascular plants, the resulting numbers are not surprising. A ratio of the naturalised species number to the total number of species in the family may be used to assess invasive success of individual families. Resulting ratios are plotted in Figure 8.2. This representation of individual plant families is very different from that in Table 8.1 with only Gramineae still among the first three families. This is certainly an extremely successful family in California. Asteraceae does not appear to be any more successful than other families. Over-representation of Cruciferae and under-representation of Cyperaceae in some introduced floras has already been noted by other authors (Rollins & Al-Shehbaz, 1986; Crawley, 1987). Over-representation of Geraniaceae and Polygonaceae may be a phenomenon specific to California. There are only 10 native species of Geraniaceae in California and introduced species are certainly much more aggressive and widespread. With the exception of several annual grasses, *Erodium cicutarium* and *E. botrys* (including *E. brachycarpum*) are very likely the most common invaders in California. Over-representation of Polygonaceae seems to be because it contains several less aggressive introduced species.

As Raven (1988) pointed out, nearly three-quarters (72 per cent) of the

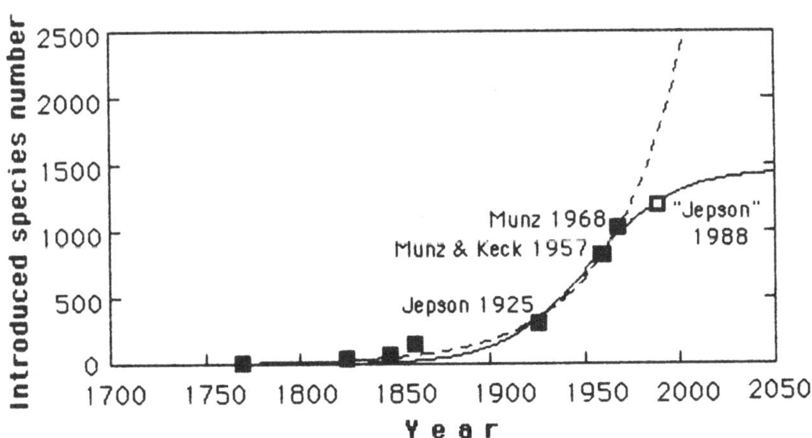

Figure 8.1. Cumulative number of introduced species of vascular plants in California during the period 1700–1988. Fitting of the data by a logistic function results in a substantially different prediction depending on whether an estimate based on incomplete material ('Jepson', 1988 – see text) is included or not.

Table 8.1. *Numbers of introduced species (N, naturalised; E, occasional escapees from cultivation; T, total) in California according to their regions of origin (NAm, North America; CAm, Central America – including Mexico; SAm, South America; Eur, Eurasia – including North Africa; SAf, southern Africa; ANZ, Australia and New Zealand; Oth, other parts of the world; Un, origin uncertain)*

Family (estimated total number of species per family)	Origin									
		NAm	CAm	SAm	Eur	SAf	ANZ	Oth	Un	Total
Gramineae	N	7	0	10	95	3	4	3	5	127
(9000)	E	2	0	2	12	3	0	2	3	24
	T	9	0	12	107	6	4	5	8	151
Asteraceae	N	16	2	18	48	4	4	3	10	105
(25000)	E	8	3	4	14	5	0	3	0	37
	T	24	5	22	62	9	4	6	10	142
Leguminosae	N	0	1	1	35	0	0	3	0	40
(17000)	E	0	1	1	18	0	0	2	2	24
	T	0	2	2	53	0	0	5	2	64
Cruciferae	N	0	0	1	37	0	0	0	0	38
(3000)	E	1	0	1	15	0	0	0	0	17
	T	1	0	2	52	0	0	0	0	55
Caryophyllaceae	N	0	0	0	19	0	0	0	0	19
(2000)	E	0	0	0	11	0	0	0	0	11
	T	0	0	0	30	0	0	0	0	30
Scrophulariaceae	N	0	0	0	12	0	1	0	1	14
(3000)	E	0	0	0	11	1	1	0	0	13
	T	0	0	0	23	1	2	0	1	27
Solanaceae	N	2	0	6	3	0	1	0	0	12
(2500)	E	5	0	8	2	0	0	0	1	16
	T	7	0	14	5	0	1	0	1	28
Chenopodiaceae	N	0	0	4	10	0	3	0	0	17
(1500)	E	0	0	0	3	0	1	0	0	4
	T	0	0	4	13	0	4	0	0	21
Labiatae	N	0	0	0	13	0	0	0	0	13
(3000)	E	0	0	1	6	1	0	0	0	8
	T	0	0	1	19	1	0	0	0	21
Polygonaceae	N	0	0	0	12	0	0	0	0	12
(750)	E	2	0	0	5	1	0	0	1	9
	T	2	0	0	17	1	0	0	1	21

Table 8.1 (*cont.*)

Family (estimated total number of species per family)	Origin									
		NAm	CAm	SAm	Eur	SAf	ANZ	Oth	Un	Total
Geraniaceae	N	0	0	0	10	2	0	0	0	12
(750)	E	0	0	0	2	6	3	0	0	11
	T	0	0	0	12	8	3	0	0	23
Aizoaceae	N	0	0	0	3	8	0	1	0	12
(2300)	E	0	0	0	0	2	0	0	0	2
	T	0	0	0	3	10	0	1	0	14
Rosaceae	N	1	0	0	3	0	1	0	0	5
(3370)	E	3	0	0	12	0	0	0	0	15
	T	4	0	0	15	0	1	0	0	20
Umbelliferae	N	0	0	0	10	0	0	0	0	10
(2700)	E	0	0	0	7	0	0	0	0	7
	T	0	0	0	17	0	0	0	0	17
Cyperaceae	N	1	0	2	5	0	0	0	0	8
(4000)	E	5	0	0	1	0	0	0	0	6
	T	6	0	2	6	0	0	0	0	14
Malvaceae	N	0	0	1	5	0	0	2	0	8
(1000)	E	0	0	0	5	0	0	0	0	5
	T	0	0	1	10	0	0	2	0	13
Boraginaceae	N	1	0	0	5	0	0	0	0	6
(2000)	E	0	0	1	7	0	0	1	0	9
	T	1	0	1	12	0	0	1	0	15
Ranunculaceae	N	0	0	0	10	0	0	0	0	10
(1800)	E	1	0	0	1	0	0	0	0	2
	T	1	0	0	11	0	0	0	0	12
Other families	N	14	2	16	143	6	6	3	2	192
	E	9	2	10	62	6	5	5	5	104
	T	23	4	26	205	12	11	8	7	296
Totals	N	42	5	59	478	23	20	15	18	660
	E	28	6	27	194	25	10	13	12	315
	T	70	11	86	672	48	30	28	30	975

Numbers derived from Munz & Keck (1957) and Munz (1968). Estimates of the total numbers of species in individual families are from Heywood (1985).

naturalised flora of California is from Eurasia and North Africa. This proportion is slightly lower (69 per cent) for all introduced plants. Most of the families show the same pattern. Notable exceptions are Aizoaceae, Solanaceae and Onagraceae; in the first two families 67 per cent of naturalised species are from southern Africa or the Americas respectively. All ten introduced species of Onagraceae are from the Americas. The fact that most of the introduced species in these three families have been found in California only recently (cf. Brewer & Watson, 1879; Jepson, 1925; Munz & Keck, 1957; Munz, 1968) indicates a partial shift in region of origin from Eurasia to other parts of the world during this century.

Genetic changes after introduction

Salisbury (1961) introduced the concept of 'infection pressure' or 'level of aggressiveness' which has to be developed before an introduced species is able to spread widely from a point of introduction. It is difficult to determine to what extent this phenomenon is just a manifestation of exponential population growth and resulting 'mass effects' and to what extent genetic processes may be involved (Shontz & Shontz, 1972; Jain & Martins, 1979; Martins & Jain, 1979, 1980; Brown & Marshall, 1981; Baker, 1986; Rejmanek, 1989). One of the first and most successful invaders in California – *Avena barbata* – is an excellent example of apparent change in the genetic structure of introduced populations.

The genetics of *Avena barbata* have been studied extensively over the last 20 years. The genetic structure of Californian populations of this species was described for allozyme genotypes by Clegg & Allard (1972), Hamrick & Allard (1972) and Miller (1977). Recently, Cluster *et al.* (1984) described polymorphism for ribosomal DNA in Californian material of *A. barbata*. We compared *A. barbata* populations of California and Eurasia (mainly the Mediterranean Basin, including northern Africa) for allozyme and ribosomal DNA markers using a combination of new material (Peters, 1988), material from Miller (1977) and from P. D. Cluster (personal communication).

Allozyme analysis has been successfully used in taxonomic and evolutionary studies for almost 30 years (Gottlieb, 1981; Richardson *et al.*, 1986). Polymorphism analysis using restriction fragments of ribosomal DNA (rDNA) is a more recent method. Ribosomal DNA (rDNA) genes are arranged in arrays of multiple copies, separated by non-transcribed regions which are believed to play a role in the regulation of transcription. The rDNA spacer fragment used in the analysis of variation in both Californian and Eurasian material is delimited by two restriction endonuclease (Eco RI) sites. This fragment includes a variable number of repetitive DNA sequences. Consequently, the Eco RI fragments obtained by enzymatic digestion vary in length. The different fragments are called spacer-length variants.

At the loci examined in this study no alleles unique to California were observed. Nine out of 23 rDNA variants present in Eurasian material were not found in Californian samples. Frequencies of alleles at the 11 allozyme loci and 18 rDNA spacer-length variants were substantially different between Eurasian regions and California, and these differences were reflected in results provided by several methods of multivariate analysis. Hedrick's coefficient of genotypic distance (Hedrick, 1983) was used in the allozyme study. This distance measure could not be used for rDNA data because the relationship between rDNA spacer-length variants and loci is not known. Therefore, Euclidean distance was used in the rDNA study. Results of ordination into two-dimensional space by non-metric multi-dimensional scaling (see, e.g. Orloci & Kenkel, 1985) are shown in Figure 8.3. Both sets of data give essentially the same results: Californian populations differ from Eurasian samples more than any two Eurasian regions differ among themselves.

P. Garcia (personal communication) provided evidence for a western

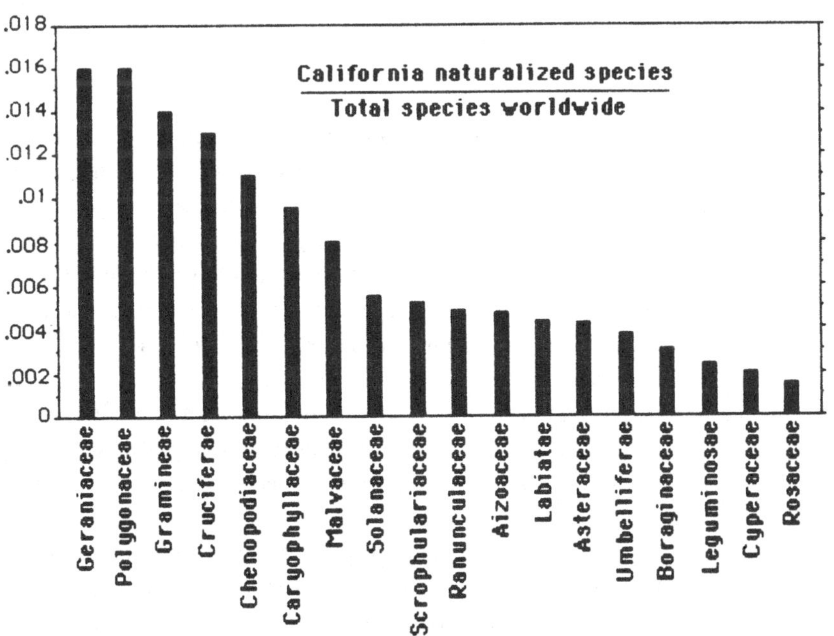

Figure 8.2. Ratios of numbers of Californian naturalised species to total numbers of species per family for the plant families with more than 10 introduced species in California.

Mediterranean origin of Californian populations of *Avena barbata*. In their recent survey of 43 Spanish populations of *A. barbata*, however, each represented by approximately 100 plants, none of the most frequent five-locus allelic associations found in California was observed. This holds as well for the present comparison of Eurasian and Californian material. Clegg & Allard (1972), Hamrick & Allard (1972) and Hamrick & Holden (1979) offered evidence for selection as a cause of the strong correlation between genotypic composition of populations in California and a degree of aridity of the environment. Two major ecotypes were recognised: the 'xeric' and the 'mesic'. Hamrick & Allard (1975) suggest that the biotic environment, particularly the degree of interspecific competition among plants, could be another important selection component associated with the evolution of these ecotypes.

Drift and local selection, under the new physical and biotic environment the species has faced in the western United States, may have provided the opportunity for evolutionary experimentation in genotypic rearrangements. Selection apparently has favoured the xeric genotype of *Avena barbata* over large areas where it was particularly adaptive, whilst a strategy based on greater polymorphism was more advantageous in mesic regions. At any rate, new selection pressures must have operated on *A. barbata* and it is likely that this is also the case with other species introduced to California.

Invaded habitats

Are some ecosystems more vulnerable to plant invasions than others? This is not an easy question to answer with the data available. Analyses of invasibility based on simple *a posteriori* observations are unsatisfactory, because in most of the cases we do not know anything about the quality and quantity of imported propagules of introduced species (Rejmanek, 1989). According to Baker (1986), some ecosystems seem to be relatively resistant to invasions, and these include dense forests, high montane ecosystems, salt marshes, and deserts. Holland & Jain (1988) suggested that the vernal pool habitat has also resisted invasions. But are these ecosystems really resistant, or do they contain lower numbers of introduced species only because these ecosystems have not been exposed to sufficient numbers of propagules of introduced species imported from similar environments?

Talbot *et al.* (1939) estimated that, in the populous San Joaquin valley of California, introduced plant species constituted 63 per cent of total herbaceous cover in grasslands, 66 per cent in woodlands, and 54 per cent in the chaparral. Baker (1986) expects that these proportions must be even greater at the present time. The only available recent analysis of proportions of introduced species over several habitats in California was published by Fiedler & Leidy (1987).

They studied quantitatively all major plant communities of Ring Mountain nature reserve in Marin County and their results are summarised in Table 8.2. They found the lowest proportions of introduced species to occur in the serpentine bunchgrass community characterised by *Stipa pulchra*, *Melica torreyana* and other native perennial grasses. The highest proportions of introduced plant species were found, not surprisingly, in the annual non-native grassland dominated by the European grasses *Lolium multiflorum* and *Briza maxima* (see also Jackson, 1985).

It is believed that openings in plant cover, generally associated with disturbance, are the most important factors promoting invasions of introduced species into natural and semi-natural communities (Peart & Foin, 1985; Baker, 1986; Rejmanek, 1989). However, under some circumstances, disturbance can also prevent successful invasion. Results of recent experiments on the effects of fertiliser addition and subsequent gopher disturbance in a serpentine annual

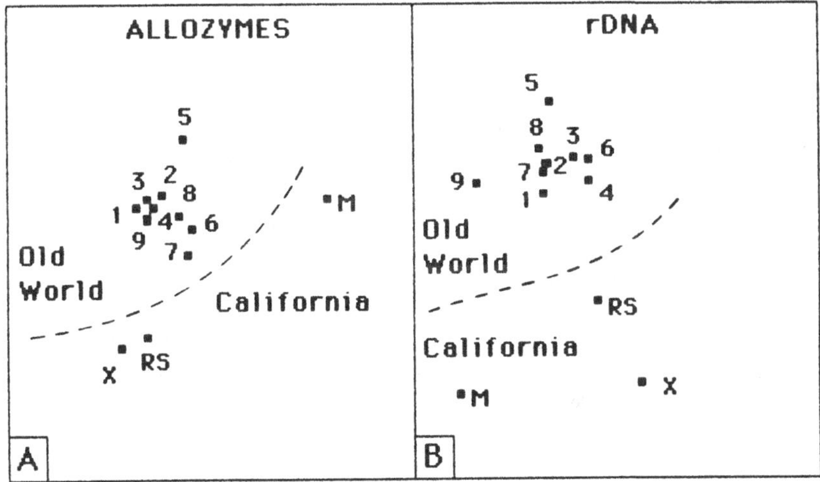

Figure 8.3. Non-metric ordination of nine Eurasian regions (numbered 1 to 9) and California (RS, random sample; M, a monomorphic 'mesic' population; X, a monomorphic 'xeric' population) based on frequency data of *Avena barbata* allozyme loci (A) and rDNA spacer-length variants (B). Ten electrophoretic zones (corresponding to 11 allozyme loci) and 18 rDNA spacer-length variants were studied. Frequency data were derived from 541 Eurasian germplasm samples and 97 Californian populations of *A. barbata* (RS). The Eurasian regions, including North Africa, in this figure are as follows: 1, Canary Islands; 2, Morocco and Algeria; 3, Tunisia and Libya; 4, Israel, Lebanon and Syria; 5, Iraq, Iran and Azerbaijan; 6, Turkey; 7, Greece, Bulgaria, Yugoslavia and Italy; 8, Crete, Cyprus, Sicily, Sardinia and Corsica; 9, France and Spain.

Table 8.2. *Introduced species in plant communities at Ring Mountain nature reserve, Marin County, California*

Plant community	Number of introduced species	% species introduced	% cover introduced
Serpentine bunchgrass (see text)			
North slope	6	12.5	2.8
South slope	5	16.7	8.9
Ridge-top	7	15.2	3.0
Non-native grassland (see text)			
North slope	9	32.1	55.5
South slope	16	44.4	84.6
Mixed broadleaf evergreen forest (*Quercus agrifolia, Lithocarpus densiflora, Umbellularia californica*)			
Tree layer	0	0.0	0.0
Shrub layer	0	0.0	0.0
Herb layer	6`	33.3	61.2
Northern coyote brush (*Baccharis pilularis* ssp. *consanguinea*)			
Upper slope	1	9.1	31.0
Mid-slope	5	50.1	49.5
Lower slope	5	38.5	41.9
Freshwater marsh (introduced *Festuca arundinacea, Carex* spp.)	7	31.6	45.7
Freshwater seep (*Carex densa, Juncus phaeocephalus*)	6	26.1	14.1

Source: Fiedler & Leidy, 1987.

grassland at Jasper Ridge, northern California (Hobbs *et al.*, 1988) illustrate this point.

Islands

Table 8.3 summarises important features of eleven major offshore islands belonging to the Californian floristic province and numbers of native and introduced vascular plant species on them. Results of multiple regression analysis revealed a significant positive dependence of both native and introduced species on the logarithm of the area and a significant negative dependence of both on the square root of the distance from the mainland (Figure 8.4). We used the equations given in Figure 8.4 to predict numbers of species in eight Californian coastal

Table 8.3. *Untransformed data on the numbers of native and introduced plants on the offshore islands of California in relation to area and distance to mainland*

Island	Area (km^2)	Distance to mainland (km)	Number of native spp.	Number of introduced spp.
Farallon Islands	0.37	32	13	23
Angel	3	2	282	134
San Miguel	36	42	171	50
Santa Rosa	218	44	370	80
Santa Cruz	244	31	477	137
Anacapa	3	21	166	40
San Nicolas	57	98	114	66
Santa Barbara	3	61	72	29
Santa Catalina	194	32	417	175
San Clemente	145	79	259	83
Guadalupe	249	253	168	38

Sources: Coulter, 1971; Ripley, 1980; Wallace, 1985.

local floras (Distance = 0) from Marin to San Diego Counties published after 1960 (Table 15.3 in Mooney *et al.*, 1986; Beauchamp, 1986). Predicted numbers of native species are on average 15 per cent lower than the published numbers, and predicted numbers of introduced species are on average 21 per cent lower than the published numbers. If the flora of San Luis Obispo County, which lists a surprisingly low number of species, is excluded, the differences are even higher: 24 per cent and 28 per cent respectively.

Insularity, therefore, has a similar effect on richness of both native and non-native floras. Islands of continental origin (Anacapa, Santa Cruz, Santa Rosa, and San Miguel) very likely lost some native species after they were cut off from the mainland (Thorne, 1969). Oceanic islands, such as Guadalupe, may never have been saturated to the level of continental floras because they completely depend on species arriving over water. Finally, some non-native species present on the mainland apparently have not had enough time to spread to all the islands.

Control of invasive plant species in California

Federal, State and County control programs have had a major influence on the distribution and abundance of certain invasive taxa in California. This applies to introduced taxa that have become widespread as well as to those that have been prevented from spreading because of suppression and eradication efforts following their initial introduction.

Biocontrol

Biological control programs have been successful using phytophagous insects for several invasive plants in the State. The first program ever attempted was for the native *Opuntia littoralis*, *O. oricola* and their hybrids on Santa Cruz island off the southern Californian coast. Overgrazing by cattle and feral goats had denuded much of the island's coastal scrub and grassland vegetation and allowed these prickly pear cacti to increase to the extent that in many areas they became the dominant cover (Goeden *et al.*, 1967). Following the release of cochineal insects (*Dactylopius* spp.) to the island and improved range management practices, considerable reductions of the target *Opuntia* spp. were observed.

The *Hypericum perforatum* program in north-western California is another example. Logging activities and overgrazing during the 1920s and 1930s had given *H. perforatum* an opportunity to expand over an estimated 1 million hectares (Sampson & Parker, 1930; Baker, 1962). Following the release of chrysomelid beetles (*Chrysolina* spp.) in heavily infested areas there was a rapid decline of *H. perforatum* to 1 per cent of its former occurrence (Huffaker & Kennett, 1959).

Attempts to control *Senecio jacobaea* with phytophagous insects have met with some success: reductions of up to 90 per cent have been obtained on well-established stands at Fort Bragg, Mendocino County (Andres *et al.*, 1976).

Other programs are underway for major weeds such as *Centaurea solstitialis*, *Carduus* spp., *Chondrilla juncea* and *Eichhornia crassipes* but more time is needed to evaluate their long-term success.

Weed detection and control/eradication programs

Long-term weed detection and control programs conducted by the California Department of Food & Agriculture (CDFA) and County Agriculture Commissioners on about 125 agricultural weeds designated as 'noxious' have resulted in the suppression, containment or eradication of many immigrant species (Thomsen, 1985; CDFA, 1987, 'noxious weed rating list'). Under this program regular surveys are conducted by State and County pest detection biologists with the purpose of detecting incipient colonies of newly arrived plants or disjunct occurrences of weeds that occur elsewhere in the State. When a weed is listed for the State as 'noxious', it is evaluated and assigned a rating of 'A', 'B', 'C' or 'Q' (Barbe, 1985). The rating of species reflects their agricultural importance, ability to spread, difficulty to control and their statewide distribution. It also establishes what control measures have to be taken on a species.

Generally, A-rated weeds have a limited distribution statewide and the policy is to either attempt eradication or contain large infestations that have the potential to spread. B-rated species may be locally common in areas but are typically not widespread throughout the state. The control policy is left to the discretion of the County Agriculture Commissioner and, unlike A-rated plants, no state funds are allotted for their control. Noxious weeds with a C-rating are those that are common and have a widespread occurrence; although they are often highly undesirable, their overall abundance make them a low priority in County control programs. Q-rated species are weeds of known agricultural importance not yet established within the State. The policy is to attempt interception of these incoming species through quarantine inspections. The following examples of A-rated noxious weeds illustrate how early control of certain weeds can result in eradication.

Alhagi pseudalhagi is a spiny and woody-stemmed, bushy legume from Asia Minor that was first noted by the CDFA in 1921 near the bank of a reservoir in Coachella Valley, Riverside County. Bottell (1933) described the root

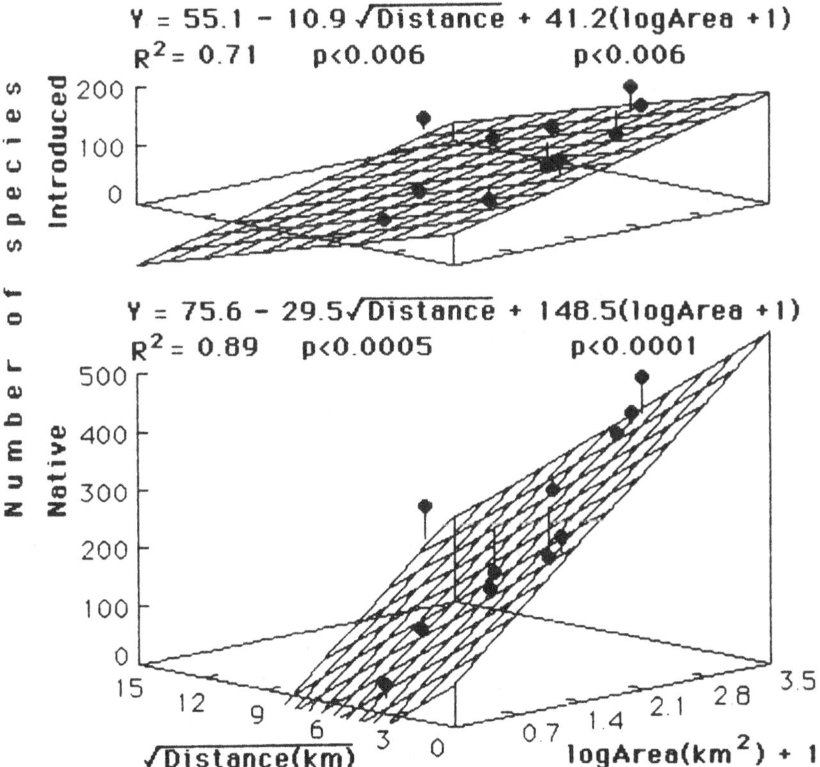

Figure 8.4. Three-dimensional plots of the combined effects of area and distance to mainland on the number of native and introduced species on 11 islands of the California floristic province (see Table 8.3).

system of *A. pseudalhagi* as the most rapidly spreading of any weed known in California, exceeding *Convolvulus arvensis* and *Sorghum halepense*. Although *A. pseudalhagi* has appeared in 17 counties in the State, a continuous control program since 1925 has reduced its present distribution to only four counties (Siebe, 1961). Complete eradication from the State was projected by 1990. *Scolymus hispanicus*, declared noxious in parts of Australia, provides another example of how early detection, followed by subsequent control, can bring about the eventual eradication of an aggressive weed that is not yet widespread. In 1968, a large infestation of *S. hispanicus* was discovered in the cattle-grazed hills of Alameda County. The infestation encompassed over 300 hectares, varying from widely scattered plants to stands of high density (Johnston, 1970). In 1972 *S. scolymus* was found in adjacent Solano County over 25 hectares of land, an indication that the species was well adapted to the environment of the San Francisco Bay region and had potential for wider distribution. At both sites, control efforts began immediately and eradication was expected to be complete by 1990. *Crupina vulgaris*, an established rangeland weed in Idaho and Montana, was found in 1976 in Sonoma County in an oak savannah community forming a dense patch of less than 1 hectare. Subsequent control action eliminated this plant entirely from that locality and it has not been observed there or elsewhere in California since that time. A total of eleven noxious weeds have been eradicated from the State. Twenty-six species are currently the subjects of eradication programs, although eight of these have populations sufficiently large that a goal of 'containment', rather than eradication, is more realistic. Many State-listed noxious weeds occur in the State as locally established infestations and are contained within a designated boundary. The goal of containment is to prevent the target species from spreading through eradication of satellite occurrences or incipient colonies outside the contained area.

The suppression and control for many weeds rated B or C is usually restricted to 'satellite' occurrences or small colonies that have not yet become well established in a particular place or county. This approach has been an effective means of preventing even further expansion of some species that are common elsewhere. A noteworthy example is that of *Centaurea solstitialis*. Known in California since 1869 (Maddox, 1985), *C. solstitialis* now occurs over an estimated 3.2 million hectares and is still spreading (Maddox, 1985; Thomsen, 1985). Whilst it is abundant in many counties in central and northern California, it is nearly absent from southern California south of the Tehachapi Range. Agriculture commissioners have been controlling satellite infestations there for over 35 years and have clearly prevented it from becoming a common weed in those areas. In some cases very costly and labour-intensive control efforts have been made for troublesome weeds that are locally common or dominant. The control programs for *Cynara cardunculus* on grazing lands in some counties near the San Francisco Bay and in southern California provide examples that have resulted in significant reductions of a robust and well-adapted rangeland weed (Thomsen *et al.*, 1986).

The examples discussed above are a few of many that exist for State-listed noxious weeds and illustrate how years of close detection, control and monitoring of invasive plants considered detrimental to agriculture can determine whether or to what extent a noxious species will be incorporated into the State's flora.

Acknowledgements. We are grateful to Douglas Barbe of the California Department of Food and Agriculture, Sacramento, and James Hickman of the Jepson Herbarium, Berkeley, who provided valuable information and access to unpublished materials. We would also like to thank Jack Major, Grady Webster, John Messina, George Robinson and Dottie Pendleton for helpful comments on the manuscript.

References

Andres, L. A., Dunn, P. H., Hawher, R. B. & Maddox, D. M. (1976). Current happenings in biological control. *Proceedings of the 28th Annual California Weed Conference*, pp. 82–3.

Aschmann, H. (1984). A restrictive definition of mediterranean climates. *Bulletin de la Société Botanique de France, 131, Actualités Botaniques*, **1984–2/3/4**, 21–30.

Baker, H. G. (1962). Weeds – native and introduced. *Journal of the California Horticultural Society*, **23**, 97–104.

Baker, H. G. (1978). Invasions and replacement in Californian and neotropical grasslands. In *Plant Relations and Pastures*, ed. J. R. Wilson, pp. 367–84. Melbourne: CSIRO Australia.

Baker, H. G. (1986). Patterns of plant invasions in North America. In *Ecology of Biological Invasions of North America and Hawaii*, ed. H. A. Mooney & J. A. Drake, pp. 44–57. New York: Springer-Verlag.

Barbe, D. (1985). The role of the California Department of Food and Agriculture. *Fremontia*, **13/2**, 13–14.

Barbour, M. G. & Major, J. (ed.) (1988). *Terrestrial Vegetation of California*, 2nd edn. Sacramento: California Native Plant Society.

Bartolome, J. W., Klukkert, S. & Barry, W. J. (1986). Opal phytoliths as evidence for displacement of native Californian grassland. *Madroño*, **33**, 217–22.

Beauchamp, R. M. (1986). *A Flora of San Diego County, California*. National City, California: Sweetwater River Press.

Bottell, A. E. (1933). Introduction and control of camel thorn. *California Department of Agriculture Bulletin*, **22**, 261–3.

Brewer, W. H. & Watson, S. (1876). *Geological Survey of California. Botany*, vol. 1. Cambridge, Mass.: Welch, Bigelow & Co.

Brown, A. H. D. & Marshall, D. R. (1981). Evolutionary changes accompanying colonization in plants. In *Evolution Today*, ed. G. G. E. Scudder & J. L. Reveal, pp. 351–63. Pittsburgh: Carnegie-Mellon University.

Burcham, L. T. (1982). *California Rangeland*, 2nd edn. Center for Archaeological Research at Davis, Publication No. 7. Davis: University of California.

Clegg, M. T. & Allard, R. W. (1972). Patterns of genetic differentiation in the slender wild oat species *Avena barbata*. *Proceedings of the National Academy of Sciences, USA*, **69**, 1820–4.

Clewell, A. F. (1985). *Guide to the Vascular Plants of the Florida Panhandle*. Tallahassee: University Press of Florida.

Cluster, P. D., Jorgensen, R. A., Bernatzky, R., Hakim-Elahi, A. & Allard, R. W. (1984). The genetics and geographical distribution of ribosomal DNA spacer-length variation in the wild oat *Avena barbata*. *Genetics*, **107**, S21.

Coulter, M. (1971). A flora of the Farallon Islands, California. *Madroño*, **21**, 131–7.

Crawley, M. J. (1987). What makes a community invasible? In *Colonization, Succession and Stability*, ed. A. J. Gray, M. J. Crawley & P. J. Edwards, pp. 429–53. Oxford: Blackwell.

Cronquist, A. (1982). Map of floristic provinces of North America. *Brittonia*, **34**, 144–5.

Edgar, E. & Shand, J. E. (1987). Checklist of panicoid grasses naturalised in New Zealand; with key to native and naturalised genera and species. *New Zealand Journal of Botany*, **25**, 343–53.

Esler, A. E. & Astridge, S. J. (1987). The naturalisation of plants in urban Auckland, New Zealand. 2. Records of introduction and naturalisation. *New Zealand Journal of Botany*, **25**, 523–37.

Fiedler, P. L. & Leidy, R. A. (1987). Plant communities of Ring Mountain Preserve, Marin County, California. *Madroño*, **34**, 173–92.

Frenkel, R. E. (1970). Ruderal vegetation along some California roadsides. *University of California Publications in Geography*, **20**, 1–163.

Garnock-Jones, P. J. (1987). Checklist of dicotyledons naturalised in New Zealand. 19. Asteraceae (Compositae) subfamily Cichorioidaea. *New Zealand Journal of Botany*, **25**, 503–10.

Gentry, A. H. (1986). Endemism in tropical versus temperate plant communities. In *Conservation Biology*, ed. M. E. Soulé, pp. 153–81. Sunderland: Sinauer.

Goeden, R. D., Fleschner, C. A. & Richer, D. W. (1967). Biological control of prickly pear cacti on Santa Cruz Island, California. *Hilgardia*, **38**, 579–606.

Gottlieb, L. D. (1981). Electrophoretic evidence and plant populations. *Progress in Phytochemistry*, **7**, 1–46.

Gray, A. (1986). Do invading species have definable genetic characteristics. *Philosophical Transactions of the Royal Society of London*, **B 314**, 655–74.

Groves, R. H. (1986). Invasion of mediterranean ecosystems by weeds. In *Resilience in Mediterranean-type Ecosystems*, ed. B. Dell, A. J. M. Hopkins & B. B. Lamont, pp. 129–45. The Hague: Junk.

Hamrick, J. L. & Allard, R. W. (1972). Microgeographical variation in allozyme frequencies in *Avena barbata*. *Proceedings of the National Academy of Sciences, USA*, **69**, 2100–4.

Hamrick, J. L. & Allard, R. W. (1975). Correlations between quantitative characters and enzyme genotypes in *Avena barbata. Evolution*, **29**, 438–42.

Hamrick, J. L. & Holden, L. R. (1979). Influence of microhabitat heterogeneity on gene frequency distribution and gametic phase equilibrium in *Avena barbata. Evolution*, **33**, 521–33.

Hedrick, P. W. (1983). *Genetics of Populations*. Boston: Science Books International.

Henderson, L. & Wells, M. J. (1986). Alien plant invasions in the grassland and savanna biomes. In *The Ecology and Management of Biological Invasions in Southern Africa*, ed. I. A. W. Macdonald, F. J. Kruger & A. A. Ferrar, pp. 109–17. Cape Town: Oxford University Press.

Heywood, V. H. (ed.) (1985). *Flowering Plants of the World*. Englewood Cliffs, New Jersey: Prentice-Hall.

Hobbs, R. J., Gulmon, S. L., Hobbs, V. J. & Mooney, H. A. (1988). Effects of fertiliser addition and subsequent gopher disturbance on a serpentine annual grassland community. *Oecologia*, **75**, 291–5.

Holland, R. F. & Jain, S. K. (1988). Vernal pools. In *Terrestrial Vegetation of California*, 2nd edn, ed. M. G. Barbour & J. Major, pp. 515–33. Sacramento: California Native Plant Society.

Howell, J. T. (1972). A statistical estimate of Munz' Supplement to *A California Flora. Wasmann Journal of Biology*, **30**, 93–6.

Huffaker, C. B. & Kennett, C. E. (1959). A ten-year study of vegetational changes associated with biological control of Klammath weed. *Journal of Range Management*, **12**, 69–82.

Jackson, L. E. (1985). Ecological origins of California's mediterranean grasses. *Journal of Biogeography*, **12**, 349–61.

Jain, S. K. & Martins, P. S. (1979). Ecological genetics of the colonizing ability of rose clover (*Trifolium hirtum* All.). *American Journal of Botany*, **66**, 361–6.

Jepson, W. L. (1925). *A Manual of the Flowering Plants of California*. Berkeley: University of California Press.

Johnston, J. (1970). Golden thistle eradication in Alameda County. *Proceedings of the 22nd Annual California Weed Conference*, pp. 75–6.

Kahler, A. L., Allard, R. W., Krzakowa, M., Wehrhahn, C. F. & Nevo, E. (1980). Associations between isozyme phenotypes and environment in slender wild oats (*Avena barbata*) in Israel. *Theoretical & Applied Genetics*, **56**, 31–47.

Knapp, R. (1981). Bibliographical review on the vegetation of California. Part I. *Excerpta Botanica, sectio B*, **21**, 121–53.

Knapp, R. (1982). Bibliographical review on the vegetation of California. Part II. *Excerpta Botanica, sectio B*, **22**, 175–88.

Long, R. W. & Lakela, O. (1976). *A Flora of Tropical Florida*, 2nd edn. Miami: Banyan Books.

Mack, R. N. (1989). Temperate grasslands vulnerable to plant invasions: characteristics and consequences. In *Ecology of Biological Invasions: A Global Perspective*, ed. J. A. Drake, H. A. Mooney, F. di Castri, R. H. Groves, F. J. Kruger, M. Rejmanek & M. Williamson, pp. 155–79. Chichester: Wiley.

Maddox, D. M. (1985). Yellow starthistle infestations are on the increase. *California Agriculture*, **39/11**, 10–12.

Major, J. & Rejmanek, M. (1988/9). Bibliographic review on the vegetation of California and its ecology. Parts III and IV. *Excerpta Botanica, sectio B*, **25**, 279–320, **26**, 1–125.

Martins, P. S. & Jain, S. K. (1979). Role of genetic variation in the colonizing ability of rose clover (*Trifolium hirtum* All.). *American Naturalist*, **113**, 591–5.

Martins, P. S. & Jain, S. K. (1980). Interpopulation variation in rose clover. *Journal of Heredity*, **71**, 29–32.

Miller, R. D. (1977). Genetic variability in slender wild oat *Avena barbata* in California. Ph.D. thesis, University of California, Davis.

Mooney, H. A., Hamburg, S. P. & Drake, J. A. (1986). The invasions of plants and animals into California. In *Ecology of Biological Invasions of North America and Hawaii*, ed. H. A. Mooney & J. A. Drake, pp. 250–72. New York: Springer-Verlag.

Munz, P. A. (1968). *Supplement to A California Flora*. Berkeley: University of California Press.

Munz, P. A. & Keck, D. D. (1959). *A California Flora*. Berkeley: University of California Press.

Myers, R. M. & Henry, R. D. (1979). Changes in the alien flora in two west-central Illinois counties during the past 140 years. *American Midland Naturalist*, **101**, 226–30.

Orloci, L. & Kenkel, N. C. (1985). *Introduction to Data Analysis with Examples from Population and Community Ecology*. Fairland: International Co-operative Publishing House.

Parish, S. B. (1920). The immigrant plants of southern California. *Bulletin of the Southern California Academy of Science*, **14**, 3–30.

Peart, D. R. & Foin, T. C. (1985). Analysis and prediction of population and community change: a grassland case study. In *The Population Structure of Vegetation*, ed. J. White, pp. 313–39. The Hague: Junk.

Peters, I. (1988). Allozyme and rDNA spacer-length variation in Mediterranean collections of *Avena barbata* Pott ex Link. Ph.D. thesis, University of California, Davis.

Raven, P. H. (1988). The California flora. In *Terrestrial Vegetation of California*, 2nd edn, ed. M. G. Barbour & J. Major, pp. 109–37. Sacramento: California Native Plant Society.

Raven, P. H. & Axelrod, D. I. (1978). Origin and relationships of the California flora. *University of California Publications in Botany*, **72**, 1–115.

Rejmanek, M. (1989). Invasibility of plant communities. In *Ecology of Biological Invasions: A Global Perspective*, ed. J. A. Drake, H. A. Mooney, F. di Castri, R. H. Groves, F. J. Kruger, M. Rejmanek & M. Williamson, pp. 369–88. Chichester: Wiley.

Richardson, B. J., Baverstock, P. R. & Adams, M. (1986). *Allozyme Electrophoresis*, London: Academic Press.

Ripley, J. D. (1980). Plants of Angel Island, Marin County, California. *Great Basin Naturalist*, **40**, 385–407.

Robbins, W. W., Bellue, M. K. & Ball, W. S. (1951). *Weeds of California*. Sacramento: California Department of Agriculture.

Rollins, R. C. & Al-Shehbaz, I. A. (1986). Weeds of South-west Asia in North America with special reference to the Cruciferae. *Proceedings of the Royal Society of Edinburgh*, **89B**, 289–99.

Salisbury, E. J. (1961). *Weeds and Aliens*. London: Collins.

Sampson, A. W. & Parker, K. W. (1930). St John's wort on rangelands of California. *California Agricultural Experiment Station Bulletin*, **503**, 1–48.

Shontz, N. N. & Shontz, J. P. (1972). Rapid evolution in populations of *Galinsoga ciliata* (Compositae) in western Massachusetts. *American Midland Naturalist*, **88**, 183–99.

Siebe, C. (1961). More about *Alhagi camelorum*. *Proceedings of the 13th Annual California Weed Conference*, pp. 67–9.

Smith, C. W. (1985). Impact of alien plants on Hawai'i's native biota. In *Hawai'i's Terrestrial Ecosystems. Preservation and Management*, ed. C. P. Stone & J. M. Scott, pp. 108–250. Honolulu: University of Hawaii Press.

Specht, R. L. (1981). Major vegetation formations in Australia. In *Ecological Biogeography of Australia*, ed. A. Keast, pp. 165–297. The Hague: Junk.

Stebbins, G. L. & Major, J. (1965). Endemism and speciation in the California flora. *Ecological Monographs*, **35**, 1–35.

Takhtajan, A. (1986). *Floristic Regions of the World*. Berkeley: University of California Press.

Talbot, M. W., Biswell, H. H. & Hormay, A. L. (1939). Fluctuations in the annual vegetation of California. *Ecology*, **20**, 349–402.

Thomsen, C. D. (1985). An assessment of noxious range weeds in California. M.S. thesis, University of California, Davis.

Thomsen, C. D., Barbe, G. D., Williams, W. A. & George, M. R. (1986). 'Escaped' artichokes are troublesome pests. *California Agriculture*, **40**, 7–9.

Thorne, R. F. (1969). The California islands. *Annals of the Missouri Botanic Gardens*, **56**, 391–408.

Wallace, G. D. (1985). *Vascular Plants of the Channel Islands of Southern California and Guadalupe Island, Baja California, Mexico*. Los Angeles: Natural History Museum of Los Angeles County.

Webb, C. J. (1987). Checklist of dicotyledons naturalised in New Zealand. 18. Asteraceae subfamily Asteroideae. *New Zealand Journal of Botany*, **25**, 489–501.

Wunderlin, R. P. (1983). *Guide to the Vascular Plants of Central Florida*. Tampa: University Press of Florida.

9

Introduction of plants into the mediterranean-type climate area of Chile

G. MONTENEGRO, S. TEILLIER, P. ARCE & V. POBLETE

The true mediterranean-climate zone of Chile can be divided physiographically into four regions: the Cordillera de Los Andes in the east with numerous peaks over 5000 metres; the Cordillera de la Costa in the west with peaks that seldom rise over 1000 metres; between these two ranges lie a series of graben known as the 'Depresion Intermedia' or Central Valley. The fourth region is the littoral fringe, a narrow coastal strip of step-like marine terraces, sea cliffs, stacks, headlands, dunes and beaches bordering the Pacific Ocean.

The natural vegetation of the arid and semi-arid zones of central Chile comprises perennials of mostly evergreen sclerophyllous shrubs and succulents, and annuals which appear after the onset of winter rains and complete their life cycles during the short wet season.

The matorral vegetation is centred in the summer-dry region of central Chile extending from 32° to 30° S latitude (di Castri, 1981) and along an altitudinal gradient from the coast to 2000 metres. The coastal scrub, occupying ocean bluffs and ocean-facing lower slopes up to 300 metres altitude, is formed mostly by chamaephytes such as *Baccharis concava*, *B. macraei*, *Haplopappus foliosus*, *Schinus polygamus* and *Senna candolleana*. Among the phanerophytes the dominant species are *Pouteria splendens*, *Peumus boldus*, *Schinus latifolius*, *Lithrea caustica* and *Flourensia thurifera* (Montenegro *et al*., 1981). The dune vegetation is characterised by the presence of *Puya chilensis*, *Eryngium paniculatum*, *Bahia ambrosioides*, *Chenopodium paniculatum*, *Centaurea chilensis*, *Scirpus nodosus*, *Poa lanuginosa*, *Festuca tunicata* and several species of *Oenothera* (V. Poblete, personal communication).

The inland matorral consists mainly of evergreen phanerophytes such as *Quillaja saponaria*, *Cryptocarya alba*, *Lithrea caustica*, *Kageneckia oblonga* and *Colliguaja odorifera*. A xeric low matorral formed by *Trevoa trinervis*, *Talguenea quinquinervia*, *Colletia spinosa*, *Baccharis linearis*, *Trichocereus*

Table 9.1. *Statistical data on the Chilean flora*

	Family	Genus	Species
Pteridophyta	27	52	157
Gymnospermae	5	10	17
Angiospermae			
Dicotyledonae	132	750	3996
Monocotyledonae	28	220	1045
Total	192	1032	5215

Source C. Marticorena & M. Quezada (1985).

chilensis and *Puya berteroniana* occurs on north-facing slopes and open areas. Alongside small water courses are found *Drimys winteri*, *Luma chequen*, *Maytenus boaria*, *Crinodendron patagua*, *Persea lingue* and *Psoralea glandulosa* (Montenegro *et al.*, 1979). The central valley is dominated by the thorny scrub *Acacia caven* and to a lesser extent by two other shrubs *Porlieria chilensis* and *Prosopis chilensis*. The general aspect of this community is that of a more or less open thicket with a rich ground cover of introduced ephemerals which appear after winter rain and a few perennial bunch grasses and forbs. According to di Castri (1968) results of studies on the soil fauna support the thesis that the 'espinal', a steppe-like community dominated by *Acacia caven*, has arisen because of human actions, because cosmopolitan populations are more abundant there than in the sclerophyllous woodlands from which it has supposedly regressed (Bahre, 1979).

The typical sclerophyllous shrubland of the Andean piedmont, which is dominated by *Colliguaja odorifera*, *Kageneckia oblonga*, *Gochnatia foliolosa* and *Porlieria chilensis*, is replaced at about 1850 metres by a montane evergreen scrub with species such as *Colliguaja integerrima*, *Kageneckia angustifolia* and *Guindilia trinervis*; this gradually gives way to a low sub-alpine scrub where predominant species are *Berberis empetrifolia*, *Chuquiraga oppositifolia* and *Tetraglochin alatum* (Aljaro & Montenegro, 1981). Communities dominated by alpine herbs and cushion plants appear at about 3000 metres (Arroyo *et al.*, 1981).

There is not a complete list of the flora of the areas of mediterranean-type climate for Chile. Recently, Marticorena & Quezada (1985) reported 5215 species for the entire country (Table 9.1), 13 per cent of which were introduced. About 3000 of these species are probably found between 30° and 40° S latitude. The herbs recorded as introduced or naturalised to the mediterranean zone of Chile are listed in Appendix 9.1 and the introduced woody species in Appendix

9.2. The life form characterising each of the species is indicated in the lists, and the community-type it is found in and its region of origin are also given.

Introduced or naturalised herbs

An extensive vegetation of annual grassland species comprises the understorey of the matorral which, unlike the native shrublands, is dominated by European species, especially species introduced from the Mediterranean Basin; these species have invaded the original vegetation during the period of European settlement. European agriculture and livestock grazing became dominant in Chile only in the last 200 to 400 years (Gulmon, 1977).

Many species have escaped from agricultural areas and become part of the wild vegetation. They have replaced or displaced native species and become well established, even dominant, in the herbaceous stratum of natural communities.

Of the invasive plants, over 50 per cent are annuals (Figure 9.1), mostly of European origin, whilst less than 30 per cent are herbaceous perennials. The particular life cycle of the annuals as drought-escaping species, together with the fact that allelopathy has been shown to be weaker in Chile than in some other regions of mediterranean climate (Keeley & Johnson, 1977; Montenegro *et al.*, 1978), have resulted in a clear numerical ascendancy of herbs over introduced shrubs or trees.

According to Navas (1973, 1976, 1978), over 25 per cent of the herbaceous plants that grow in the Santiago 'basin' are introduced species at present naturalised. Among these species with a high percentage cover are *Erodium cicutarium*, *E. moschatum*, *Vulpia myuros*, *V. bromoides*, *Capsella bursa-pastoris*, *Anthemis cotula*, *Lolium multiflorum* and *Stellaria media*.

Mature communities of central Chile have few or no invasive shrubs (Figure

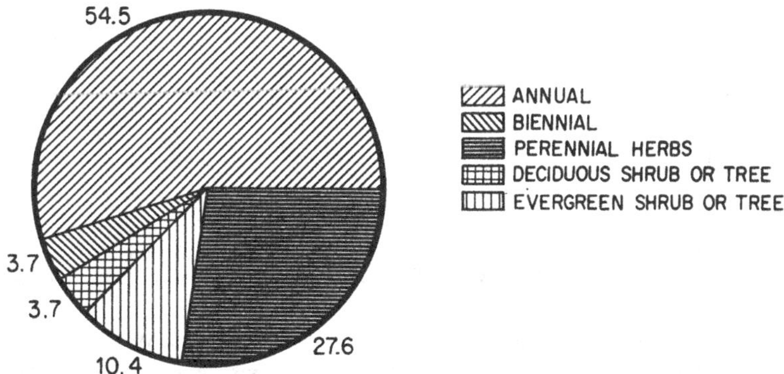

Figure 9.1. Percentages of different life forms of plant species introduced to the mediterranean-climate zone of central Chile.

9.2) as compared to herbs. Herbs are the life form that shows a significantly greater frequency in the sclerophyllous matorral, in the espinal and in marshy areas, as compared to other communities in central Chile.

The level of invasion of some herb species is closely correlated with the degree of disturbance. *Eschscholzia californica*, the California poppy, has become naturalised and spread widely in the country, being particularly abundant alongside railway tracks. Annuals such as *Poa annua*, *Fumaria capreolata*, *Marrubium vulgare*, *Verbascum virgatum*, *Conium maculatum* and *Foeniculum vulgare*, are usually found in specialised habitats maintained by humans, such as roadsides, paths and cultivated fields.

Introduced woody plants

Life forms of invasive woody plants are shown in Figure 9.2. Contrasting with the herb species, woody species do not seem to be aggressive enough to have become naturalised within the natural communities, whether forest or matorral.

Results of a survey by the authors have shown a total of 19 woody species (47 per cent trees and 53 per cent shrubs) which were introduced either for ornamental purposes (such as *Acacia dealbata*) or for timber (such as *Pinus radiata*

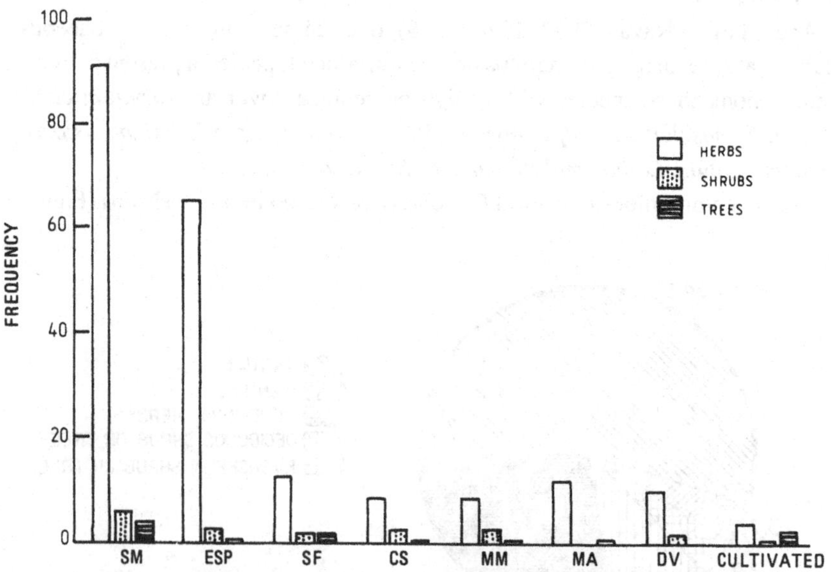

Figure 9.2. Life form frequencies of plants introduced to different plant communities of central Chile. SM, sclerophyllous matorral; ESP, espinal; SF, sclerophyllous forest; CS, coastal scrub; MM, montane matorral; MA, marshy areas; DV, dune vegetation.

and *Eucalyptus globulus*). To a lesser extent, *Atriplex semibaccata* has been cultivated in the mediterranean arid area for livestock grazing.

Rubus ulmifolius was introduced to Chile in 1860 by German colonists, to be used as field hedges. Because of its invasive character and efficient vegetative reproduction, it escaped from cultivated areas and became the most difficult invasive species to control, it being particularly deleterious to agriculture and livestock grazing. At present it has invaded about 2 million hectares, mostly in the south of the country. In the central zone, it flourishes in humid places, especially along streams, associated with other introduced species such as *Ulmus carpinifolia* and *Foeniculum vulgare*.

Rosa moschata grows spontaneously in the humid part of the mediterranean-climate zone along with other naturalised species of *Rosa*, such as *R. rubiginosa* and *R. canina* (Marticorena & Quezada, 1985). *Acacia dealbata*, an ornamental tree from south-eastern Australia, is abundant all over the mediterranean-climate zone where it occupies open habitats created by human activity. Even on mechanically disturbed sites *A. dealbata* has been noted to recruit young plants. *Ulex europaeus*, a thorny legume, was introduced as a hedge plant and is now very common in the southern part of the country where it grows in association with another legume, *Teline monspessulana*.

Chile has almost a million hectares covered by *Pinus radiata*. This is 37 per cent of the total area planted in the world. The communities most affected by

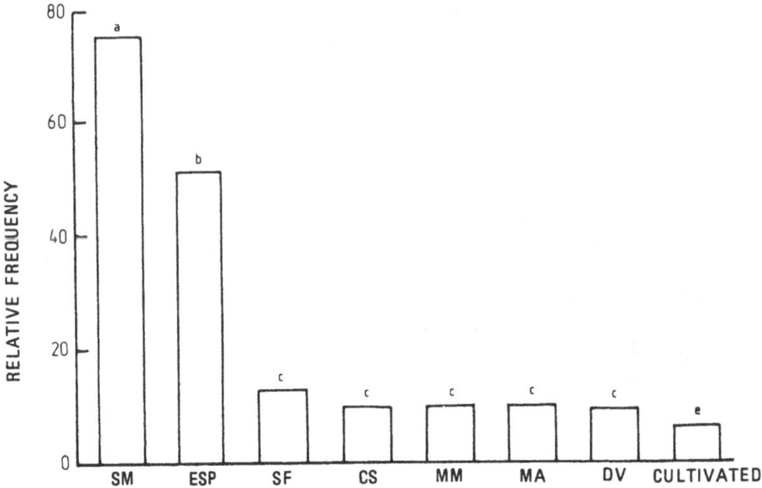

Figure 9.3. Percentages of introduced species in each plant community of central Chile. Different letters indicate statistically significant differences ($P < 0.001$). See Figure 9.2. for key to abbreviations.

reafforestation by *P. radiata* and the planting of trees as a source of timber are the deciduous native forests of *Nothofagus obliqua* and *N. glauca* in the south, and the native sclerophyllous forests and matorral near the coast in the semi-arid central zone. *Eucalyptus globulus* has also been cultivated extensively in Chile, covering about 50000 hectares, particularly on the coastal terraces and in the Andean foothills of the central zone of Chile.

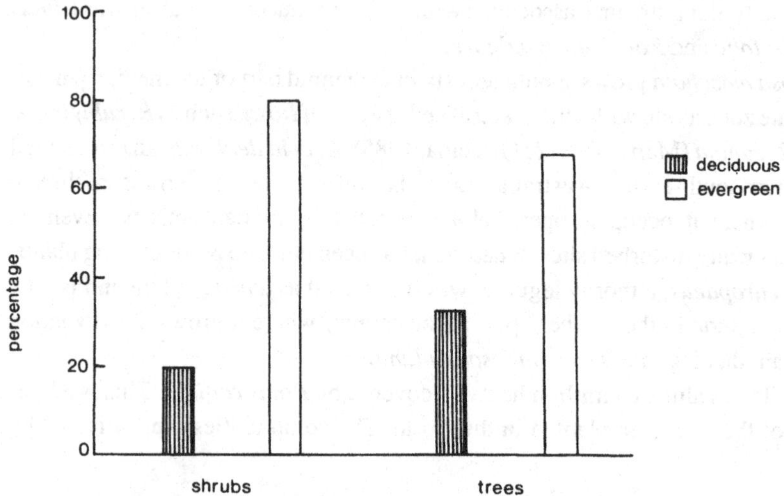

Figure 9.4. Percentages of evergreen and deciduous woody species introduced to the mediterranean-climate zone of central Chile.

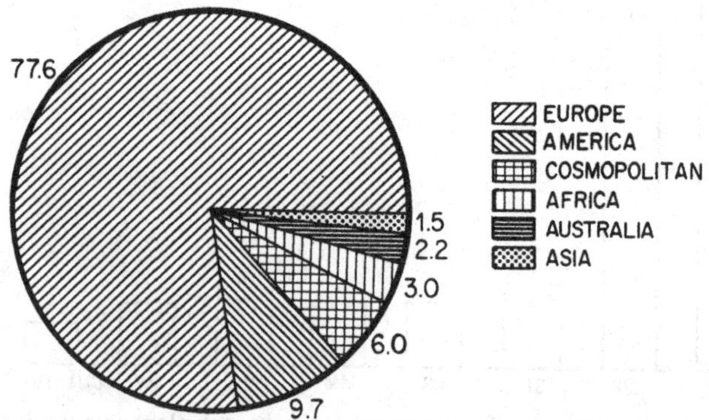

Figure 9.5. Percentages of regions of origin of plants introduced to central Chile as a percentage of the total introductions.

Conclusions

The introduced invasive flora of Chile is clearly dominated by herbaceous taxa, a trend similar to that found in California (Mooney *et al.*, 1986). In general, introduced species show significantly greater proportions in the matorral and espinal of the coastal range (Figure 9.3), the matorral being the ecosystem most extensively invaded by introduced plants. Introduced woody species are mostly evergreen (Figure 9.4) and used mainly as timber and fuel sources. The origin of most introduced species is Europe (Figure 9.5) with a significant proportion coming from North and Central America.

Acknowledgements. The writing of this chapter was funded through the grants FONDECYT 199/88 and DIUC 89/87 to Professor G. Montenegro. We are grateful to all who participated by returning questionnaires and to Mr Miguel Gómez for drawing the figures.

Appendixes

Appendix 9.1. *The introduced or naturalised herbs of the mediterranean-climate zone of Chile surveyed in this chapter, their life form and communities in which they have been recorded*[a] *and their regions of origin*

Species	Life form	Communities recorded[a]	Region of origin
Aira caryophyllea	Annual grass	SM, ESP	Europe
Ambrosia chamissonis	Perennial	DV	North America
Anthemis cotula	Annual	SM, ESP	Europe
Anthriscus caucalis	Annual	SM, ESP, SF	Europe
Argemone mexicana	Annual	SM	Central America
Atriplex patula	Annual	DV, SM	Europe
Avena barbata	Annual grass	SM, ESP	Europe
A. sterilis	Annual grass	SM, ESP	Europe
Bellis perennis	Perennial	MA	Europe
Bidens pilosa	Annual	MA	Asia
Briza minor	Annual grass	SM, ESP	Europe
Bromus hordeaceus	Annual grass	SM, ESP	Europe
B. rigidus	Annual grass	DV	Europe & Africa
Calendula officinalis	Annual	MA	Europe
Calystegia sepium	Perennial	SM	Europe
Capsella bursa-pastoris	Annual	SM, ESP	Europe & Asia
Cardamine hirsuta	Annual	SM, SF	Europe & Asia
Cardaria draba	Perennial	SM, ESP	Europe
Carduus pycnocephalus	Annual	SM, ESP	Europe
Carthamus lanatus	Annual	SM, ESP	Europe
Centaurea melitensis	Annual	SM, ESP	Europe
C. solstitialis	Annual	SM, ESP	Europe
Cerastium arvense	Perennial	MM	Europe & Asia
Chenopodium ambrosioides	Perennial	SM, MM	Central America
Chrysanthemum coronarium	Annual	DV	Europe & Asia
C. parthenium	Annual	SM, MM	Europe & Asia
Cichorium intybus	Biennial	SM, ESP	Europe & Asia
Conium maculatum	Annual	SM, ESP, SF	Europe & Asia
Convolvulus arvensis	Perennial	SM, ESP	Europe
Cotula coronopifolia	Perennial	MA	Africa
Critesion murinum	Annual grass	SM, ESP	Europe
Cymbalaria muralis	Annual	MM	Europe
Cynara cardunculus	Perennial	SM, ESP	Europe
Cynodon dactylon	Perennial grass	DV, CS	Europe & Asia
Cynoglossum creticum	Biennial	SM, MM	Europe
Dactylis glomerata	Perennial grass	ESP	Europe & Asia
Datura stramonium	Annual	SM, ESP	Central America
Dichondra sericea	Perennial	SM	America

Appendix 9.1. (*cont.*)

Species	Life form	Communities recorded[a]	Region of origin
Digitaria sanguinalis	Perennial grass	CS, SM	Europe
Echinochloa crus-galli	Annual grass	CS, SM, MA	Asia
Echium vulgare	Perennial	SM, MM	Europe & Asia
Elodea canadensis	Perennial	MA	North America
Erodium botrys	Annual	SM, ESP	Europe
E. cicutarium	Annual	SM, ESP	Europe
E. malacoides	Annual	SM, ESP	Europe
E. moschatum	Annual	SM, ESP	Europe
Eschscholzia californica	Perennial	SM, ESP	North America
Euphorbia peplus	Annual	SM, ESP, SF	Europe
Filago gallica	Annual	SM, ESP	Europe
Foeniculum vulgare	Perennial	SM, ESP	Europe
Fumaria agraria	Annual	SM, ESP, SF	Europe
F. capreolata	Annual	SM, ESP, SF	Europe
F. officinalis	Annual	SM, ESP, SF	Europe
F. parviflora	Annual	SM, ESP, SF	Europe
Galium aparine	Annual	SM, ESP, SF	Europe & Asia
G. murale	Annual	SM, ESP	Europe
Gastridium ventricosum	Annual grass	SM, ESP	Europe
Geranium robertianum	Annual	SM, SF	Europe
Hypochoeris glabra	Annual	DV, SM, ESP	Europe
H. radicata	Perennial	DV, CS, SM	Europe
Kickxia elatine	Annual	MM	Europe
Lactuca serriola	Annual	SM, MM	Europe
Lobularia maritima	Perennial	CS	Europe
Lolium multiflorum	Perennial grass	SM, ESP	Europe
L. perenne	Perennial grass	SM, ESP	Europe
L. temulentum	Perennial grass	DV, SM	Europe
Malva nicaensis	Annual	MS, ESP	Europe
Marrubium vulgare	Perennial	MS, ESP	Europe
Medicago arabica	Annual	MS, ESP	Europe
M. lupulina	Annual	SM, ESP	Europe & Asia
M. minima	Annual	SM, ESP	Europe
M. polymorpha	Annual	ESP	Europe
Melilotus alba	Biennial	SM, ESP	Europe & Asia
M. indica	Annual	SM, ESP	Europe
M. officinalis	Biennial	SM, ESP	Europe & Asia
Mentha piperita	Perennial	SM	Europe

Appendix 9.1. (*cont.*)

Species	Life form	Communities recorded[a]	Region of origin
Nasturtium officinale	Perennial	SM	Europe & Asia
Nothoscordum inodorum	Bulbous perennial	SM, ESP	North America
Phalaris amethystina	Annual grass	SM, ESP	Europe
Picris echioides	Annual	SM	Europe
Plantago lanceolata	Perennial	SM, ESP	Europe & Asia
Poa annua	Annual grass	SM, ESP	Europe
Polycarpon tetraphyllum	Annual	MS	Europe
Polygonum persicaria	Annual	MA	Europe
Polypogon monspeliensis	Annual grass	MA	Europe
Ranunculus muricatus	Perennial	MS	Europe
Raphanus sativus	Annual	CS, MS, ESP	Europe
Rumex acetosella	Perennial	MS	Europe & Asia
Ruta graveolens	Perennial	MS	Europe
Setaria geniculata	Perennial grass	MS	Europe & Asia
Silene gallica	Annual	MS, ESP	Europe
Silybum marianum	Perennial	MS, ESP	Europe
Sisymbrium officinale	Annual	MS, ESP	Europe
Solanum nigrum	Annual	MS, ESP, SF	Europe
Sonchus asper	Annual	CS, SM	Europe
S. oleraceus	Annual	CS, SM	Europe
Sorghum halepense	Perennial grass	SM	Europe
Spergula arvensis	Annual	SM	Europe
Spergularia media	Annual	SM	Europe
Stellaria media	Annual	SM, ESP, SF	Europe
Taraxacum officinale	Perennial	MS, ESP, MA	Europe
Torilis nodosa	Annual	MS, ESP	Europe
Trifolium glomeratum	Annual	MS, ESP	Europe
Trigonella monspeliaca	Annual	MS, ESP	Europe
Tropaeolum majus	Annual	MA	South America
Urtica urens	Annual	MS, ESP, SF	Europe
Verbascum thapsus	Perennial	MS, MM	Europe & Asia
V. virgatum	Annual	MS	Europe
Veronica anagallis-aquatica	Perennial	MA	Europe
Vicia sativa	Annual	MS, ESP	Europe
Vulpia bromoides	Annual grass	DV, MS, ESP	Europe
V. myuros	Annual grass	DV, MS, ESP	Europe
Xanthium spinosum	Annual	MS, ESP	Central America

[a] DV, dune vegetation; CS, coastal scrub; SM, sclerophyllous matorral; ESP, espinal; SF, sclerophyllous forest; MM, montane matorral; MA, marshy areas.

Information on regions of origin from Bailey (1951), Holm *et al.* (1977) and Tutin *et al.* (1964–1980).

Appendix 9.2. *The woody species introduced to the mediterranean-climate zone of Chile surveyed in this chapter, their life form*[a] *and communities in which they have been recorded*[b] *and their regions of origin*

Species	Life form[a]	Communities recorded[b]	Region of origin
Acacia dealbata	Ev Tree	SM, SF	Australia
Atriplex semibaccata	D Shrub	Cultivated	North America
Crataegus monogyna	Ev Tree	SM	Africa
Eucalyptus globulus	Ev Tree	Cultivated	Australia
Lavatera assurgentiflora	Ev Shrub	DV, CS	North America
Nicotiana glauca	D Shrub	SM	South America
Pinus radiata	Ev Tree	Cultivated	North America
Populus nigra var. *italica*	D Tree	Cultivated	Europe
Ricinus communis	Ev Tree	CS	Africa & Asia
Rosa canina	Ev Shrub	SM, ESP, MM	Europe
R. moschata	Ev Shrub	SM, ESP, MM	Europe
R. rubiginosa	Ev Shrub	SM, ESP, MM	Europe
Rubus ulmifolius	Ev Shrub	SM, SF	Europe
Salix babylonica	D Tree	SM, MA	Asia
Schinus molle	Ev Tree	ESP, MM	South America
Senecio mikanioides	Ev Shrub	DV, CS	Africa
Teline monspessulana	Ev Shrub	CS, SM	Europe
Ulex europaeus	Ev Shrub	SF	Europe
Ulmus carpinifolia	Ev Tree	SM, SF	Europe

[a] D, deciduous; Ev, evergreen

[b] DV, dune vegetation; CS, coastal scrub; SM, sclerophyllous matorral; ESP, espinal; SF, sclerophyllous forest; MM, montane matorral; MA, marshy areas.

Information on regions of origin from Bailey (1951), Holm *et al.* (1977) and Tutin *et al.* (1964–1980).

References

Aljaro, M. E. & Montenegro, G. (1981). Growth of dominant Chilean shrubs in the Andean Cordillera. *Mountain Research & Development*, **1**, 287–91.

Arroyo, M. T. K., Armesto, J. & Villagran, C. (1981). Plant phenological patterns in the high Andean Cordillera of central Chile. *Journal of Ecology*, **69**, 203–23.

Bahre, C. J. (1979). Destruction of the natural vegetation of north-central Chile. *University of California Publications in Geography*, **23**, 1–116.

Bailey, L. H. (1951). *Manual of Cultivated Plants*. New York: Macmillan.

di Castri, F. (1968). Esquisse ecologique du Chile. In *Biologie de l'Amérique Australe*, vol. 4, ed. C. Delamere Deboutteville & E. Rapaport, pp. 7–52. Paris: CNRS.

di Castri, F. (1981). Mediterranean-type shrublands of the world. In *Ecosystems of the World*, vol. 11, *Mediterranean-type Shrublands*, ed. F. di Castri, D. W. Goodall & R. L. Specht, pp. 1–52. Amsterdam: Elsevier.

Gulmon, S. L. (1977). A comparative study of the grasslands of California and Chile. *Flora*, **166**, 261–78.

Holm, L. G., Plucknett, D. L., Pancho, J. V. & Herberger, J. P. (1977). *The World's Worst Weeds: Distribution and Biology*. Honolulu: University of Hawaii Press.

Keeley, S. C. & Johnson, A. W. (1977). A comparison of the pattern of herb and shrub growth in comparable sites in Chile and California. *American Midland Naturalist*, **97**, 120–32.

Marticorena, C. & Quezeda, M. (1985). Catálogo de la Flora Vascular de Chile. *Gayana*, **42**, 5–157.

Montenegro, G., Rivera, O. & Bas, F. (1978). Herbaceous vegetation in the Chilean matorral. *Oecologia*, **36**, 237–44.

Montenegro, G., Aljaro, M. E. & Kummerow, J. (1979). Growth dynamics of Chilean matorral shrubs. *Botanical Gazette*, **140**, 114–19.

Montenegro, G., Aljaro, M. E., Walkowiak, A. & Saenger, R. (1981). Seasonality growth and net-productivity of herbs and shrubs of the Chilean matorral. In *Proceedings of the Symposium on Dynamics and Management of Mediterranean-Type Ecosystems, San Diego, 1981*, ed. C. E. Conrad & W. Oechel, pp. 129–35. USDA Forest Service General Technical Report PSW-58.

Mooney, H. A., Hamburg, S. P. & Drake, J. A. (1986). The invasion of plants and animals into California. In *Ecology of Biological Invasions of North America and Hawaii*, ed. H. A. Mooney & J. A. Drake, pp. 250–72. New York: Springer-Verlag.

Navas, L. E. (1973). *Flora de la Cuenca de Santiago de Chile*, tomo 1. Santiago: Ediciones Andres Bello.

Navas, L. E. (1976). *Flora de la Cuenca de Santiago de Chile*, tomo 2. Santiago: Ediciones Andres Bello.

Navas, L. E. (1978). *Flora de la Cuenca de Santiago de Chile*, tomo 3. Santiago: Ediciones Universidad de Chile.

Tutin, T. H., Heywood, V. H., Burges, N. A., Moore, D. M., Valentine, D. H., Walters, S. M. & Webb, D. A. (ed.) (1964–1980). *Flora Europaea*, 5 vols. Cambridge: Cambridge University Press.

10

Introduced plants of the fynbos biome of South Africa

M. J. WELLS

This chapter reviews information on introduced plants of the fynbos biome in the context of invasions by introduced plants in the southern African subcontinent as a whole. Southern Africa is defined as that part of the continent lying south of latitude 22° S. It covers a land area of approximately 2077700 km^2, in which seven biomes are represented: fynbos, succulent karoo, nama karoo, desert, grassland, savanna and forest (Rutherford & Westfall, 1986; Figure 10.1). The fynbos biome occurs only within the boundaries of the Republic of South Africa.

The southern Africa flora region, which extends north of latitude 22° S to the northern borders of Namibia, covers an area of approximately 2573000 km^2, from which about 20000 indigenous species (Gibbs Russell, 1985) and 1000 naturalised introduced species have been recorded.

Mediterranean-type ecosystems in southern Africa

The winter-rainfall region of southern Africa, treated as a single climatic habitat (Figure 10.2), occupies an area of approximately 160000 km^2, i.e. 7.8 per cent of the area of the subcontinent. It stretches from the south-western Cape, where rainfall on mountain peaks may exceed 3000 mm per year, northwards for approximately 1100 km, ending in Namibia, where the mean annual rainfall may be as low as 20 mm. The northern extremity of the region, not shown in Figure 10.2, approximates to that of the succulent karoo biome in Figure 10.1. Of the winter-rainfall region, an estimated 31 per cent (49000 km^2) is covered by the fynbos biome, and 69 per cent (111000 km^2) by the succulent karoo biome (Rutherford & Westfall, 1986).

Macdonald & Richardson (1986) noted that it is the lower mountain slopes in the west of the area of the fynbos biome that experience a true mediterranean-type climate with cold wet winters and dry hot summers (Koppen's climatic type Cs). Most accounts (including those of Deacon, Bigalke & Pepler, and Brooke

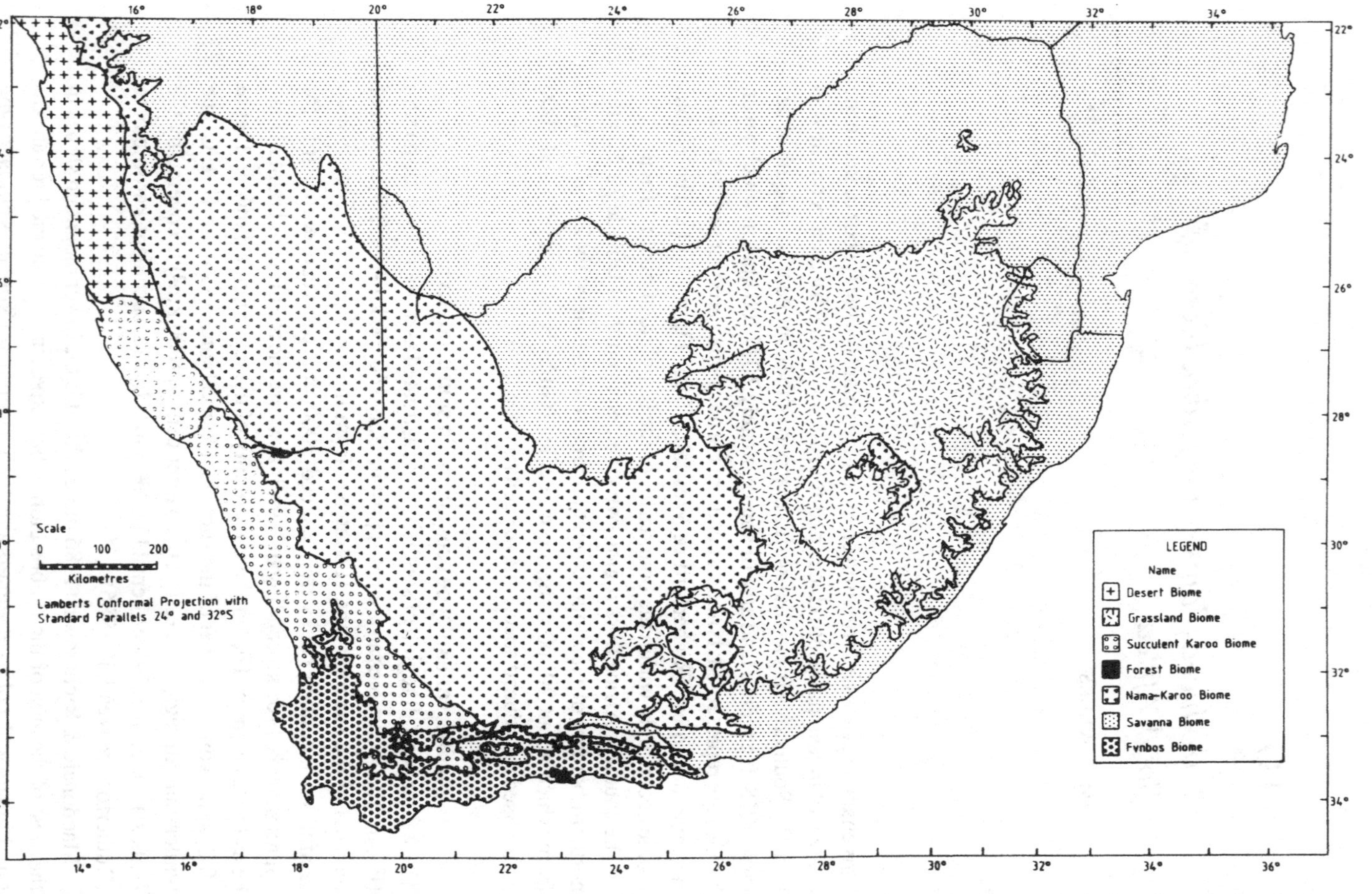

Figure 10.1. The distribution of the fynbos biome in southern Africa in relation to some of the other major biomes, including the succulent karoo biome at a scale of 1:10000000. After Rutherford & Westfall (1986).

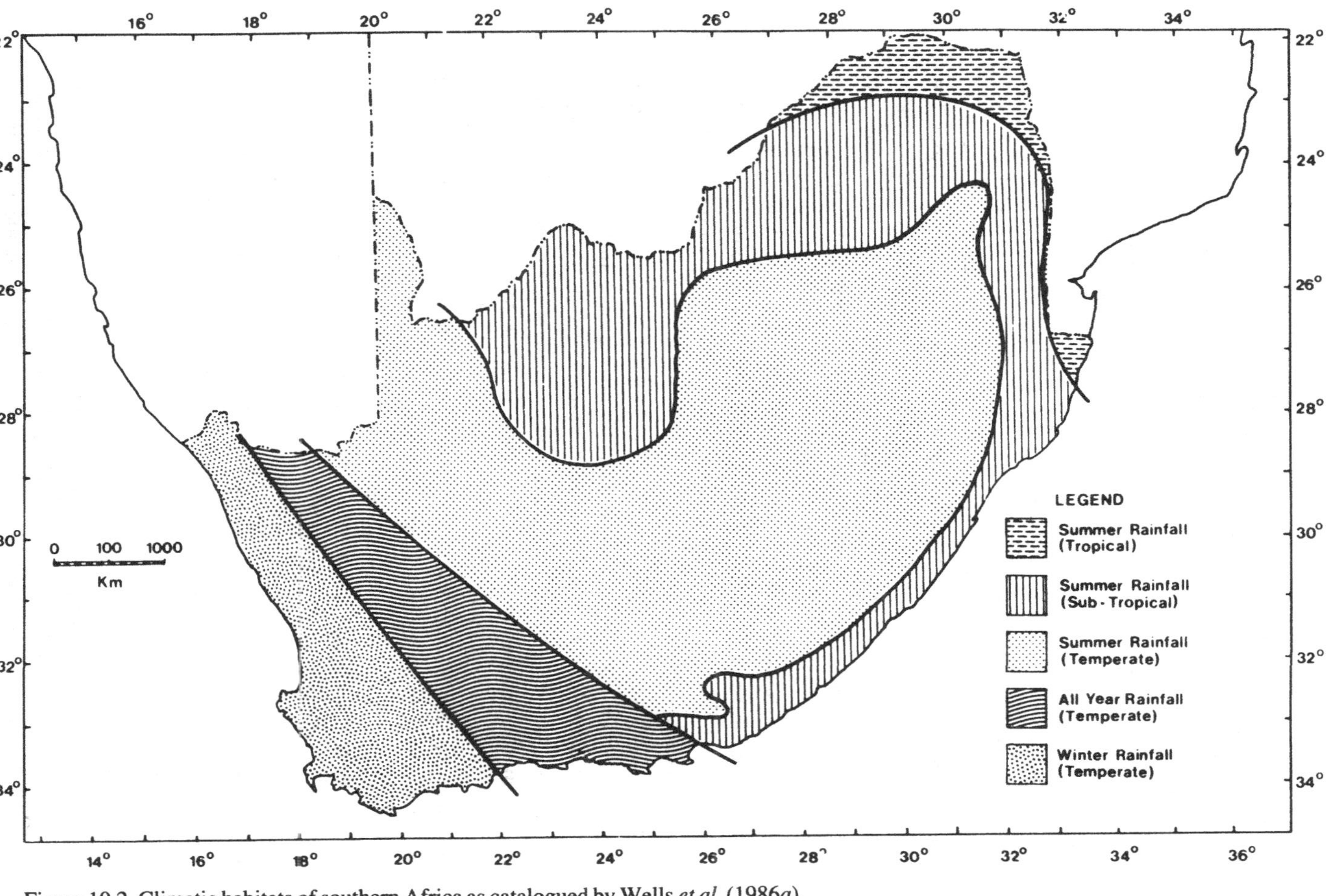

Figure 10.2. Climatic habitats of southern Africa as catalogued by Wells *et al*. (1986*a*).

& Siegfried, this volume) of mediterranean-type ecosystems in southern Africa deal with the entire fynbos biome, including those parts that fall outside the winter-rainfall area, whilst they exclude the wholly winter-rainfall succulent karoo biome.

The prevailing exclusion of the succulent karoo from consideration as a mediterranean-type ecosystem, despite its similarity to arid ecosystems in the Mediterranean Basin in regions such as northern Israel and the Persian Gulf (M. C. Rutherford, personal communication), is based partially on climatic definitions (e.g. Aschmann, 1973) that set upper temperature limits for 'mediterranean climates'. It is also based on the widely held opinion that the succulent karoo is biogeographically more closely allied to nama karoo than to fynbos. For instance, White (1983) placed the arid winter-rainfall vegetation of southern Africa in the Karoo-Namib regional centre of endemism. In a recent re-analysis of biome classifications of southern Africa, Rutherford & Westfall (1986) clearly distinguished the succulent karoo biome of the arid winter-rainfall region from the even-rainfall and summer-rainfall nama karoo biome. The close relationship of the two winter-rainfall biomes and their dissimilarity from the summer-rainfall biomes are strongly indicated by the results of recent floristic analyses (Gibbs Russell, 1987).

The two winter-rainfall biomes are linked by the occurrence of three large families, viz. Proteaceae, Oxalidaceae and Campanulaceae, that are not important in summer-rainfall biomes. They are also linked by the relatively low ranking of Poaceae, as well as by the unimportance of eleven families that are generally important in summer-rainfall biomes, viz. Acanthaceae, Malvaceae, Lamiaceae, Curcurbitaceae, Amaranthaceae, Rubiaceae, Convolvulaceae, Anacardiaceae, Solanaceae, Capparaceae and Boraginaceae. Such floristic similarities have resulted in support for the consideration of the concept of a single 'winter-rainfall biome' in southern Africa, as first suggested by Bayer (1984). Be that as it may, fynbos is generally accepted as the mediterranean-type ecosystem of South Africa (see di Castri & Mooney, 1973; Kruger *et al.*, 1983; Dell *et al.*, 1986). Thus, despite justification for its consideration as a mediterranean-type ecosystem, succulent karoo is not included further in this review. An appropriate balance between this chapter and other South African chapters in this volume is thereby retained.

Fynbos biome

Fynbos is fine-leaved evergreen sclerophyllous shrub vegetation occurring in the south-western Cape. It is also characterised by the co-dominance of phanerophytes, chamaephytes and hemicryptophytes, and is made up mainly of elements of the temperate floral kingdom 'Capensis' of Good (1964). The term 'fynbos' is further discussed, especially its status as a heathland, by Moll & Jarman (1984*a*, *b*).

Life form and species dominance and the height and structure of the community vary both with local habitat and with age after fire (van Wilgen, 1981). The prevalence of high-intensity fires occurring at intervals of greater than 5 to 10 years may be the most important single environmental determinant of fynbos (Macdonald & Jarman, 1984). The fynbos biome is largely distinguished from other biomes in which fires are common by the preponderance of obligate seeding plants, which are particularly vulnerable to unnaturally frequent (human-induced) fires that deplete their seed banks (Rutherford & Westfall, 1986).

The fynbos biome approximates the Cape floristic kingdom. This flora, together with forest and other inclusions, covers an area of 90000 km^2 and contains 152 families, 986 genera and 8504 species (Bond & Goldblatt, 1984). It has a species/square kilometre ratio of 0.0094, compared to 0.0057 for the rest of the southern Africa flora area (Gibbs Russell, 1985) and 68 per cent of its species are endemic to the Cape flora area (Bond & Goldblatt, 1984).

The fynbos biome, in the sense of Rutherford & Westfall (1986), covers an area of about 70000 km^2 and is bounded by the Atlantic and Indian oceans to the west and south, and by karoo basins in the interior. Its topography is dominated by the Cape folded mountain belt, attaining a maximum elevation of 2325 m above sea level, with undulating lowlands between the mountains and the coast. Most of the area receives a mean annual rainfall of 300–600 mm, with areas in the north-western portion receiving as little as 210 mm and some mountain peaks in the south-west over 3000 mm per year. Variations in climate and drainage over the rugged terrain, together with a wide range in soils, make for a great variety of habitats, some of which may be very localised.

Approximately 40000 km^2 (70 per cent) of the area covered by the fynbos biome (see Figures 10.1 & 10.2) receives most of its rainfall in winter, whilst the remaining area of approximately 21000 km^2 (30 per cent) covered by the biome falls in the all-year-(even-) rainfall area. There are also outlier communities between Port Elizabeth and Grahamstown, i.e. within the summer-rainfall area, whose floristics and stratigraphy correlate well with that of the fynbos biome, but which are placed in the savanna biome on the basis of life form combinations and moisture matrix relations (Kruger, 1979; Rutherford & Westfall, 1986).

Perspective of research

Early scientific interest in the diversity, origin and affinities of both the fynbos vegetation and the Cape flora (see Bond & Goldblatt, 1984) was soon followed by concern for its future (Wicht, 1945).

An extensive survey of vegetation types in southern Africa recognised and described four 'fynbos veld types' (Acocks, 1953). Subsequently, semi-intensive ecological surveys of the fynbos resulted in descriptions of the Cape of Good Hope Nature Reserve (Taylor, 1969), the lowland fynbos (Boucher, 1983) and the fynbos as a whole (Taylor, 1978; Boucher & Moll, 1981). Such surveys in turn led to studies of major invasive plants (Fugler, 1979), fynbos structure (Campbell, 1985) and, currently, productivity (M. C. Rutherford, personal communication).

The South African Department of Environmental Affairs is responsible for conserving catchment areas; since the 1970s it has built on the work of Wicht and others in initiating studies of fynbos dynamics and the control of introduced plants in fynbos (see, for example, Kruger & Bigalke's (1984) study of the effect of fire in fynbos, and Donald's (1982) study of the control of *Pinus pinaster* in fynbos). The Cape Provincial Department of Nature and Environmental Conservation published a study of information on the major plant invasions (Stirton, 1978). Recently, and especially since the initiation of the Fynbos Biome Project of the National Programme for Environmental Sciences in 1979, efforts have been made to co-ordinate and stimulate interdisciplinary research, for instance, on birds as dispersal agents of invasive plants (Glyphis *et al.*, 1981). In synthesising such results Macdonald & Richardson (1986) had access to over 200 references relevant to plant invasions in the fynbos, 70 per cent of which post-date 1970 (Macdonald & Jarman, 1984). Research results synthesised included intensive studies of introduction and establishment (Shaughnessy, 1980), the distribution of particular species, and of introduced plants of particular areas (e.g. Hall, 1961; Cowan, 1977; Mclachlan *et al.*, 1980; Fugler, 1982; Bruwer, 1983; Richardson & Brown, 1986). Lacking then and now are experimentally based comparable data on the distribution and status of plant invaders throughout the biome. Thus, whilst control actions proceed, much basic plant geographic and ecological data still require to be gathered.

In the absence of status evaluations for all invasive vascular plants, they will be reviewed at two levels: firstly, all naturalised introduced plants, i.e. all those species that have taken the initial invasive step of growing in competition with the indigenous flora (the so-called 'flora weeds' of Wells *et al.*, 1986*a*); and secondly, the ten most important invasive plant species (the so-called 'transformer species' of Wells *et al.*, 1986*b*).

Naturalised plants: possible introductions

Southern Africa has a long history of settlement by migrating African pastoralists and agriculturalists, and of contact with Asian traders. Tropical crops, almost certainly accompanied by weeds, were introduced to the subcontinent, where

Table 10.1. *Number of species and percentage of species from each region of origin in the temperate winter-rainfall region in relation to the total for southern Africa*

	Temperate winter-rainfall		Total for southern Africa	
Region of origin	Number	%	Number	%
Europe and Asia	149	61	243	100
South America	53	38	139	100
North America	15	33	46	100
Australia	30	79	38	100
Elsewhere in Africa	3	15	20	100
Pantropical	4	25	16	100
Central America	3	25	12	100
Americas (region indeterminate)	4	44	9	100
Other	2	28	7	100
Total	263	50	530	100

their naturalisation would have been favoured by the practice of shifting agriculture (see also Deacon, this volume).

The approximately 20 per cent non-endemic species of the southern African flora region (Goldblatt, 1978) includes several hundred that are often regarded as indigenous, but have world-wide or tropical distributions, as well as uses or weedy attributes which suggest that they could have been introduced long before the existence of adequate botanical records (Wells *et al.*, 1986*b*). Most of these plants are species of subtropical, summer-rainfall areas, but some of them penetrate the winter-rainfall region. They are mainly herbaceous agrestal weeds, such as *Cyperus rotundus, Eleusine indica* ssp. *indica* and *Sorghum bicolor* ssp. *arundinaceum*, or burweeds such as *Juncus bufonius* and *Setaria verticillata*. They are most often found in disturbed areas and are not referred to in the literature as being a threat to indigenous species or communities.

Naturalised plants: certain introductions

Herbarbium and literature records show that all of the 263 naturalised plant species recorded from the temperate winter-rainfall habitat (Table 10.1) occur in the fynbos biome. Further analysis of the species list from the fynbos biome core (Gibbs Russell, 1987) confirms the presence of 288 naturalised vascular plant species and infra-specific taxa, and a combination of listings yields 330 taxa. This result represents a ratio of at least 1 naturalised species per 212 km^2 and per 26 indigenous species in the biome.

Table 10.2. *Life forms of introduced naturalised and 'important' (in parentheses) invasive vascular plants in southern Africa*

	Percentage occurrence of life forms amongst species[a]							
Biome/area	Tree	Tree/ shrub	Shrub	Climber	Dwarf shrub	Erect herb	Sprawling herbs	Other
Fynbos[b] 263 sp. (+10)	5(30)	9(70)	3	1	1	62	13	6
Southern Africa 530 sp.	6	8	5	5	1	57	15	3

[a] Catalogued (Wells *et al.*, 1986*a*).
[b] Important species (Macdonald & Jarman, 1984), column Ac in Table 10.3, this chapter.

That the occurrence of various life forms amongst naturalised plants of the fynbos biome does not differ markedly from that for naturalised plants in the subcontinent as a whole (see Table 10.2) is not surprising in view of a demonstrable overlap of introduced floras between the various climatic habitats. The presence of these introduced species in flora listings, without an indication of their success, may mask the true state of plant invasions in the biome. It is nevertheless of note that tree (large phanerophyte) species and small tree/large shrub (phanerophyte) species are no more common, and erect and sprawling herb (hemicryptophyte) species no less common amongst the introduced plants of the fynbos biome than they are in the subcontinent as a whole.

The naturalised plants are drawn from 52 plant families, with over 50 per cent coming from the five largest families (Poaceae, 19 per cent; Asteraceae, 11 per cent; Fabaceae, 11 per cent; Brassicaceae, 6 per cent; and Solanaceae, 4 per cent). This composition differs only marginally from the order for the subcontinent as a whole (Wells *et al.*, 1986*b*). Amongst the grasses the ratio of C3 to C4 photosynthetic pathways is 55:13 (*fide* Ellis) with most of the C4 grasses occurring in the all-year-(even-) rainfall region, or in irrigated areas.

Important invasive plants

Hall *et al.* (1984) recorded that the fynbos biome contains 68 per cent of the threatened and rare plants of the entire southern African region, and that invasive plants contributed significantly to the threat to 44 out of 70 of the threatened plants that were investigated. Most recent publications on plant invaders of the fynbos biome are directly concerned with the urgent priority of conserving indigenous ecosystems and species. Estimates of the extent and degree of invasions (Hall & Boucher, 1977; Stirton, 1978; Macdonald & Jarman, 1984) are usually limited to invaders such as woody phanerophytes and chamaephytes that have a highly visible impact on the fynbos. Often choice or ranking of species is influenced by our ability to control them. As an instance of this, since its successful biological control, *Hypericum perforatum* has virtually disappeared from the South African literature. With a few exceptions, such as *Stipa trichotoma*, it is seldom that hemicryptophytes such as 'Mediterranean grasses' (Macdonald *et al.*, 1985) are mentioned, other than in a local survey or in a flora context, e.g. Bond & Goldblatt (1984). From these sources and flora listings it is clear that a large number of herbaceous species (such as *Fumaria muralis, Lolium perenne, Polypogon monspelianus, Salsola kali, Silene gallica, Spergula arvensis* and *Spergularia media*) are widespread within the biome and are not limited to disturbed areas. Many others, like *Brachypodium distachyon, Briza maxima, Digitaria sanguinalis, Polygonum aviculare* and *Rumex angiocarpus* thrive on disturbance. But at present too little is known of the distributions of these and other invasive herbs to use their success (or lack of it) to help define the range or degree of invasiveness to which the biome may be susceptible.

The parameters used by various authors to estimate the importance of invasive plants in the fynbos biome are varied (see footnotes to Table 10.3), as are the amounts and methods of sampling. Although not quantitatively comparable, these estimates provide an indication of the most successful invasive species in the biome and region.

Pockets of forest less than 20 km across, i.e. biome outliers unmappable at a biome scale of 1:10 million (Rutherford & Westfall, 1986) are present in mesic situations of winter-, all-year- and summer-rainfall biomes. Important invaders of these forest patches are shared by the biomes in which they occur (Geldenhuys *et al.*, 1986). Similarly, riverine habitats, also unmappable at biome level, and their invaders are present in all biomes. Important invasive plants of forest and riverine habitats in the fynbos biome (e.g. *Acacia mearnsii, A. melanoxylon, Eucalyptus* spp., *Nicotiana glauca, Populus* × *canescens*, and *Sesbania punicea*) also tend to be shared by several biomes and regions. It is not suggested that the introduced plants of these special habitats (and others such as sand dunes) are not characteristic of the biomes in which they occur. Their exclusion from biome descriptions would result in large areas of, for instance, the grassland biome on the high veld, being credited with no woody invaders at all. But many of these invasive plants are not diagnostic of particular biomes and, unless they vary significantly in kind and extent of invasion, they can be excluded from comparative studies.

Results of such investigations suggest that, whilst some species may differ (with *Paraserianthus lophantha* and *Acacia longifolia* replacing *Salix*

Table 10.3. *The most important[a] invasive vascular plants in the fynbos biome of southern Africa*

Species	Parameter[b] Aa	Ab	Ac	Ad	Ae
Hakea sericea	1	4.5	4	1	+
Acacia cyclops	2.5	2	6	10	4
Acacia saligna	2.5	1	1	+	4
Pinus pinaster	4	7	2	2	11
Acacia longifolia	5	3	5	10	4
Acacia mearnsii	6	7	+	5.5	4
Mediterranean grasses[c]	7	+	+	+	+
Pinus radiata	8	+	7.5	3.5	11
Leptospermum laevigatum	9	7	3	7	+
Pinus halepensis	10	+	7.5	3.5	11
Paraserianthus lophantha	+	4.5	9	+	4
Acacia melanoxylon	+	9	+	10	4
Sesbania punicea	+	10	+	5.5	+
Hakea gibbosa	+	+	10	8	+
Pinus canariensis	+	+	+	+	11
Pinus pinea	+	+	+	+	11
Nicotiana glauca	–	+	–	+	–
Opuntia ficus-indica	+	–	+	–	–
Lantana camara	+	–	–	+	–
Rubus cuneifolius	+	–	–	–	–
Solanum mauritianum	+	–	–	+	+
Acacia dealbata	+	–	+	+	+
Eucalyptus spp.	+	–	+	–	–
Populus × *canescens*	+	–	+	–	+

[a] Rated 1–10 if this is indicated by the various authors and workshop groups, or if it can be deduced from the statistics presented. Marked with a '+' if probably within the top 40 but not top 10 (see text).

[b] As in Macdonald & Jarman (1984). Parameters used by various sources are: a, current area of invasion of the biome as a whole; b, the sum of importance values based on current area of invasion in vegetation types (treated equally although their areas are unequal) within a biome or region; c, potential area of invasion in biome as a whole; d, potential rate of spread based on seed production, dispersal mechanisms, etc.; e, impact, i.e. potential area of invasion, number of species that would be displaced, impact on hydrological cycles, aesthetic values etc.

[c] *Avena*, *Briza* and *Lolium* spp.

babylonica and *Acacia dealbata* in the winter-rainfall area), and with others such as *Acacia mearnsii* and *Populus* × *canescens* being shared, streambank invasions in the fynbos can hardly exceed the almost total replacement of indigenous streambank vegetation (Henderson & Musil, 1984) in large parts of the summer-rainfall area.

If forest and streambank invaders are excluded, there is very little sharing of the most important introduced species between the fynbos biome and other biomes. Five species stand out as exemplifying invasion in the main body of the fynbos: *Acacia cyclops*, *A. longifolia*, *A. saligna*, *Hakea sericea* and *Pinus pinaster*. Macdonald *et al.* (1985) record that the most recent estimate of area under *Hakea* and *Pinus* throughout the biome is 7592 km^2, and *Acacia* and other thicket-forming woody plants are estimated to have invaded some 8962 km^2. There can be no doubt as to the seriousness of this invasion in terms of the depleted area of surviving fynbos and the wealth of indigenous species that stands to be lost. Whether the fynbos biome is any more liable to invasion than other biomes is a moot point. Macdonald (1984) concluded that 'although there is as yet no firm indication that the fynbos biome has a disproportionate number of invasive alien species, there exists a reasonable body of data to suggest that alien species that have invaded the fynbos have often been more successful than have those in other South African biomes'. Even this conclusion may be questioned on the grounds that invaders of other biomes, e.g. *Chromolaena*, *Opuntia* and *Stipa* species, could have been just as successful as invaders of the fynbos if their spread had not been controlled, or limited by agricultural development.

Savanna biome communities in the eastern Cape, characterised by a summer-rainfall pattern and which resemble fynbos in floristics, stratigraphy and flammability (but not in life form and moisture matrix), have been invaded by *Pinus pinaster* and Australian *Acacia* and *Hakea* species in much the same way as fynbos communities in the heart of the winter-rainfall area. Also the prevalence of Australian *Acacia* species in streambank habitats throughout the subcontinent suggests that it is their ability to take advantage of more than adequate soil water at some season (and to survive the intervening drought) that favours them, i.e. rather than the season in which the rainfall occurs. A pertinent question is perhaps: would the fynbos biome, subject as it is to fierce fires and severe summer aridity, and less subject than most biomes to selective grazing and browsing by domestic stock, have been at all susceptible to invasion by anything other than a select assemblage of fire- and drought-adapted, mainly Australian, plants? Comparison of the home and invasive ranges of invading plant species might provide a predictive tool, as well as throw light on the mechanics of invasion (Neser, 1984).

Invasive plants in lowland and mountain fynbos

Fynbos is most frequently subdivided into either lowland or mountain fynbos (Moll & Bossi, 1984). The former covers an area of 28 508 km^2, of which at least 68 per cent has been transformed, whilst the latter covers an area of 42 064 km^2, of which only 10 per cent of the main block had been transformed by 1981 (Moll & Bossi, 1984).

By far the most important invasive plants of lowland fynbos are *Acacia cyclops* and *A. saligna* (Macdonald & Richardson, 1986). Within further subdivisions of lowland fynbos (Moll & Bossi, 1984) the following invasive plants are also important (Macdonald & Jarman, 1984): *Acacia longifolia*, especially on lime and sand in coastal fynbos, and in valley/riverine situations; *Acacia mearnsii*, *Nerium oleander*, *Nicotiana glauca*, *Paraserianthus lophantha* and *Sesbania punicea* in valley/riverine habitats; *Hakea suaveolens* in coastal renosterveld; *Leptospermum laevigatum* in coastal fynbos, especially on lime and sand (including dunes); *Myoporum serratum* on dunes; and *Pinus pinaster* on sand. All these species are woody trees or shrubs and most of them have hard seed dispersed by birds and mammals (Macdonald & Richardson, 1986).

The three most important invasive plants in mountain fynbos are the serotinous *Hakea sericea* and *Pinus pinaster*, and *Acacia longifolia* whose hard seeds are dispersed by both birds and water (Macdonald & Richardson, 1986). Within further subdivisions of mountain fynbos (Moll & Bossi, 1984) the following invasive plants may also be important (Macdonald & Jarman, 1984): *Acacia cyclops* in xeric mountain habitats, especially on the Cape Peninsula; *Acacia mearnsii* and *Paraserianthus lophantha* in mesic mountain and riverine habitats; *Acacia melanoxylon*, *Pittosporum undulatum* and *Solanum mauritianum* in forest and riverine habitats; and *Populus* × *canescens* and *Sesbania punicea* in riverine habitats only.

Discussion

The fynbos biome of South Africa is not the only mediterranean-type ecosystem on the subcontinent, but it is particularly important in terms of species diversity, endemism, rare and threatened species, and the threat posed to these species by introduced plants.

Invasion by large shrubs and trees, notably *Acacia*, *Hakea* and *Pinus* species, occurs not only in winter, but also in all-year- (even-) rainfall fynbos and fynbos-like communities of the summer-rainfall region, and seems to be more closely linked to vegetation type, fire regime and fire adaptation of the invaders than to moisture matrix. The contribution of Australian trees and shrubs and of European herbs is a feature of the invader assemblage. Many of the data presented in this chapter, however, are the products of workshops rather than of surveys. An objective assessment both of the invaders and of the susceptibility of the fynbos and succulent karoo to invasion would require far more, and comparable, data both on the invaders and the invasions.

References

References cited in the text are listed below. For more complete listings on introduced plants in the fynbos biome the reader should consult Macdonald & Jarman (1984) and in southern Africa generally, Moran & Moran (1982).

Acocks, J. P. H. (1953). Veld types of South Africa. *Memoirs of the Botanical Survey of South Africa*, No. 28.

Aschmann, H. (1973). Distribution and peculiarity of mediterranean systems. In *Mediterranean-type Ecosystems: Origin and Structure*, ed. F. di Castri & H. A. Mooney, pp. 11–19. Berlin: Springer-Verlag.

Bayer, M. B. (1984). The Cape flora and the karoo. *Veld & Flora*, **70**, 17–19.

Bond, P. & Goldblatt, P. (1984). Plants of the Cape flora – a descriptive catalogue. *Journal of South African Botany*, Supplement No. 13.

Boucher, C. (1983). Floristic and structural features of the coastal foreland vegetation south of the Berg River, western Cape Province, South Africa. *Bothalia*, **14**, 669–74.

Boucher, C. & Moll, E. J. (1981). South African mediterranean shrublands. In *Ecosystems of the World*, vol. 11, *Mediterranean-type Shrublands*, ed. F. di Castri, D. W. Goodall & R. L. Specht, pp. 233–48. Amsterdam: Elsevier.

Bruwer, J. P. (1983). Besmetting van *Sesbania punicea* en ander onkruide in die lope van sekere riviere in Wes-Kaap. Unpublished report, Department of Agriculture, Elsenburg.

Campbell, B. M. (1985). A classification of the mountain vegetation of the fynbos biome. *Memoirs of the Botanical Survey of South Africa*, No. 50.

Cowan, G. (1977). An investigation of the encroachment of exotic tree species within the Howison's Poort catchment area. B.Sc. Hons thesis, Department of Geography, Rhodes University, Grahamstown.

Dell, B., Hopkins, A. J. M. & Lamont, B. B. (ed.) (1986). *Resilience in Mediterranean-type Ecosystems*. The Hague: Junk.

di Castri, F. & Mooney, H. A. (ed.) (1973). *Mediterranean-type Ecosystems. Origin and Structure*. Berlin: Springer-Verlag.

Donald, D. G. M. (1982). The control of *Pinus pinaster* in the fynbos biome. *South African Forestry Journal*, **123**, 3–7.

Fugler, S. R. (1979). Some aspects of the autecology of three *Hakea* species in the Cape Province, South Africa. M.Sc. thesis, University of Cape Town.

Fugler, S. R. (1982). Infestations of three Australian *Hakea* species in South Africa and their control. *South African Forestry Journal*, **120**, 63–8.

Geldenhuys, C. J., Le Roux, P. J. & Cooper, K. H. (1986). Alien invasions in indigenous evergreen forest. In *The Ecology and Management of Biological Invasions in Southern Africa*, ed. I. A. W. Macdonald, F. J. Kruger & A. A. Ferrar, pp. 119–31. Cape Town: Oxford University Press.

Gibbs Russell, G. E. (1985). Analysis of the size and composition of the southern African flora. *Bothalia*, **15**, 613–29.

Gibbs Russell, G. E. (1987). Preliminary floristic analysis of the major biomes in southern Africa. *Bothalia*, **17**, 213–28.

Glyphis, J. P., Milton, S. J. & Siegfried, W. R. (1981). Dispersal of *Acacia cyclops* by birds. *Oecologia*, **48**, 138–41.

Goldblatt, P. (1978). Analysis of the flora of southern Africa: its characteristics, relationships and origins. *Annals of the Missouri Botanical Garden*, **65**, 369–436.

Good, R. (1964). *The Geography of the Flowering Plants*, 4th edn. London: Longman.

Hall, A. V. (1961). Distribution studies of introduced trees and shrubs in the Cape Peninsula. *Journal of South African Botany*, **27**, 101–10.

Hall, A. V. & Boucher, C. (1977). The threat posed by alien weeds to the Cape flora. In *Proceedings of the Second National Weeds Conference of South Africa*, pp. 35–45. Cape Town: Balkema.

Hall, A. V., De Winter, B., Fourie, S. P. & Arnold, T. H. (1984). Threatened plants in southern Africa. *Biological Conservation*, **28**, 5–20.

Henderson, L. & Musil, K. J. (1984). Exotic woody plant invaders of the Transvaal. *Bothalia*, **15**, 297–313.

Kruger, F. J. (1979). South African heathlands. In *Ecosystems of the World*, vol. 9A, *Heathlands and Related Shrublands*, ed. R. L. Specht, pp. 19–80. Amsterdam: Elsevier.

Kruger, F. J. & Bigalke, R. C. (1984). Fire in fynbos. In *Ecological Effects of Fire in South African Ecosystems*, ed. P. de V. Booysen & N. M. Tainton, pp. 67–114. Berlin: Springer-Verlag.

Kruger, F. J., Mitchell, D. T. & Jarvis, J. U. M. (ed.) (1983). *Mediterranean-type Ecosystems: The Role of Nutrients*. Berlin: Springer-Verlag.

Macdonald, I. A. W. (1984). Is the fynbos biome especially susceptible to invasion by alien plants? *South African Journal of Science*, **80**, 369–77.

Macdonald, I. A. W. & Jarman, M. L. (ed.) (1984). Invasive alien organisms in the terrestrial ecosystems of the fynbos biome, South Africa. Pretoria: South African National Scientific Programmes, Report No. 85, CSIR.

Macdonald, I. A. W., Jarman, M. L. & Beeston, P. M. (ed.) (1985). Management of invasive alien plants in the fynbos biome. Pretoria: South African National Scientific Programmes Report No. 111, CSIR.

Macdonald, I. A. W. & Richardson, D. M. (1986). Alien species in terrestrial ecosystems of the fynbos biome. In *The Ecology and Management of Biological Invasions in Southern Africa*, ed. I. A. W. Macdonald, F. J. Kruger & A. A. Ferrar, pp. 77–91. Cape Town: Oxford University Press.

Mclachlan, D., Moll, E. J. & Hall, A. V. (1980). Resurvey of the alien vegetation in the Cape Peninsula. *Journal of South African Botany*, **46**, 127–46.

Moll, E. J. & Bossi, L. (1984). A current assessment of the extent of the natural vegetation of the fynbos biome. *South African Journal of Science*, **80**, 355–8.

Moll, E. J. & Jarman, M. L. (1984*a*). Clarification of the term fynbos. *South African Journal of Science*, **80**, 351–2.

Moll, E. J. & Jarman, M. L. (1984*b*). Is fynbos a heathland? *South African Journal of Science*, **80**, 352–5.

Moran, V. C. & Moran, P. M. (1982). Alien invasive vascular plants in South African natural and semi-natural environments: bibliography from 1830. Pretoria: South African National Scientific Programmes Report No. 65, CSIR.

Neser, S. (1984). Theories on why some introduced plants become aggressive invaders. Unpublished paper, presented at the Sixth National Weeds Conference (of South Africa), Nelspruit.

Richardson, D. M. & Brown, P. J. (1986). Invasion of mesic mountain fynbos by *Pinus radiata*. *South African Journal of Botany*, **52**, 529–36.

Rutherford, M. C. & Westfall, R. H. (1986). Biomes of southern Africa – an objective categorization. *Memoirs of the Botanical Survey of South Africa*, No. 54.

Shaughnessy, G. L. (1980). Historical ecology of alien woody plants in the vicinity of Cape Town, Ph.D. thesis, University of Cape Town.

Stirton, C. H. (ed.) (1978). *Plant Invaders; Beautiful but Dangerous*. Cape Town: Department of Nature and Environmental Conservation of the Cape Provincial Administration.

Taylor, H. C. (1969). The vegetation of the Cape of Good Hope Nature Reserve. M.Sc. thesis, University of Cape Town.

Taylor, H. C. (1978). Capensis. In *Biogeography and Ecology of Southern Africa*, ed. M. J. A. Werger, pp. 171–229. The Hague: Junk.

van Wilgen, B. W. (1981). Some effects of fire frequency on fynbos plant community composition and structure at Jonkershoek, Stellenbosch. *South African Forestry Journal*, **18**, 42–55.

Wells, M. J., Balsinhas, A. A., Joffe, H., Engelbrecht, V. M., Harding, G. & Stirton, C. H. (1986*a*). A catalogue of problem plants in southern Africa. *Memoirs of the Botanical Survey of South Africa*, No. 53.

Wells, M. J., Poynton, R. J., Balsinhas, A. A., Musil, K. J., Van Hoepen, E. & Abbott, S. K. (1986*b*). The history of introduction of invasive alien plants to southern Africa. In *The Ecology and Management of Biological Invasions in Southern Africa*, ed. I. A. W. Macdonald, F. J. Kruger & A. A. Ferrar, pp. 21–35. Cape Town: Oxford University Press.

White, F. (1983). *Vegetation Map of Africa*. Paris: The UNESCO Press.

Wicht, C. L. (1945). *Report of the Committee on the Preservation of the Vegetation of the South-west Cape*. Cape Town: The Royal Society of South Africa.

11

Invasive plants of southern Australia

P. M. KLOOT

The native flora of southern Australia has links with that of the other southern continents, reflecting their common origins in the Gondwanaland palaeoflora (Nelson, 1981). It also has later links with the floras of the landmasses to the north of Australia which developed after the Australian plate drifted northwards and collided with the Pacific plate to the north-east and the Eurasian plate to the north-west (Powell, Johnson & Veevers, 1981). As a result of these influences the areas of mediterranean climate in southern Australia developed a flora physiognomically similar to, but genetically distinct from, that of other areas of mediterranean-type climate (Raven, 1973).

Southern Australia, however, now shares many species with other mediterranean areas of the world. This sharing has been a dual process. One aspect of the process is slow and largely confined to a particular environment, whereas the other has been rapid and affects all environments. The first process is the acquisition of species as a result of bird migration between wetlands and surrounding areas of the northern hemisphere and similar environments in the southern hemisphere (Kloot, 1984). The propagules may adhere to migrating birds (and other creatures) or they may be small seeds caught in mud which escaped preening. Whilst this transportation is both rare and erratic it has been going on annually for some 70 million years. The large number of birds involved each year over such a long period ensured that even extremely rare events could have occurred a number of times. Consequently, apart from the species themselves that have been introduced in this way, various genotypes of individual species could also have been introduced. Incidentally, it would still be necessary for the plant to be self-fertile (Baker, 1974) because the chances of two representatives of an outcrossing species successfully being introduced and established, and at the same time sufficiently close to be able to cross successfully, are many magnitudes less than for a single successful establishment. As the birds return to the northern hemisphere each year with propagules from their southern ranges, not only would they move southern hemisphere species to the north, but there would also be a small but continuing return of material originating in the north. The flow of genetic material in both directions would tend to inhibit the formation of new species. Although the Australian populations of such plants are indeed part of the Australian flora, they did not evolve there. I have termed these 'acquired' species (Kloot, 1984). The so-called 'cosmopolitan' species are in this category although I would differentiate between the temperate species, which are being discussed here, and which have formed disjunct populations with their native ranges in the northern hemisphere, and the tropical cosmopolitan species that form a continuous band through equatorial land masses.

The second process is the introduction of species from elsewhere through human agency, both intentional and unintentional. This process has been rapid; for example, over 900 species of higher plants have been naturalised in South Australia in less than 200 years (Kloot, 1986*a*) compared to only about 60 species acquired over millions of years by the first process (Kloot, 1984). The origin of these 900 naturalised species and their routes of introduction to one political region of southern Australia, namely South Australia, will be discussed at length below.

The early botanists in southern Australia were unaware of the process of acquisition or the rapidity of spread and establishment of introduced species and thus misunderstood the presence of these species in the early years of European settlement. They thought that the species were native to Australia; many of these misconceptions have persisted until very recently. Biogeographers relying on such botanical information have either devised complicated explanations or abandoned attempts to do so for the distribution patterns of these species which were created by human transportation well beyond their original range (Kloot, 1984).

This chapter will examine various attributes of the naturalised flora of only part of the mediterranean zone of southern Australia, namely the State of South Australia. The data are derived from a major investigation of the introduced flora of that State (Kloot, 1985). It is believed that the South Australian situation is at least indicative of that prevailing over the entire region of southern Australia with a mediterranean-type climate but confirmation must await the completion of similar studies that, so far, have been proposed only for Victoria (W. T. Parsons, personal communication).

Geographical origins

Some clear trends are revealed by classifying the naturalised flora of South Australia according to its geographical origin (Kloot, 1987*b*). The years 1855,

Table 11.1. *The origins of the naturalised flora of South Australia at three different periods*

Origin	1855 No. of species	1855 (%)	1909 No. of species	1909 (%)	1984 No. of species	1984 (%)
Mediterranean	25	25	128	32	284	31
Europe[a]	50	49	142	35	232	26
Eurasia[a]	9	9	28	7	41	5
Asia	–	–	2	+	11	1
Eastern Asia	–	–	–	–	11	1
Old world tropics	–	–	3	+	8	+
California	2	2	2	+	8	+
North America[a]	2	2	9	2	51	6
Central America	1	1	4	+	19	2
South America	4	4	18	5	63	7
South Africa	8	8	49	12	132	15
East Africa	–	–	1	+	9	1
Western Australia	–	–	1	+	5	+
South Australia	–	–	3	+	5	+
Eastern Australia and New Zealand	–	–	4	+	15	2
Garden origin	–	–	3	+	10	1
Totals	101	100	397	100	904	100

\+ Less than 1 per cent of total
[a] Excluding that area immediately preceding in the list.
Source: Kloot, 1987*b*.

1909 and 1984 divide the historical period of South Australia from 1802 to the present into convenient phases and there is substantial information available for each of the years selected. By 1855 the early flurry of botanical activity had ended; in 1909, J. M. Black published his *Naturalised Flora of South Australia* and by 1984 this author had completed a major investigation of the naturalised flora of South Australia (Kloot, 1985) in which the data available for 1855 and 1909 were reviewed, supplemented and modified; it is these modified data that are presented and discussed here.

Initially, as indicated by the figures for 1855 (Table 11.1), a high proportion of the naturalised flora originated in Europe outside the Mediterranean Basin, or Eurasia. Proportionately, only a few species were from the Mediterranean Basin itself and from South Africa. By 1909 some trends were already apparent. The proportion of species from Europe/Eurasia had fallen and that from the mediterranean areas was increasing. This trend continued and is even more marked in the data for 1984, although the proportion from the Mediterranean Basin has hardly varied since 1909. The proportion originating from South Africa has risen consistently. The proportion of plants from America has risen but the number and proportions from South America have, so far, always exceeded those from North America.

Southern Australia was settled by northern Europeans largely migrating directly from that region. It is understandable therefore that the introduced plants originating from the same areas would have had the best opportunity for early transportation to the new settlement. With time, plants originating from areas more similar environmentally to South Australia found their way here and became established, thereby altering the proportions shown in Table 11.1. Apart from South African species that reached South Australia by way of Europe, the stopover of many ships at South African ports on the way to Australia, particularly prior to the opening of the Suez Canal, facilitated the movement of local plants to Australia. This could have occurred by contamination of fodder loaded there, e.g. *Emex australis*, *Pentzia suffruticosa* and probably *Cyperus tenellus*, by adherence to animals or humans, e.g. *Cotula* species and *Arctotheca calendula*. Intentional movement is also implicated, particularly in the case of some ornamentals (e.g. species of the family Aizoceae) or potentially useful plants (e.g. *Ehrharta* spp.).

The same argument applies to South America, although not to Chile. Many ships stopped at South American ports, particularly Rio de Janeiro and Buenos Aires, before heading for the Cape of Good Hope, or sailing further south and making directly for Australia (Charlwood, 1981). Also, sailing ships returning to Europe often went around Cape Horn and then called at South American ports. These ships eventually returned to Australia and some contamination leading to the transport of propagules is at least theoretically possible.

Conversely, the proportion of North American species has always been low. There was no regular direct link between there and South Australia. Almost all of the North American species listed for South Australia are also found in Europe (Tutin *et al.*, 1964–1980), which suggests that these plants reached Australia by way of Europe. The movement of fodder from North America, which on a large scale at least was rather erratic, could have been responsible for the arrival of some species that became successfully naturalised afterwards. *Solanum elaeagnifolium* appears to be such an example, as it is believed to have been brought, to South Australia at least, in hay imported from North America during the 1914 drought.

Willis (1972) remarked that no North American species of *Trifolium* had been introduced (become established?) in Victoria. It should be noted that none of these species has become established in Europe either (Tutin *et al.*, 1964–1980). As these species have not been commercialised, there appears to be no intentional movement of propagules and they do not seem to have dispersal mechanisms to facilitate their movement.

Everist (1959) demonstrated convincingly that settlers' origins affected the composition of the introduced flora. He showed that the introduced flora of Queensland in 1959 was predominantly temperate in origin, although the local environment is basically subtropical. Subsequently, the development of agricultural systems based on subtropical pastures and crops necessitated the importation of large quantities of seeds from climatically similar environments. Consequently, the proportion of subtropical species established in Queensland has risen sharply only recently (Kleinschmidt & Johnson, 1977).

The settlement of other mediterranean areas of the world demonstrates the same influence of ethnic origin of the settlers. South Africa (Wells & Stirton, 1982) and southern Australia were both settled by northern Europeans of Dutch, English or German origin. The naturalised floras of these regions show the same general trends in time in the change from a high proportion of European species to those of more specifically Mediterranean origins. California and Chile were both settled initially by the Spanish. These regions have similar naturalised floras which always had a high proportion of species from the Mediterranean Basin (Gulmon, 1977) and which, for California at least, were specifically noted as originating chiefly in Spain (Naveh, 1967).

Because of environmental limitations the species that are most likely to establish successfully are those that originate from similar mediterranean environments or those that possess a 'general-purpose' genotype (Baker, 1974) and do not require a specific environment in which to reproduce. Furthermore, in the agricultural areas of southern Australia, to succeed in areas that are cultivated regularly, the species must have an annual life cycle or be a deep-rooted perennial that is not checked by cultivation (Kloot, 1986*b*).

The first species to become naturalised successfully in South Australia were largely those possessing 'general-purpose' genotypes. They had come from, but not necessarily originated in, northern Europe, particularly Great Britain and Germany, so they were unlikely to have been adapted specifically to a mediterranean environment. Those species that had originated elsewhere but came by way of northern Europe may have been far better suited to southern Australian conditions than to those of northern Europe, e.g. *Oxalis pes-caprae* and *Medicago* spp.

In time the numbers and proportions of naturalised plants originating from other mediterranean areas increased and at present about half of the naturalised species originated from such regions. The mediterranean-climate areas of the west coast of the Americas are still under-represented, however, and a gradual accretion of species from these areas may be predicted. Furthermore, there must still be many species in South Africa and the Mediterranean Basin itself that would be adapted to southern Australian conditions.

Routes of introduction to South Australia

The present South Australian introduced flora originated predominantly in other mediterranean regions with which, historically, there was very little direct contact apart from South Africa. It is clear therefore that apart from a limited number of species that were imported directly as potential fodder plants, and those generally only in the last fifty years or so, the vast majority of these plants must have reached South Australia by a circuitous route. The species concerned must have been transported, intentionally or otherwise, to a third region from which they were then moved again, on purpose or accidentally, to South Australia. The same argument applies for other regions from which members of the introduced flora originated, e.g. China and East Africa.

Because southern Australia was settled from northern Europe, and in particular from Britain, it is reasonable to assume that those localities would be the staging posts from which plants were moved to southern Australia. It is remarkable that of the 904 introduced species recorded for South Australia, at least 765 were native to Britain or had been introduced and grown there by the 1830s (Loudon, 1830). This fact is not conclusive proof that all these plants were actually introduced to South Australia from Britain; it would, however, certainly apply to most intentional introductions such as the ornamental bulbs from South Africa and probably even such Australian ornamentals as *Sollya heterophylla*, *Pittosporum undulatum* and *Albizia lophantha*, which had all been introduced to the British horticultural trade before the colonisation of South Australia (Loudon, 1830).

The introduced flora may be categorised as to its route of introduction as follows:

1. Plants intentionally introduced or native to Britain where they were used for one or more purposes and then introduced intentionally to South Australia as ornamentals, crop or fodder plants. In some cases, such plants escaped and became naturalised in Britain and then did so in South Australia. For instance, *Briza maxima* and *Lobularia maritima* from the Mediterranean Basin, *Fuchsia magellanica* and *Bromus unioloides* from South America and *Mimulus moschatus* and *Helianthus annuus* from North America, are all naturalised in Britain (Clapham *et al.*, 1962) and in South Australia.

2. Plants unintentionally introduced, or native to Britain and introduced generally unintentionally to South Australia, e.g. *Amaranthus retroflexus*,

Coronopus didymus, *Medicago polymorpha*. Those species referred to as 'cosmopolitan' would also be included here. Some plants believed to have been introduced directly to South Australia may also have come from Britain. For instance, both *Cyperus tenellus* (Kloot, 1979) and *Solanum elaeagnifolium* were believed to have reached South Australia in contaminated fodder from South Africa and North America respectively, but both species were being grown in Britain by 1830 (Loudon, 1830).

3. Plants intentionally moved directly to South Australia from their origin, e.g. *Pentzia virgata* introduced from South Africa, *Paspalum dilatatum* from South America, *Ehrharta* spp. from South Africa and *Medicago rugosa* from the Mediterranean Basin as potential fodder plants.

4. Plants unintentionally moved directly to South Australia from their native origin as fodder or ballast contaminants or attached to implements or other tools, e.g. *Cyperus arenarius* from southern Asia; *Scirpus hamulosus* from central Asia in camel fodder or harnesses; *Eragrostis curvula* from South Africa, apparently as a contaminant of *Ehrharta* seed; *Emex australis* from South Africa as a fodder contaminant; *Galenia* spp. from South Africa and *Suaeda aegyptiaca* from Europe in ballast. It is a feature of such species that they have never been recorded from Britain or other parts of north-western Europe with which South Australia has historical ties. Consequently, very few such plants are of temperate northern origin, but rather from either the Mediterranean Basin or sub-tropical regions.

Botanical relationships

Throughout the south-eastern Australian States (Table 11.2) there is a remarkable similarity between the number of species and rankings of the major families of the naturalised flora. In Western Australia there is an unusual reversal between the Compositae and the Leguminosae but otherwise the ranking is somewhat similar. Because north-eastern New South Wales is an almost sub-tropical environment, the Solanaceae and Cyperaceae, both being families more typically tropical, are more heavily represented in that state. It would be expected that a similar effect would be found in Western Australia which also extends to the subtropics but the fact that it does not may be the result of less land development in that region. Similarly, subtropical grasses not found in either South Australia or Victoria markedly enhance the number of Gramineae in New South Wales and, to a lesser extent, Western Australia. Conversely, the naturalised species of the family Iridaceae originating from South Africa appear to have found the mediterranean-climate region of South Australia and (south-western) Western Australia more congenial and are more numerous there.

At the generic level, there is also much similarity between the States' respective floras (Table 11.3). *Solanum* and *Cyperus* being more subtropical in

Table 11.2. *The number of introduced species in the most numerous families in four Australian states*

Family	South Australia	Victoria	New South Wales	Western Australia
Dicotyledons				
Compositae	123	101	153	91
Leguminosae	83	65	110	103
Cruciferae	53	41	49	40
Caryophyllaceae	34	30	34	21
Solanaceae	29	32	38	27
Rosaceae	26	24	45	12
Labiatae	25	17	24	18
Scrophulariaceae	24	25	33	16
Monocotyledons				
Gramineae	142	136	208	175
Iridaceae	43	20	26	44
Liliaceae	27	4	19	18
Cyperaceae	14	9	21	9

See Kloot (1986*a*) for South Australia; Ross (1976), Todd (1979, 1981, 1985) for Victoria; Jacobs & Pickard (1981) for New South Wales; and Green (1985) for Western Australia.

distribution are better represented in New South Wales and to a lesser extent in Western Australia. The case with *Crotalaria* is even more striking. Introduced species of this genus are not found at all in South Australia and Victoria. The distributional data provided by Jacobs & Pickard (1981) show that *Crotalaria* is largely confined to the northern coast of New South Wales where the environment tends to be subtropical. The considerably higher numbers of *Bromus* for New South Wales and *Rubus* for New South Wales and Victoria, probably reflect more intensive taxonomic research in those particular genera in local institutions, thereby leading to recognition of more species. The situation with *Oenothera* is not so certain. Whilst detailed investigations may be responsible for some of the extra species, there may be some biological reason for the large number of species in New South Wales; alternatively, there may even be an historical explanation, in that Sydney would probably have been the most common port of call for ships travelling from the west coast of North America.

The two pasture legume genera *Trifolium* and *Medicago*, apart from being so prominent at present in all four States, have headed the genera list for South

Table 11.3. *Numbers of introduced species in the most numerous genera in four Australian states*

Genus	South Australia	Victoria	New South Wales	Western Australia
Trifolium	26	20	22	28
Medicago	14	10	11	12
Solanum	13	12	16	10
Oxalis	12	9	12	11
Opuntia	12	6	5	2
Euphorbia	12	7	12	9
Amaranthus	10	9	10	8
Cyperus	9	8	13	7
Bromus	9	9	17	7
Juncus	7	9	12	8
Rubus	7	11	13	4
Oenothera	4	4	12	6
Crotalaria (introduced)	—	—	11	3

References as in Table 11.2.

Australia consistently since 1909 (Kloot, 1987*a*). It seems likely that they have attained those positions because they have been encouraged for their desirable fodder qualities.

The genera *Solanum*, *Oxalis* and *Opuntia* are documented as having a majority of intentionally introduced species. *Euphorbia* may have been also, because of its characteristic flowers and the attractive foliage of some species, but this is not as well documented.

Throughout southern Australia, *Amaranthus* seems to have been the most successful invasive genus. This may be because seed of the genus was a common contaminant of imported garden seed in the past; when sown, its seeds could germinate in a favourable environment.

Manner of introduction

How were all these invasive plants introduced? Intensive investigation (Kloot, 1987*b*) has revealed that the majority (57 per cent) of naturalised plants were introduced intentionally (Table 11.4). The largest group were introduced as ornamentals, but escaped fodder and culinary plants are also potent sources of successful invaders. The documentation for accidentally introduced species is very scattered, but contaminated stock, seed and ballast are likely to have been the most important means of introduction.

Table 11.4. *The manner of introduction of the naturalised species of South Australia*

Intentionally introduced			
	Documented	Suspected	Total
Ornamentals	319	40	359
Fodder plants	58	17	75
Culinary plants	43	1	44
Hedges	14	—	14
Medicinals	8	5	13
Other	9	1	10
Total	451	64	515
Unintentionally introduced			
	Confirmed	Possible[a]	Total
Attached to stock	4	88	92
Contaminated seed	16	41	57
Ballast plants	7	36	43
Contaminated footwear	—	11	11
Contaminated fodder	3	3	6
Others	5	—	5
Total	35	179	214
No information			175
Grand total			904

[a] Based on overseas information.
Source Kloot, 1987*b*.

Rate of naturalisation

Specht (1981) assembled from a range of sources the number of naturalised species recorded in four States of southern Australia between 1876 and 1978. By plotting these values against time he showed that overall the rate of increase in the number of naturalised species was 5.86 per annum. This value is, in this author's opinion, only a rough estimate of the situation because it is based on inaccurate data. For a number of reasons (Kloot, 1987*a*), the number of naturalised plants at any time is consistently underestimated by contemporary observers. Furthermore, apart from such systematic errors, there is great variability in the extent of botanical activity in the area. Thus in South Australia the apparent rate of increase in naturalised species has varied between 0.7 and 27.4 species per year at different periods (Kloot, 1987*a*). I suggest that this is not a real effect but an artefact caused by variation in the level of botanical activity. Over the long term, however, i.e. between the year of first European contact with South Australia in 1802 and 1984, the average is 6.1 species per year. For Victoria, Ross (1976) calculated that the figure was about six species per year. From the data provided by Green (1985) for Western Australia a slightly lower value of 5.4 species per year may be derived. The Australian data suggest a linear relationship between the number of naturalised plants with time. There does not seem to be any suggestion of a geometric increase as found by Frenkel (1970) for California.

Groves (1986) questioned how long it would take for the rate of naturalisation to slow as a result of contemporary quarantine services. In view of the large number of introduced species growing in Australia, it seems highly likely that they will provide sufficient candidates for naturalisation for the foreseeable future. It may take many years for a given species to find its way, accidentally or otherwise, to an environment in which it will thrive unaided. Indeed, for many species such an environment might not exist locally. By reviewing the likely course of events, however, it seems predictable that many introduced species now grown in Australia will eventually become naturalised over a greater or lesser area. An example may be *Phalaris aquatica*, a species which has been widely adopted as a perennial forage grass in southern Australia. Its establishment and persistence were once regarded as difficult but highly desirable and many landholders expended much time and effort in planting it on their properties. It is now a significant invasive plant of southern Australian environments where it thrives on roadsides and other uncultivated areas and crowds out other species, particularly remnants of low-growing native vegetation. Once established *P. aquatica* is very difficult to control in non-arable situations, and it is a significant invasive plant far beyond the areas for which it was recommended initially.

The ever-increasing speed and frequency of international travel also increases the chance of an accidental or (illegal) intentional introduction, not detected by the quarantine service.

References

Baker, H. G. (1974). The evolution of weeds. *Annual Review of Ecology & Systematics*, **5**, 1–24.

Charlwood, D. (1981). *The Long Farewell*. Ringwood: Allen Lane.

Clapham, A. R., Tutin, T. G. & Warburg, E. F. (1962). *Flora of the British Isles*, 2nd edn. Cambridge: Cambridge University Press.

Everist, S. L. (1959). Strangers within the gates. *Queensland Naturalist*, **16**, 49–60.

Frenkel, R. E. (1970). Ruderal vegetation along some California roadsides. *University of California Publications in Geography*, **20**, 1–163.

Green, J. W. (1985). *Census of the Vascular Plants of Western Australia*, 2nd edn. Perth: Western Australian Herbarium.

Groves, R. H. (1986). Plant invasions of Australia: an overview. In *Ecology of Biological Invasions: An Australian Perspective*, ed. R. H. Groves & J. J. Burdon, pp. 137–49. Canberra: Australian Academy of Science.

Gulmon, S. L. (1977). A comparative study of the grassland of California and Chile. *Flora*, **166**, 261–78.

Jacobs, S. W. L. & Pickard, J. (1981). *Plants of New South Wales*. Sydney: Government Printer.

Kleinschmidt, H. E. & Johnson, R. W. (1977). *Weeds of Queensland*. Brisbane: Government Printer.

Kloot, P. M. (1979). The native and naturalised *Cyperus* species in South Australia. *Journal of the Adelaide Botanic Gardens*, **1**, 333–41.

Kloot, P. M. (1984). The introduced elements of the flora of southern Australia. *Journal of Biogeography*, **11**, 63–78.

Kloot, P. M. (1985). Studies in the alien flora of the cereal rotation areas of South Australia. Ph.D. thesis, University of Adelaide.

Kloot, P. M. (1986*a*). A review of the naturalised alien flora of the cereal areas of South Australia. *Department of Agriculture, South Australia, Technical Paper*, **12**.

Kloot, P. M. (1986*b*). Checklist of the introduced species naturalised in South Australia. *Department of Agriculture, South Australia, Technical Paper*, **14**.

Kloot, P. M. (1987*a*). The naturalised flora of South Australia. 2. Its development through time. *Journal of the Adelaide Botanic Gardens*, **10**, 91–8.

Kloot, P. M. (1987*b*). The naturalised flora of South Australia. 3. Its origin, introduction, distribution, growth forms and significance. *Journal of the Adelaide Botanic Gardens*, **10**, 99–111.

Loudon, J. C. (1830). *Loudon's Hortus Brittanicus*. London: Longman.

Naveh, Z. (1967). Mediterranean ecosystems and vegetation types in California and Israel. *Ecology*, **43**, 445–59.

Nelson, E. C. (1981). Phytogeography of southern Australia. In *Ecological Biogeography of Australia*, vol. 1, ed. A. Keast, pp. 733–59. The Hague: Junk.

Powell, C. M., Johnson, B. D. & Veevers, J. J. (1981). The Early Cretaceous break-up of Eastern Gondwanaland, the separation of Australia and India, and their interaction with South-east Asia. In *Ecological Biogeography of Australia*, vol. 1, ed. A. Keast, pp. 15–29. The Hague: Junk.

Raven, P. H. (1973). The evolution of Mediterranean floras. In *Mediterranean-type Ecosystems: Origin and Structure*, ed. F. di Castri & H. A. Mooney, pp. 213–24. Berlin: Springer-Verlag.

Ross, J. H. (1976). An analysis of the flora of Victoria. *Muelleria*, **3**, 169–76.

Specht, R. L. (1981). Major vegetation formations in Australia. In *Ecological Biogeography of Australia*, vol. 1, ed. A. Keast, pp. 163–297. The Hague: Junk.

Todd, M. A. (1979). A conspectus of new records and nomenclature for vascular plants in Victoria during the period 1970–1977. *Muelleria*, **4**, 173–9.

Todd, M. A. (1981). A conspectus of new records and nomenclature for vascular plants in Victoria. 2. 1978–early 1980. *Muelleria*, **4**, 429–38.

Todd, M. A. (1985). A conspectus of new records and nomenclature for vascular plants in Victoria. 3. Early 1980–early 1984. *Muelleria*, **6**, 59–78.

Tutin, T. G., Heywood, V. H., Burges, N. A., Moore, D. M., Valentine, D. H., Walters, S. M. & Webb, D. A. (ed.) (1964–1980). *Flora Europaea*, 5 vols. Cambridge: Cambridge University Press.

Wells, M. J. & Stirton, C. H. (1982). South Africa. In *Biology and Ecology of Weeds*, ed. W. Holzner & M. Numata, pp. 339–43. The Hague: Junk.

Willis, J. H. (1972). *A Handbook to Plants in Victoria*, vol. 2. Melbourne: Melbourne University Press.

12

Life cycles of some Mediterranean invasive plants

I. OLIVIERI, P.-H. GOUYON & J.-M. PROSPERI

In this chapter we will consider some life history traits, such as longevity and fecundity, dispersal in space and in time, and the type of reproductive system (inbreeding or outbreeding). Some of the factors which may act on these traits, as reviewed by Stearns (1976, 1977) and Charlesworth (1980), can be summarised as:

1. Perennial types (polycarpic or monocarpic) (Hart, 1977) are to be favoured if, because of environmental conditions, juvenile survival is low compared to adult survival (Michod, 1979), or alternatively, if the environmental conditions are less predictable during the juvenile stage than later in the life cycle (Murphy, 1968).

2. Perennial genotypes are also favoured in species in which population sizes are usually stable or decreasing, whilst annual types are favoured in species in which populations are increasing most of the time. As pointed out by Caswell (1982), however, no population is always increasing or stable, so that the result for a given genotype depends on the relative importance of phases of increase and decrease.

3. High dispersal rates are selected against within each population and selected for at the establishment of new populations. Van Valen (1971) and others (e.g. Slatkin & Wade, 1979; Olivieri & Gouyon, 1985) emphasised the importance of the rate of extinction on the evolutionary stable dispersal rate. We shall see subsequently that there is another important environmental factor which should also be considered.

4. Inbreeding versus outbreeding is more difficult to summarise. Jain (1976) reviewed most of the ecological factors which may explain the existence of inbreeders, viz. shortage of pollinators (at establishment or after (Baker, 1963)), mechanism of genetic isolation (Antonovics, 1968), etc.

More recent reviews, especially those of geneticists, have attempted to explain the existence of outbreeding species (Lloyd, 1979; Wells, 1979; Feldman & Christiansen, 1984; Lande & Schemske, 1985; Holsinger, 1986). Indeed, from the point of view of population genetics, in the absence of any force favouring outbreeding, inbreeding should always be favoured (Fisher, 1930); this in turn is related to the general problem of the cost of meiosis (Maynard Smith, 1978). Briefly summarised, the amount of inbreeding depression, the availability of pollinators in entomophilous species and the importance of heterozygosity at establishment (which allows a coloniser to produce genetically variable off-spring) and subsequently (co-evolution with pathogens, for instance) will determine the outcome of selection on the reproductive regime.

The general patterns and trends of life history attributes of invasive species can be found in Baker & Stebbins (1965), and especially in Lewontin (1965) and Bazzaz (1986). Since mediterranean-type ecosystems are probably more disturbed than most others because of the long history of fire and heavy grazing, some characteristic features linked to disturbance are likely to be enhanced in colonising and invasive species of mediterranean-type ecosystems. Classically, colonising species are supposed to be annual polyploid self-crossers with a high fecundity, and having efficient dispersal mechanisms. But invasive plants, especially some originating in the Mediterranean Basin, do not always fit this generalisation.

In this chapter, we shall discuss some examples of Mediterranean plants – *Medicago* species, *Trifolium subterraneum*, the slender thistles *Carduus pycnocephalus* and *C. tenuiflorus*, and *Thymus vulgaris* – which show diverse life history traits in relation to their biogeography and their invasive abilities.

Biogeography of the genus *Medicago*

We shall firstly compare invasive and non-invasive species of the genus *Medicago* by using results from both the literature and from our own recent biogeographic study on the distribution of *Medicago* species in Spain and Corsica.

Of 55 recognised species of *Medicago*, of diverse ploidy levels ($2n = 16$ to $6x = 48$), 33 are selfing annuals, of which six species are cultivated as annual forage species in cereal rotations in the ley-farming systems of mediterranean-climate regions of southern Australia. A further 18 species are outcrossing perennials (including *Medicago sativa* – lucerne or alfalfa – the important forage crop). Three species are selfing perennials (namely *Medicago suffruticosa*, *M. hybrida* (Lesins & Lesins, 1979) and *M. marina* (Olivieri, unpublished data) and one species is a selfing facultative annual, biennial or perennial (*M. lupulina*).

A complete description of the genus and its evolution may be found in Heyn (1963) and Lesins & Lesins (1979). According to Heyn (1963), the genus

Table 12.1. *Number of species of annual (Section Spirocarpos) and perennial (Section Falcago) species of* Medicago *occurring in different regions, from different floras*

Region	No. *Medicago* species	No. perennials	No. annuals
Italy and adjacent islands	27	4 (15%)	23 (85%)
Eastern Mediterranean countries	34	12 (35%)	22 (65%)
All parts of the USSR	33	18 (55%)	15 (45%)

Source: Heyn, 1963.

Medicago is native to western Asia and the Mediterranean Basin. She wrote that:

> Kousnetzoff (1926) who investigated the area of four perennial species of *Medicago* (*M. sativa*, *M. falcata*, *M. platycarpa* and *M. lupulina*) regards the genus *Medicago* as Mediterranean. In the case of the annual species of *Medicago* difficulties arise when defining the areas of distribution of the species. The main reason for the spurious distribution of the annual species is to be sought in the extensive trade in raw wool not cleaned of attached spiny fruits. Species or varieties of *Medicago* with soft spiny pods are of course the most affected. This explains the world-wide distribution of such species as *M. minima*, *M. arabica*, and *M. polymorpha* . . . The majority of annual species of *Medicago* is distributed throughout the northern Mediterranean from the Iberian Peninsula in the west to Palestine in the east, their number largest at the two extremes. Unlike section Spirocarpos (annual), the species of section Falcago (perennial) are mostly Central and Western Asiatic.
>
> The Mediterranean region is considered by several authors to have provided a refuge for plants from adjacent regions especially Central and Western Asia, in periods when climatic conditions worsened.

From data collected from different floras, Heyn showed that perennial species of *Medicago* are much more strongly represented in the USSR and eastern Mediterranean countries than in western Mediterranean countries (Table 12.1). This pattern could be related to the geographic origin of perennial species (Iran, Turkey) and to the high level of juvenile mortality found in northern countries (USSR), where only perennial species can establish, because of the cold winter. Annual species are supposed to have originated from perennial species at the

Table 12.2. *Life history traits of* Medicago *species found in four regions of the Mediterranean Basin*

	Species	Lo	P	Spi	Re	Sa	Co	Sp	Al
L	*M. lupulina**	P	16, 32	N	S	—	@	@	—
L	*M. secundiflora*	A	16	N	S	—	@	@	2%
OC	*M. carstiensis*	P	16	N	O	—	—	—	—
OP	*M. platycarpa*	P	16	N	O	—	—	—	—
OP	*M. ruthenica*	P	16	N	O	—	—	—	—
OO	*M. orbicularis*	A	16	N	S	21%	64%	33%	26%
OHy	*M. radiata*	A	16	N	S	—	—	—	—
OHE	*M. heyniana*	A	16	N	S	—	—	—	—
OC	*M. cretacea*	P	16	N	O	—	—	—	—
MFF	*M. falcata*	P	16, 32	N	O	—	—	—	—
MFF	*M. sativa**	P	16, 32	N	O	—	@	55%	—
MFF	*M. glomerata*	P	16	N	O	—	—	—	—
MFF	*M. glutinosa*	P	32	N	O	—	—	—	—
MFF	*M. prostrata*	P	16, 32	N	O	—	—	—	—
MFR	*M. rhodopea*	P	16	N	O	—	—	—	—
MFR	*M. saxatilis*	P	48	N	O	—	—	—	—
MFR	*M. rupestris*	P	16	N	O	—	—	—	—
MFR	*M. cancellata*	P	48	N	O	—	—	—	—
MFD	*M. daghestanica*	P	16	N	O	—	—	—	—
MFD	*M. pironae*	P	16	N	O	—	—	—	—
MFP	*M. dzawakhetica*	P	32	N	O	—	—	—	—
MFP	*M. papillosa*	P	16, 32	N	O	—	—	—	—
MA	*M. arborea**	P	32, 48	N	O	—	—	—	§
MM	*M. marina*	P	16	N	S	—	—	@	—
MS	*M. suffruticosa*	P	16	N	S	—	—	1%	—
MS	*M. hybrida*	P	16	N	S	—	—	—	—
SR	*M. rotata*	A	16	N	S	—	—	—	—
SR	*M. bonarotiana*	A	16	N	S	—	—	—	—
SR	*M. noeana*	A	16	N	S	—	—	—	4%
SR	*M. shepardii*	A	16	N	S	—	—	—	—
SR	*M. rugosa**	A	32	N	S	—	—	!	1%
SR	*M. scutellata**	A	32	N	S	—	—	3%	3%
SP	*M. soleirolii*	A	16	N	S	—	—	!	1%
SP	*M. tornata**	A	16	N	S	—	—	3%	1%
SP	*M. littoralis**	A	16	Y	S	—	2%	21%	2%
SP	*M. truncatula**	A	16	Y	S	11%	47%	29%	27%
SP	*M. rigidula*	A	14	Y	S	—	25%	1%	1%
SP	*M. murex*	A	14, 16	Y	S	21%	33%	2%	4%
SP	*M. constricta*	A	14	Y	S	—	—	—	—
SP	*M. turbinata*	A	16	N	S	7%	—	5%	—
SP	*M. doliata* (= *aculeata*)	A	16	Y	S	—	1%	27%	21%

Table 12.2 (*cont.*)

	Species	Lo	P	Spi	Re	Sa	Co	Sp	Al
SL	*M. sauvagei*	A	16	Y	S	—	—	—	—
SL	*M. laciniata*	A	16	Y	S	—	—	—	1%
SL	*M. minima*	A	16	Y	S	9%	27%	37%	12%
SL	*M. praecox*	A	14	Y	S	—	23%	1%	—
SL	*M. coronata*	A	16	Y	S	—	—	!	—
SL	*M. polymorpha**	A	14	Y	S	54%	88%	48%	46%
SL	*M. arabica*	A	16	Y	S	36%	65%	2%	1%
SL	*M. lanigera*	A	16	N	S	—	—	—	—
SL	*M. disciformis*	A	16	Y	S	—	—	!	—
SL	*M. tenoreana*	A	16	Y	S	—	—	!	—
SI	*M. intertexta*	A	16	Y	S	6%	—	1%	4%
SI	*M. ciliaris*	A	16	Y	S	—	—	5%	20%
SI	*M. muricoleptis*	A	16	Y	S	—	—	—	—
SI	*M. granadensis*	A	16	N	S	—	—	!	—

The initials in the first column represent the subgenus, the section and the subsection to which the species belongs according to Lesins & Lesins (1979). Lo, longevity; P, chromosome number (which indicates the ploidy level, $n = 7$ or 8); Spi, presence (Y) or absence (N) of spines on pods; Re, reproductive system (S, selfer; O, outcrosser); *, used by humans as forage species; — species not found; !, species not found, but is given as present in Heyn (1963); @, only a few individuals of the species found; §, species introduced recently; Sa, Sardinia; Co, Corsica; Sp, Spain; Al, Algeria.

time when the Strait of Gibraltar was closed, as a result of successive flooding and drying of the Mediterranean Sea (Lesins & Lesins, 1979).

On the smaller scale of the western Mediterranean region, Table 12.2 shows, along with some life history traits, the frequency of occurrence (per cent of collection sites) of each of the 54 species of the genus in Sardinia (data from 208 sites from Piano *et al.*, 1980), Corsica (data from 91 sites of Prosperi *et al.*, 1989b), Spain (data from 190 sites of Prosperi *et al.*, 1989a) and Algeria (data from 202 sites of A. Abdelguerfi *et al.*, 1988). These data indicate the success of establishment of the different species in western Mediterranean regions. Among the annual species which were not found in these four regions, most of them are well represented in more eastern countries, some being endemic there (e.g. *M. noeana* and *M. shepardii* in Turkey and Iraq, *M. lanigera* in central Asia, *M. muricoleptis* in Italy and *M. sauvagei* in Morocco).

In the western Mediterranean region, annual species are more successful than perennial species (Table 12.2). Among the perennials, even *M. sativa*, although cultivated all over the world, is not frequently found as natural populations. There is no other perennial species of *Medicago* which can be considered a successful invader. For instance, *M. marina* is found only on the littoral on sand dunes and *M. glomerata*, a species which is intermediate between septentrional *M. falcata* and Mediterranean *M. sativa*, may be considered an endangered species; we could find it only near Nice in southern France.

Most annuals which are successfully invasive have spiny pods. An exception is *M. orbicularis*, which has big but smooth pods. We do not agree with Heyn (1963) that it might be dispersed by wind. Rather, the main reason for its success could be the high number of seeds per pod (about 20, instead of the four or five for most other species) and, moreover, the high percentage of hard (dormant) seeds.

Some annual species have ceased being invasive because the environmental conditions have changed, although they were initially well adapted to the western Mediterranean conditions. An example of this group is *M. scutellata* which used to be frequent in southern France until the last century where it was associated with vineyards and cereal rotations. Because of changes in land use in the region in this century (see Lepart & Debussche, this volume), one can now hardly find populations of this species. In addition to changes in land use its pods are not spiny, so that it would be difficult for this species to establish in new habitats.

Because most annual species of the genus *Medicago* are inbreeders, the relationship between reproductive system and invasive status is obvious, but it is not necessarily linked to selection for invasive ability. Also, most annuals are diploid (only two are tetraploid), whilst nine out of 21 perennials are polyploid (tetraploid or hexaploid). This difference could be explained by radiation in a diploid group followed by the development of polyploidy in some species. In *Medicago* there is, therefore, an association between a low degree of ploidy and invasive ability, the reverse of what we expected.

Changes after introduction or colonisation

Trifolium subterraneum *(subterranean clover)*

Trifolium subterraneum is a forage species which was accidentally introduced to Australia in the last century (Gladstones, 1966; Rossiter, 1966*a*, *b*, 1977). The species rapidly colonised the mediterranean-climate regions of southern Australia. Subsequently, it was recognised as a useful forage species and an artificial genealogical selection program was begun using material collected mainly from natural populations in Corsica and Portugal. The species is peculiar in that it buries its seeds in the soil just after or even prior to self-fertilisation, i.e. it may be cleistogamous. The degree of seed dormancy is very high and dispersal in space (and therefore colonisation) seems to have occurred only because of the attachment of the lanigerous pods to the wool of sheep.

In a recent study, we compared two Australian cultivars (Mount Barker and Clare) and eight Corsican ecotypes, all grown as 50 spaced plants per population in a randomised block design. We could distinguish two types of plants, with the two cultivars being extreme for most of the characters measured. The first type was the Mount Barker type which was characterised by many stolons, short internodes, many branches, a very high level of green biomass, but very few pods. The mean dry matter allocated to sexual reproduction was very low (30 per cent), whilst the length of the growth cycle was the longest. The other type of response was the Clare type which was characterised by very few but longer stolons, with no branching and many flowers. The reproductive parts represented about 52 per cent of the total dry matter, and the length of the growth cycle was the shortest of all ten populations. Corsican populations were intermediate between the two cultivars for most characters, except for the reproductive biomass, which in some populations represented an average of 61 per cent of the total dry matter.

The two extremes, which are the result of artificial selection after colonisation, may be seen as representative of two different strategies. The 'Clare' pattern is probably more successful when colonising empty sites (enhanced dispersal on a small scale), whilst the 'Mount Barker' pattern can prevent the species from being eliminated from a successional sequence (higher competitive ability).

Carduus pycnocephalus *and* C. tenuiflorus *(slender thistles)*

Carduus pycnocephalus and *C. tenuiflorus* are Mediterranean species which have invaded Australia and the western USA. Although they occur in pastures, they have, however, no forage value and occur as spiny-leaved weeds. We showed (Olivieri, 1984) that Australian populations were much less resistant to a pathogen naturally present in southern France than plants collected near the place where the pathogen was located. We also showed (Olivieri, 1985) that in California the two species seemed to be more genetically separated than in France, since the seed set of the hybrids was much lower there than that in southern France. These two results may be interpreted as consequences of invasion.

At the intra-specific level we also showed (Olivieri & Gouyon, 1985) that the evolution of dispersal in colonising species has to be studied at the 'metapopulation' level (Couvet *et al.*, 1985). We define a metapopulation as a group of populations which are founded one by the others, and then evolve independently until extinction through the ecological processes of succession or stochastic disturbances. In particular, thistles produce two types of seeds (Olivieri *et al.*,

1983): dispersed seeds with a pappus and little dormancy, and non-dispersed seeds without a pappus and with a high dormancy. The ratio of dispersed seeds is partially genetically determined. We observed (Olivieri & Gouyon, 1985) a decrease of the ratio of dispersed seeds with increasing age of the population, a result which is easily explained if one considers that firstly, new populations are founded by migrants and that there is a strong selection at establishment for those genotypes which produce a higher proportion of seeds with a pappus, and secondly, that dispersed seeds are lost within a given site.

Because thistles are poor competitors, the maximum lifespan of a given population is likely to be short (less than 10 years) through the process of ecological succession. Thus the extinction of each population is ineluctable because of either stochastic disturbance or species replacement. Two measures of landscape – the degree of disturbance and the speed of species replacement in the succession – will determine the evolutionarily stable dispersal rate which is selected at the metapopulation level.

The percentage of seeds with a pappus in slender thistles is very high (about 80 per cent) compared to the evolutionarily stable dispersal rate that can be predicted from theoretical considerations (Olivieri & Gouyon, 1985). This finding suggests that the migration rate, which is calculated as the ratio of dispersed seeds, is overestimated; indeed, most seeds with a pappus do not migrate very far. It further suggests that another factor is acting on this ratio, such as escape from competition within each population. Selection for short-distance dispersal would then be the primary factor maintaining a high proportion of seeds with a pappus. Nevertheless, only those seeds which migrate far can lead to the continued establishment of new populations and, thus, the maintenance of the species in a given landscape. In other words, the high colonising ability of slender thistles with their high capacity for dispersal may be the result of selection acting either at the metapopulation or at the population level. In the former case the mechanism may be by escape to stochastic extinctions or, at the inter-specific level, escape from competition through long-distance dispersal (replacement by other species) or at the intra-specific level, by the establishment of new populations. In the latter case, the high colonising ability of the slender thistles may arise by escape from competition through short-distance dispersal.

Thymus vulgaris *(thyme)*

Thymus vulgaris is a gynodioecious Mediterranean species. Although perennial, this species is very successful at colonisation during the early stages of succession. The species is becoming invasive in the drier inland region of the province of Otago, New Zealand (Wilkinson *et al.*, 1979).

Thymus vulgaris has been shown to have a particular pattern to its reproductive system. In disturbed areas, the populations contain a majority of females (60–95 per cent), the remainder of the population being hermaphrodite. These populations are thus highly outcrossed (Dommée *et al.*, 1978, 1983). On the other hand in stable environments, populations contain high proportions of hermaphrodites and are likely to be more inbred. Dommée & Jacquard (1985) observed that young populations (more frequent on disturbed sites) contain more females than older populations (which are more likely to be found in stable environments). Since females produce more seeds than hermaphrodites (because of a re-allocation of resources previously used for the male function) and since they are outcrossed, they probably help *T. vulgaris* to colonise new habitats. Until recently, the main explanation for the success of thyme was thus its ability to regulate its selfing rate (Dommée *et al.*, 1983).

The difference between young and old populations of *Thymus vulgaris* was explained by Gouyon *et al.* (1983), Gouyon & Couvet (1985) and Couvet *et al.* (1986) as a result of differences in the dynamics of nuclear and cytoplasmic genes involved in the determination of sex. Male sterility is induced by cytoplasmic factors, which are transmitted by female gametes only. Specific nuclear genes can restore male fertility. Hermaphrodites may therefore be considered as 'restored cytoplasms' and females as 'non-restored cytoplasms'. Different cytoplasms are usually found in gynodioecious species, with specific nuclear restorer genes at different loci. By a founder effect, new sites are colonised by females and hermaphrodites which do not have the nuclear restorer genes corresponding to female cytoplasm. Because females produce more seeds, patches of females will grow faster than patches of hermaphrodites, so that females will be more frequent in young populations. Eventually, pollen carrying the specific restorer genes will appear through mutation or immigration of new hermaphrodites or because the different patches will be getting closer. They will invade the population, which will then become mainly composed of restored cytoplasms, i.e. of hermaphrodites. Indeed, the fitness of a nuclear gene restoring male fertility to a patch of females is likely to be very high. Then most individuals (females on the one hand and hermaphrodites coming from restored cytoplasm on the other) will have the same cytoplasm, so that the genetic determination of sex will be nuclear. In this case, it is much more difficult to maintain a high proportion of females, for it can be shown that the observed female advantage of seed production allows the maintenance of a low percentage of females only.

Even if it were useful for *T. vulgaris* to be outcrossed in disturbed habitats, the proximate cause of the reproductive pattern just described probably originates from nucleocytoplasmic interactions, and not from adaptive causes. In other words, it might be that thyme is a successful coloniser because of its genetic system, but not because of selection acting on life history traits linked to colonisation. The pattern resulting, however, seems to allow *T. vulgaris* to be very efficient at invading new sites and remaining present in later stages of succession. *Thymus vulgaris*, despite its Mediterranean habitat, is not characteristic of any association mentioned by Braun-Blanquet *et al.* (1952).

Its colonising ability could be the reason why *Thymus vulgaris*, despite its low migration rate (Belhassen *et al.*, 1987; Gouyon *et al.*, 1987) , is the only woody species of *Thymus* to have recolonised southern France after the last glaciation from among the 40 species remaining in Spain on the protected side of the Pyrenees.

Conclusion

Comparisons between related species of *Medicago* have confirmed the usual predictions linking life history traits with invasive abilities (semelparity, dispersal in space and time, inbreeding). Within a single (and moreover, annual) Mediterranean species – namely *Trifolium subterraneum* – one can find very different life histories. These arise as a result of artificial selection where humans have modified the species towards extreme strategies compared to what is observed in natural populations.

Models of breeding systems may predict different patterns according to the factor being considered. For instance, ease of selfing may be favoured at establishment, but being outbred at colonisation may provide a higher potential to adapt to local conditions through a higher heterozygosity of the founders. Obviously, different solutions have been adapted by the different species, selfing in invasive species of *Medicago* and outcrossing in invasive genotypes of *Thymus vulgaris*. To understand these different patterns, it is necessary to study the mechanisms involved in the determination of the genetic system. The example of *Thymus vulgaris* shows that being a successful coloniser may be a consequence of selection at a different level. This could be a very general case.

The evolution of dispersal and of the breeding system of invasive plants can be best understood by considering a level above that of the population, namely that of the metapopulation. Models at this level were developed to account for the evolution of dispersal in the slender thistles (Olivieri & Gouyon, 1985). They are currently being extended to allow study of the evolution of other life history traits. The basic concept underlining these studies is the disequilibrium status of most populations (Gouyon, 1990; Olivieri *et al.*, 1990).

References

Abdelguerfi, A., Chapot, J. Y. & Conesa, A. P. (1988). Contribution à l'étude de la répartition des luzernes annuelles spontanées en Algérie selou certains facteurs du milieu. *Fourrages*, **113**, 89–106.

Antonovics, J. (1968). Evolution in closely adjacent populations. V. Evolution of self-fertility. *Heredity*, **23**, 219–38.

Baker, H. G. (1963). Evolutionary mechanisms in pollination biology. *Science*, **139**, 877–83.

Baker, H. G. & Stebbins, G. L. (1965). *The Genetics of Colonizing Species*. New York: Academic Press.

Belhassen, E., Dockes, A. C., Gliddon, C. & Gouyon, P. H. (1987). Dissémination et voisinage chez une espèce gynodioique: le cas de *Thymus vulgaris*. *Génétique, Sélection, Evolution*, **19**, 307–20.

Bazzaz, F. A. (1986). Life history of colonizing plants: some demographic, genetic and physiological features. In *Ecology of Biological Invasions of North America and Hawaii*, ed. H. A. Mooney & J. A. Drake, pp. 96–110. New York: Springer-Verlag.

Braun-Blanquet, J., Negre, R. & Roussine, N. (1952). *Les Groupements Végétaux de la France Méditerranéenne*. Paris: CNRS.

Caswell, H. (1982). Life history theory and the equilibrium status of populations. *American Naturalist*, **120**, 317–39.

Charlesworth, B. (1980). *Evolution in Age-structured Populations*. Cambridge: Cambridge University Press.

Couvet, D., Gouyon, P. H., Kjelleberg, F., Olivieri, I., Pomente, D. & Valdeyron, G. (1985). De la métapopulation au voisinage: la génétique des populations en déséquilibre. *Génétique, Sélection, Evolution*, **17**, 407–17.

Couvet, D., Bonnemaison, F. & Gouyon, F. (1986). The maintenance of females among hermaphrodites: the importance of nuclear–cytoplasmic interactions. *Heredity*, **57**, 325–30.

Dommée, B. & Jacquard, P. (1985). Gynodioecy in thyme, *Thymus vulgaris* L.: evidence from successional populations. In *Genetic Differentiation and Dispersal in Plants*, ed. P. Jacquard, G. Heim & J. Antonovics, pp. 141–64. Berlin: Springer-Verlag.

Dommée, B., Assouad, M. W. & Valdeyron, G. (1978). Natural selection and gynodioecy in *Thymus vulgaris* L. *Botanical Journal of the Linnean Society*, **77**, 17–28.

Dommée, B., Guillerm, J. L. & Valdeyron, G. (1983). Régime de reproduction et hétérozygotie des populations de Thym, *Thymus vulgaris* L. *Comptes Rendus Académie Sciences Paris, III*, **296**, 111–14.

Feldman, M. W. & Christiansen, F. B. (1984). Population genetic theory and the cost of inbreeding. *American Naturalist*, **123**, 642–53.

Fisher, R. A. (1930). *The Genetical Theory of Natural Selection*. Oxford: Clarendon Press.

Gladstones, J. S. (1966). Naturalized subterranean clover (*Trifolium subterraneum* L.) in Western Australia: the strains, their distribution, characteristics and possible origins. *Journal of the Australian Institute of Agricultural Science*, **23**, 302–7.

Gouyon, P. H. (1990). Invaders and disequilibrium. In *Biological Invasions in Europe and the Mediterranean Basin*, ed. F. di Castri, A. J. Hansen & M. Debussche, pp. 365–9. Dordrecht: Kluwer.

Gouyon, P. H. & Couvet, D. (1985). Selfish cytoplasm and adaptation: variations in the reproductive system of thyme. In *Structure and Functioning of Plant Popu-*

lations, vol. 2, ed. J. Haeck & J. W. Woldendorp, pp. 299–320. Amsterdam: North Holland.

Gouyon, P. H., Lumaret, R., Valdeyron, G. & Vernet, P. (1983). Reproductive strategy and disturbance by man. In *Ecosystems and Disturbance*, ed. H. A. Mooney, pp. 213–25. Berlin: Springer-Verlag.

Gouyon, P. H., King, E. B., Bonnet, J. M., Valdeyron, G. & Vernet, P. H. (1987). Seed migration and the structure of plant populations. *Oecologia*, **73**, 92–4.

Hart, R. (1977). Why are biennials so few? *American Naturalist*, **111**, 792–9.

Heyn, C. C. (1963). *The Annual Species of* Medicago. Jerusalem: Scripta Hierosolymitana.

Holsinger, K. E. (1986). Dispersal and plant mating systems: the evolution of self-fertilization in subdivided populations. *Evolution*, **40**, 405–13.

Jain, S. K. (1976). The evolution of inbreeding in plants. *Annual Review of Ecology & Sytematics*, **7**, 469–95.

Kousnetzoff, V. A. (1926). Areas of the geographical distribution of the most important forage species of clover and alfalfa. *Bulletin of Applied Botany and Plant Breeding*, **16**, 55–88 (in Russian with English summary).

Lande, R. & Schemske, D. W. (1985). The evolution of self-fertilization and inbreeding depression in plants. I. Genetic models. *Evolution*, **39**, 24–40.

Lesins, K. A. & Lesins, I. (1979). *Genus* Medicago *(Leguminosae). A Taxonomic Study*. The Hague: Junk.

Lewontin, R. C. (1965). Selection for colonizing ability. In *The Genetics of Colonizing Species*, ed. H. G. Baker & G. L. Stebbins, pp. 77–94. New York: Academic Press.

Lloyd, D. G. (1979). Some reproductive factors affecting the selection of self-fertilization in plants. *American Naturalist*, **113**, 67–79.

Maynard Smith, J. (1978). *The Evolution of Sex*. Cambridge: Cambridge University Press.

Michod, R. E. (1979). Evolution of life-histories in response to age-specific mortality factors. *American Naturalist*, **113**, 531–50.

Murphy, G. E. (1968). Patterns in life-history phenomena and the environment. *American Naturalist*, **102**, 52–64.

Olivieri, I. (1984). Effect of *Puccinia cardui-pycnocephali* on slender thistles (*Carduus pycnocephalus* and *C. tenuiflorus*). *Weed Science*, **32**, 508–10.

Olivieri, I. (1985). Comparative electrophoretic studies of *Carduus pycnocephalus* L., *C. tenuiflorus* Curt. (Asteraceae) and their hybrids. *American Journal of Botany*, **72**, 715–18.

Olivieri, I., Couvet, D. & Gouyon, P. H. (1990). The genetics of transient populations: research at the metapopulation level. *Trends in Ecology & Evolution*, **5**, 207–10.

Olivieri, I. & Gouyon, P. H. (1985). Seed dimorphism for dispersal: theory and implications. In *Structure and Functioning of Plant Populations*, ed. J. Haeck & J. W. Woldendorp, pp. 77–90. Amsterdam: North Holland.

Olivieri, I., Swan, M. & Gouyon, P. H. (1983). Reproductive system and colonizing strategy of two species of *Carduus* (Compositae). *Oecologia*, **60**, 114–17.

Piano, E., Sardara, M. & Pesceddu, S. (1980). Observations on the distribution and ecology of subterranean clover and other annual legumes in Sardinia. *Rivista di Agronomia*, **3**, 273–83.

Prosperi, J. M., Delgado Enguita, I. & Angevain, M. (1989*a*). Prospection du genre *Medicago* en Espagne et en Portugal. *Plant Genetic Resources Newsletter*, **78 (179)**, 27–9.

Prosperi, J. M., Gensollen, Y., Olivieri, I. & Mansat, P. (1989*b*). Observations sur la répartition et l'écologie de luzernes annuelles et de trèffe souterrain en Corse. *Proceedings XVI Congrès International des Herbages, Nice, France, 1989*.

Rossiter, R. C. (1966*a*). Ecology of the mediterranean annual-type pasture. *Advances in Agronomy*, **18**, 1–56.

Rossiter, R. C. (1966*b*). The success or failure of strains of *Trifolium subterraneum* L. in a mediterranean environment. *Australian Journal of Agricultural Research*, **17**, 425–46.

Rossiter, R. C. (1977). What determines the success of subterranean clover strains in south-western Australia? *Proceedings of the Ecological Society of Australia*, **10**, 76–88.

Slatkin, M. & Wade, M. J. (1979). Group selection on a quantitative character. *Proceedings of the National Academy of Science, USA*, **75**, 3531–4.

Stearns, S. C. (1976). Life history tactics: a review of the ideas. *Quarterly Review of Biology*, **51**, 3–47.

Stearns, S. S. (1977). The evolution of life history traits: a critique of the theory and a review of the data. *Annual Review of Ecology & Systematics*, **8**, 145–71.

Van Valen, L. (1971). Group selection and the evolution of dispersal. *Evolution*, **25**, 591–8.

Wells, H. (1979). Self-fertilization: advantageous or deleterious? *Evolution*, **33**, 252–5.

Wilkinson, E. L., Dann, G. M. & Smith, G. J. S. (1979). *Thyme in Central Otago*. Christchurch, New Zealand: Tussock Grasslands & Mountain Lands Institute.

13

Invasion processes as related to succession and disturbance

J. LEPART & M. DEBUSSCHE

Interest in biological invasions has quickened recently (see Groves & Burdon, 1986; Kornberg & Williamson, 1986; Macdonald *et al.*, 1986; Mooney & Drake, 1986; di Castri *et al.*, 1990), both in terms of revising the theoretical framework for the subject and the addition of supplementary data. Two main questions underlie this awakened interest: firstly, what are the biological features that determine whether a species will become an invader or not, and secondly, what are the site properties that determine whether an ecological system will be relatively prone to or resistant to invasion? The answers are many and relate to biological characteristics such as a high reproductive rate, high dispersal ability, inbreeding capacity, disturbance frequency, lack of predators or disease, climatic matching, vacant niche, etc. Disturbance attributable to human actions is a factor in many cases, but the very multiplicity of interpretations indicates that there is neither a simple cause nor a few obvious characteristics which are able to explain biological invasions (Elton, 1958). As Crawley (1987) said, we are 'unable to predict whether a particular introduction will be successful', but we guess that many communities are invasible and that a number of common and widespread plants may behave as invaders.

In the Mediterranean Basin, invaders are comparatively few and they generally have little economic impact, contrary to the situations in other areas of mediterranean climate (Guillerm, this volume). This contrast probably originates from the fact that agriculture, farming and associated disturbances have developed in the Mediterranean Basin from 9000 to 10000 years BP (e.g. Struever, 1971; Camps, 1982), before being exported throughout the world. The causality of invasion processes should have an important historical component even though it may be difficult to evaluate. The ways by which invasions take place are in many cases easier to investigate.

Mack (1985) described a species as invasive if it 'enters a territory in which it has never before occurred, regardless of circumstances'. Gray (1986) added that 'a successful invader is one that rapidly expands from its founding colony or colonies'. Bazzaz (1986) depicted the colonisation process as consisting of different phases and these are summarised in Figure 13.1.

Invasion and succession processes seem *a priori* to be closely similar. However, three differences may be suggested:

1. Genetic drift and other evolutionary aspects could be more important in invasion processes than in succession processes, considering the limited size of the populations concerned and the lack of pre-existing adaptations to biotic and abiotic conditions.

2. The very nature of 'safe' sites could be different, as is the case when succession follows the facilitation model or when it is linked to resource changes (e.g. Clements, 1928; Odum, 1969; Connell & Slatyer, 1977; Tilman, 1982). But it may not be so different when emphasising senescence phenomena and the importance of disturbance to population recruitment (e.g. Watt, 1947; Oliver & Stephens, 1977; Pickett & White, 1985).

3. Population spread generally occurs over short distances in the case of succession and over long distances when invasion processes are considered, although such a difference could strongly depend on landscape heterogeneity.

In this chapter we shall discuss the invasion process from three different aspects by considering firstly, dispersal and landscape, secondly, disturbance, establishment and survival; and thirdly, succession. To illustrate this approach we have chosen five woody plant species growing in the Mediterranean region of France, which show a gradient from obvious invasion to successional spread. The species are *Ailanthus altissima*, which invades a climatic region to which it is alien, *Fraxinus ornus*, which invades areas in its native climatic region from which it was formerly absent, *Phillyrea angustifolia*, which extends its distribution by colonising newly created habitats, *Pinus halepensis*, which establishes extensively in old fields arising as a result of rural depopulation, and finally, *Genista scorpius*, which locally forms dense stands typical of some post-cultural successional stages.

Expansion of woody species in a historic context

Recent and present human impact on vegetation

As elsewhere in the Mediterranean Basin, human influence is evident in the French Mediterranean region. For millennia people have lived there and continuously changed the landscape, thereby causing the extension, retreat or fragmentation of plant distribution. The landscape and its patchily distributed plants reflect, with various delays in time, rises and falls in the economy (see Barry, 1952, 1960; Le Roy Ladurie, 1966). The highest density of population occurred in the middle of the nineteenth century; this period was accompanied by intensive wood cutting and sheep and goat production with flocks of thousands of grazing and browsing animals. At this time, each piece of arable land, even a plot of only a few square metres, was cultivated. Paintings and photographs show us that, until the beginning of the twentieth century, it was a stony landscape with few trees and people everywhere. Rural depopulation, starting at the end of the nineteenth century, increased during the twentieth. After attack by *Phylloxera*, viticulture has continuously encountered difficulties and vineyards have reverted to old fields. Simultaneously, the number of sheep and goat flocks has drastically decreased. Also at the same time, reafforestation on a large scale and the introduction of alien tree species were promoted, mainly in the hilly hinterland. Clear-cutting and coppices of native oaks (*Quercus ilex* and *Q. pubescens*) lasted more or less unchanged until the 1940s. Nowadays, clear-cutting for firewood still occurs, but on a very much smaller scale. In the context of this continual decrease in human population density in rural areas, an epizootic caused by myxomatosis occurred. Introduced deliberately to France in 1952, the virus invaded extremely rapidly, leading to decimation of populations of the rabbit (*Oryctolagus cuniculus*) (Giban *et al.*, 1956). Though less drastic, the disease still rages presently. This large reduction in rabbit numbers has clearly favoured the establishment and growth of seedlings of woody plants (Bourlière, 1956; Morel, 1956; Thomas, 1960, 1963). Finally, this landscape, long influenced by human actions, shows some obvious recent changes (Debussche *et al.*, 1987) and a large range of successional stages (Escarre, 1979; Escarre *et al.*, 1983). As with several other woody plants (trees and shrubs), the five species we have chosen as examples have thrived in this changing situation.

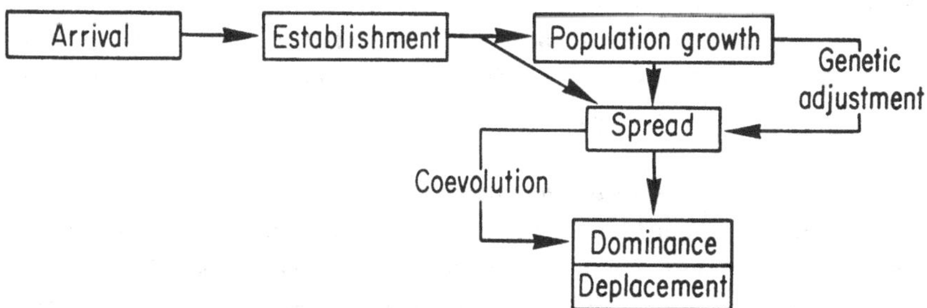

Figure 13.1. The phases of population spread of colonising species (after Bazzaz, 1986).

Ailanthus altissima*: invasion of roadsides*

Up to 20 m high, with a slender trunk and large pinnate leaves, *Ailanthus altissima* is an Asian species, introduced from China to Europe in 1751 as an ornamental tree. Occasionally in the nineteenth century it was introduced as food for silk worms (*Samia cynthia*) (Dupuis, 1862). It is extensively naturalised in central, southern and western Europe. In France, it is mainly observed on roadsides and adjacent places, abandoned quarries, ruins and sometimes on stream banks. This tree seldom forms natural woodlots but grows in small groups, patches or rows.

Such an expansion of its distribution over a period of two centuries and the habitats where it thrives, both may be related particularly to its rapid growth, efficient vegetative reproduction and early seed production (Héricourt, 1861; Dupuis, 1862; Clair-Maczulajtys, 1984). It can attain a height growth of 1 metre per year when young. Many root-sprouts are produced freely on its superficial rooting system as far as 20 or more metres from the trunk and a patch of suckers can spread 1 to 2 metres per year. When buried, a piece of stem can take root and grow in the same manner as a root piece, thereby producing a new individual. The species is able to sprout vigorously when cut or burnt. Seed formation occurs early, when the plant is only 5 to 10 years old and the wind-dispersed seeds are produced abundantly; they are able to germinate in a few days, both in light and in dark.

It is obvious that this species is very resistant to the usual disturbances, and that each time soil containing fragments of stems or roots is moved it spreads. In the same manner floods may distribute the plant along riversides. In the Languedoc region of France the plant is a common invader of roadsides and appears as small trees a few metres high, often in a row of small clumps originating from a single planted tree. A sandy or gravelly soil, as often occurs along roadsides, railway lines, in quarry deposits and in some stream banks, favours the development of a superficial root system. Suckers are not usually successful if they enter dense garrigue or closed oak forests, probably because of soil changes and competition for water. Seedling establishment in natural habitats is very unusual in this region; this quasi-absence may be related to a high moisture level required for germination (Clair-Maczulajtys, 1984).

Fraxinus ornus*: invasion of streamsides*

Widespread throughout Mediterranean and southern central Europe this tree, which may grow up to 20 m high, is doubtfully native to France (Tutin *et al.*, 1972). It was introduced mainly for ornamental purposes and only seldom for limited reafforestation, mainly during the nineteenth and the beginning of the twentieth centuries. In south-eastern France, it is observed along streams and on hill slopes covered by coppices of *Quercus ilex*, *Q. pubescens* and *Ostrya carpinifolia* (Loisel, 1976), whereas in Languedoc it is noticed as a local escape from parks or from small planted woodlots. An obvious example of invasion is along the Hérault river where this species was planted in an arboretum in 1920. The rate of spread from the site of introduction along the river banks has been about 1 kilometre per year downstream (dispersal by water), whereas it is only about 20 or more metres per year upstream (dispersal by wind) (Figure 13.2). Around the site of the arboretum and for a few kilometres downstream, a very high density of mature *F. ornus* trees (up to 800 individuals per kilometre; Figure 13.2) and saplings is observed on the banks of the river which largely dominate native species such as *Alnus glutinosa*, *Populus nigra*, *Fraxinus angustifolia* and *Salix* spp. It is also very abundant on hillsides as an understorey to *Pinus nigra* plantations or mixed with open coppices of *Quercus pubescens*. But away from the site of introduction, the species is restricted to the area flooded by the stream (Thébaud & Debussche, 1990).

In relation to this pattern of spread, Thébaud (1986) and Thébaud &

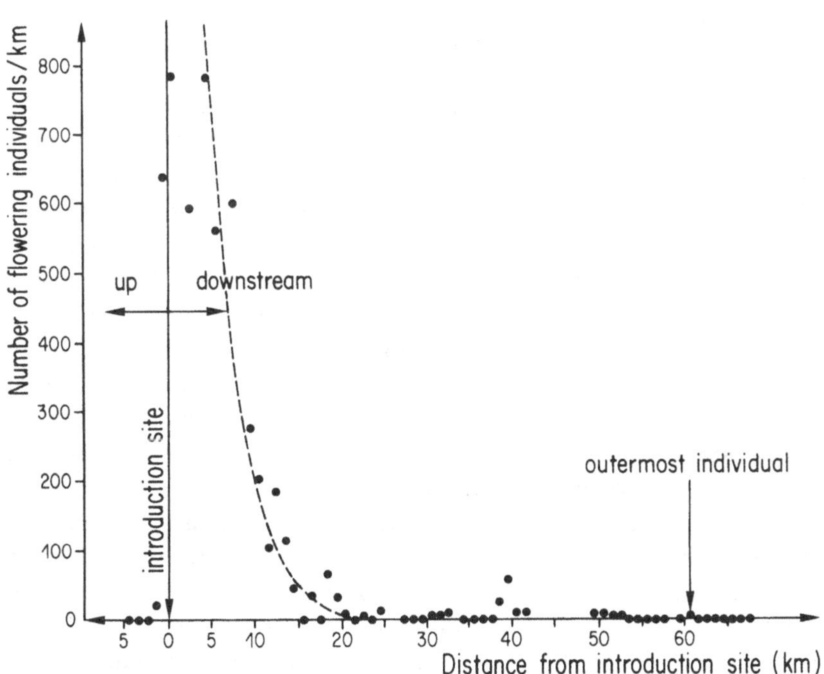

Figure 13.2. Dispersal success of *Fraxinus ornus* on the Vis and Hérault riversides 66 years after its introduction; each dot represents a sampled kilometre (after Thébaud & Debussche, 1990).

Debussche (1990) underline the importance of sprouting and seed characteristics. When felled by the effects of floods, trees are able to produce sprouts from the exposed roots and from the trunk. Treelets may produce seeds when about 12 years old. Seeds are wind-dispersed, usually only over short distances of a few tens of metres, although some seeds are carried much further. Seeds float and are easily carried by running water. Furthermore, germination probably occurs on exposure to light after a dormant period of about one year (see e.g. Villiers & Wareing, 1960, in the case of germination of *Fraxinus excelsior*).

It is obvious from this example that dispersal by water is more efficient than dispersal by wind, even though the seed shows features typical of wind-dispersed species. Water transport is probably an important process for plant invasion, however different the morphology of seeds may be (see Eriksson *et al.*, 1983, on the role of non-adaptive features). Dispersal in this way by water may explain the rapid rates of migration shown by some tree species in Europe after glaciation (Firbas, 1949; Huntley & Birks, 1983). If naturalisation of *Fraxinus ornus* is certain on river and stream banks, because of its sprouting ability after heavy floods and also because of its floating seeds, it is less certain to be found, and then only occasionally, in open woodlands of *Quercus pubescens*; it never occurs in closed oak forest or throughout dry garrigue shrublands.

Phillyrea angustifolia*: a recent and localised coloniser*

This evergreen native shrub, which may grow to 4 m high, is widespread in warm situations generally near the coast and throughout the western part of the Mediterranean Basin. In France, it is frequently observed in areas of Mediterranean climate and at a few sites along the Atlantic coastline. As with many other shrubs and trees, it has taken advantage of rural depopulation, the decrease in sheep grazing and the decimation of rabbits (Tallon, 1955). It is of little importance in the landscape, except in the Camargue region of the Rhône delta, where its expansion has been dramatic (Tallon, 1931; Molinier & Tallon, 1965; Strasberg, 1987; Strasberg & Lepart, unpublished observations).

The Rhône delta has gradually changed its shape over the last ten millennia because of the shifting of the course of the river and because of the transgression of the sea (Pons *et al.*, 1979). A continuous and secure levee bank constructed in the nineteenth century now prevents the Camargue from flooding and from further incursion of the sea. This construction, by modifying the hydrological network, creates new habitats, which are generally drier and more salty than were present prior to construction (Tallon, 1954*a*). Changes in agricultural policy, with different regimes for water management, have also modified habitats considerably (Tallon, 1954*b*; Heurteaux, 1969).

Aerial photographs show the success of *Phillyrea angustifolia* since 1942 at

'La Tour du Valat', a property with typical Camargue habitats (Figure 13.3). The area covered by *P. angustifolia* increased by 3.3 times between 1942 and 1985, with a much higher density of individuals; the different patterns of expansion are limited by local edaphic constraints, such as the presence of temporarily inundated zones and a superficial salt water table, and to changing flock management (Strasberg, 1987).

The main life history traits of this shrub (see Sébastian, 1956; Strasberg, 1987; Strasberg & Lepart, personal observations) include the fact that it sprouts vigorously when cut or burnt, the sprouts arising from dormant buds on the semi-buried lignotuber. Young plants may flower when they are 3 years old and sprouts show flowers the year after cutting or burning. Drupes ripen in autumn and are consumed by birds which disperse the seeds usually over short distances (Debussche *et al.*, 1985); however, common birds such as the European starling (*Sturnus vulgaris*) and thrushes (*Turdus* spp.) may disperse the seeds also over long distances (Debussche & Isenmann, 1985). Germination occurs a few months after dispersal; dormancy is absent.

Although very scattered, or even absent, in the Camargue in the past (as found from pollen records, see Triat-Laval, 1978), the shrub is nowadays present in many swards and post-cultural stages, except for marshy or highly salinised land. Establishment usually occurs after cessation of cultivation at sites with low levels of herbaceous cover, or in places where a dense plant cover is temporarily destroyed by cattle trampling or rabbit scraping (Strasberg, 1987). High recruitment success may follow a sequence of a summer drought followed by increasing mortality of herbaceous plants and a wet winter and spring. *Phillyrea angustifolia* spreads by invading swards close to the existing populations and increases in density around remote isolated individuals (Strasberg, 1987). When established, this vigorously sprouting shrub easily copes with fire and browsing. It is able to survive for a long time, although in a more or less sterile form, as an understorey species in open forest (mostly of *Pinus* spp., *Quercus* spp., *Populus alba* and *Ulmus minor*) or in cleared woodlots.

Pinus halepensis*: a post-disturbance pioneer*

Pinus halepensis is a native and widely distributed tree at low and medium altitudes in the western part of the Mediterranean Basin. In France, it occupies a wide coastal range near the Mediterranean. This range is known to have fluctuated, however, since the last glaciation because of climatic changes and human activities (Triat-Laval, 1978; Reille *et al.*, 1980; Vernet, 1980; Pons, 1984). Evidence for the recent extension of the distribution of *P. halepensis* has been gained from forest surveys spanning a century (Acherar, 1981). During the period 1878–1904 the area covered by this species in Languedoc and Provence increased

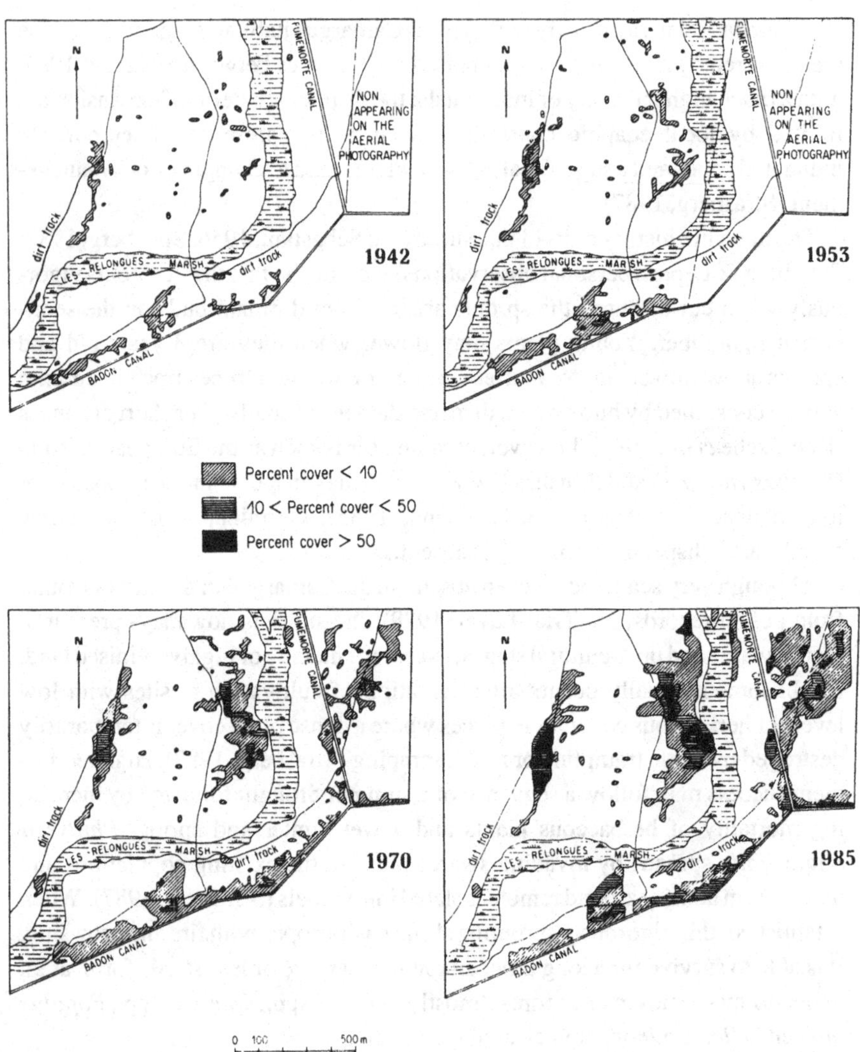

Figure 13.3. Invasion of *Phillyrea angustifolia* at the property 'La Tour du Valat' in the Camargue since 1942 (Strasberg & Lepart, unpublished data).

threefold, whereas it increased by 2.6 times in Languedoc and by 1.6 times in Provence between 1904–1908 and 1971–1978. Invasion by this species is frequently observed in old fields (Acherar, 1981; Acherar *et al.*, 1984) and less often after fires (see Trabaud *et al.*, 1985; Trabaud, this volume). The starting points very probably consist of scattered woodlots on rocky and marly outcrops near the sea, clumps of preserved trees in parks and/or newly reafforested areas.

Several life history traits, mainly characteristics of life span, dispersal and germination, are important in relation to the invasion of *P. halepensis* (see Boudy, 1950; Nahal, 1962; Acherar, 1981). The species is a short-lived tree (about 100 years) unable to sprout when cut or burnt. Cones are borne by young trees as early as 6 years from germination, with viable seeds produced by trees about 12 years old. The seeds are winged and produce abundantly, giving seed densities of up to 25 seeds per m^2 per year at the edge of a mature forest. Seed dispersal occurs mainly over short distances (Figure 13.4), but, as for other wind-dispersed species (see e.g. Ridley, 1930), some seeds may be dispersed a long distance. Germination occurs rapidly and abundantly in light and there is no dormant period. A heavy litter layer and dense vegetative cover (mainly of herbs) strongly limit germination success.

Pinus halepensis occurs in early successional stages of vegetation following cessation of cultivation; it can colonise zones of bare ground, so that in time grass cover becomes sparse (Figure 13.5). Areas of land adjacent to the seed source are colonised mainly (Figure 13.6), whereas isolated trees, surrounded rapidly by

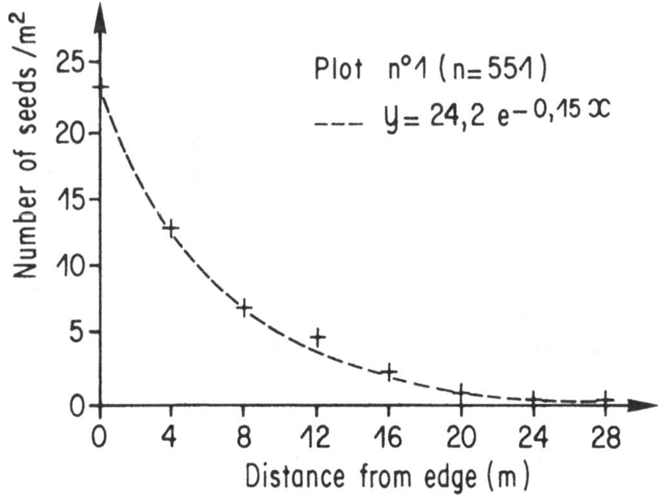

Figure 13.4. Change in the seed rain of *Pinus halepensis* with distance from the forest edge (after Acherar *et al.*, 1984).

saplings, create new foci for invasion. Such trees may be observed as far as several kilometres from any woodlot. In many cases, *P. halepensis* is a transient species, unable to reproduce in its own understorey overgrown with shrubs, and is usually replaced by oaks (*Quercus ilex* and *Q. pubescens*) (Acherar *et al.*, 1984). Just as frequent woodcutting and fires eliminate the species, a lack of disturbance is also unfavourable for its continued presence over a large part of its present range.

Genista scorpius: *pulses of dominance in rangeland and old fields*

Genista scorpius is a silvery green shrub up to 1.5 m tall with thick spines on its twigs; it is native to open habitats in Algeria, the Iberian peninsula and southern France. In this latter area, it is a common component of extensive rangelands and abandoned farmlands throughout limestone regions. This shrub probably assumed importance centuries ago, following deforestation and the development of grazing flocks, because its occitan name ('arjalas') in French Languedoc is found frequently as the name of people, hamlets and villages.

The species is characterised by the following life history traits (Escarre, 1979; Debussche *et al.*, 1980; Debussche, personal observations). If not burnt, its life span is short (about 30 years). Sprouting occurs rapidly after fires. Seeds are dispersed over short distances by the sudden opening of the dry pods and by ants, and over long distances by sheep carrying them in their wool. The seeds may remain viable in the seed bank for several years, and probably for at least 20 years. Dense formations are almost monospecific and impenetrable with a litter of accumulated dry twigs.

Human activities, mainly sheep grazing and local burning, have favoured the

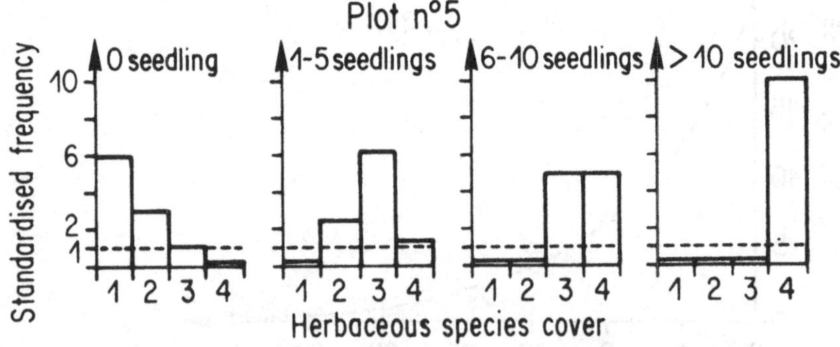

Figure 13.5. Influence of cover of herbaceous species on the density of seedlings of *Pinus halepensis* (4 m^2 sample size); with 1 > 90 per cent, 2 between 90 and 75 per cent, 3 between 75 and 25 per cent and 4 < 25 per cent (after Acherar *et al.*, 1984).

spread of the shrub from its primary habitat, which was probably openings in oak forests maintained by wild herbivores and lightning fires. Human activities may increase its dominance, resulting in a dense formation which can regularly form again at the same place, according to fire frequency (Figure 13.7). If not disturbed for about 30 years, the dense formations of *G. scorpius* senesce and gaps form in the canopy, which enable colonisation by rosaceous shrubs which then rapidly overtop *G. scorpius* (Escarre, 1979). The species is completely eliminated from the understorey when a forest of dense oak (mainly *Quercus pubescens*) has taken its place, but its seeds remain for a long time in the seed bank in the soil.

Discussion

Dispersal and landscape

The species described above are dispersed by several agents (wind, water, animals) which can explain to a certain extent their patterns of distribution and spread. Each of the five species described combine two kinds of dispersal: one, over short distances from the seed source, which generally concerns most of the seeds; whereas the other kind promotes long-distance dispersal of fewer seeds. For *Fraxinus ornus*, wind and water are concerned, for *Phillyrea angustifolia* small- and medium-sized passerine birds are the dispersal agents and for *Genista scorpius* pod opening and dispersal by ants are complemented by longer distance dispersal by sheep. Even if the same vector is responsible for both kinds of dispersal, their origins are different: as in the case of *Pinus halepensis*, with more

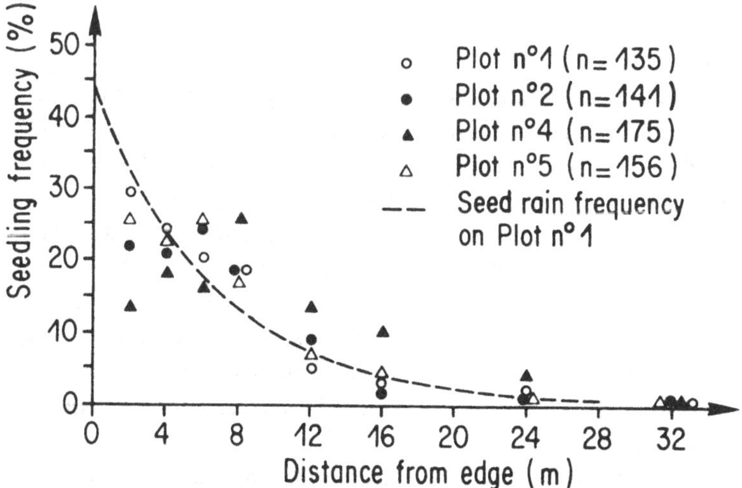

Figure 13.6. Change in the frequency of seedlings of *Pinus halepensis* with distance from the forest edge (after Acherar *et al.*, 1984).

usual wind speeds having a different effect from those of storms. Similarly, the suckers of *Ailanthus altissima* provide dispersal on a different spatial scale from the transport, by humans or by floods, of living pieces of its root and trunk. These two kinds of dispersal are also illustrated morphologically by some species of the family Compositae bearing seeds without and with a pappus which results respectively in short and longer distance dispersal from the parent plant (Venable & Lawlor, 1980; Venable & Levin, 1983).

Studies on succession and on native plant populations generally emphasise dispersal over short distances (e.g. Harper, 1977; Acherar *et al.*, 1984; Debussche *et al.*, 1982, 1985). Studies on invasions, on the other hand, have shown for a long time the importance of long-distance dispersal (e.g. Darwin, 1859; Ridley, 1930; Bannister, 1965; Mack, 1986; Mooney *et al.*, 1986). In fact, the importance given to each of these two types of dispersal seems to be related to landscape heterogeneity; as Harper (1977) has said: 'they are not contrasting, but parts of the same system'. When a species is widely distributed over extensive areas, as in the case of many native species, long-distance dispersal generally goes unnoticed and its biological impact is often considered negligible. In contrast, when a species is absent from large areas to which it is a potential coloniser, as in the case of introduced species, long-distance dispersal is the most obvious and significant aspect of the species' spread. It is likely that many plant species show both short- and long-distance dispersal and that on this point invasive plants are not different from successional plants. Such a conclusion is also suggested by the post-glaciation process of those tree species which used to be successional species in another context (Davis, 1981; Huntley & Birks, 1983).

The role of disturbance

Establishment and population growth and spread largely depend on the availability of 'safe' sites for germination and seedling survival (Harper, 1977; Johnstone, 1986). If 'it seems rational to picture at least part of the success or failure of an invading species as a function of the frequency of suitable sites for germination' (Harper, 1965), it is likely that invasive species are usually able to occupy the more widespread safe sites, i.e. patches of bare ground of various sizes resulting from multi-scale disturbances, such as drought, fire, flood, cultivation, overtrampling, scraping, roadside maintenance, etc. In these safe sites competition from resident species is absent for a time. The frequently observed connection between abundance of invasive species and disturbance frequency supports this assumption.

It is often observed in the case of successful establishment that disturbances are not so frequent, their timing fitting in with the biology of the plant, especially its reproductive features. Successive periods with a different intensity of disturbance, such as a wet year after a drought, a year with an average water level after a high flood, or a period of low grazing intensity after one of overgrazing, all seem to particularly favour invasion.

When focusing on species survival, relations between invasion and disturbance are complex. They depend on types and frequency of disturbance, environmental constraints and the biology of the particular species concerned. A large range of situations is observed, with either negative, neutral or positive impacts. Frequent or intensive disturbance can limit survival of plants – for instance *Pinus halepensis* will disappear if fires occur consecutively within a period of less than 12 years, which is the time necessary for a plant to produce a large number of viable seeds. On the other hand, absence of disturbance for several decades may favour the exclusion of species less competitive than some others; for instance, *Pinus halepensis* is currently being replaced by oaks in areas protected from fire. Disturbance is probably necessary for the maintenance of *Ailanthus altissima* and of *Fraxinus ornus*, both showing moreover a high capacity for vegetative regeneration. In other cases, disturbance may have no, or little impact on long-term survival: *Phillyrea angustifolia* sprouts easily when burnt, over-browsed or after a severe drought and, in the Camargue, no other woody plant can lead to its exclusion by competing with it successfully.

In short, the relationship between invasion and disturbance cannot be

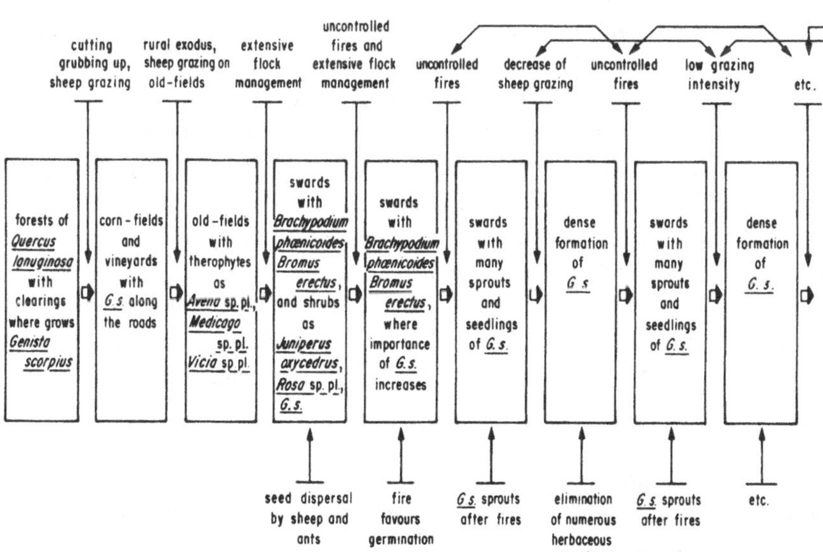

Figure 13.7. Origin and spread of *Genista scorpius* communities (after Debussche *et al.*, 1980).

summarised in a simple and comprehensive model (Ehrlich, 1986) but it may be more effectively elucidated by following a dialectic approach centred on the biology of each species.

Invasion and succession

There is no obvious difference, for the characteristics we have discussed, between invasive plants and those plants thriving during the early and intermediate stages of succession. We suggest that these plants are capable of being invasive since environmental conditions and the type and frequency of disturbance fit in with their life history traits.

The above statement *a priori* does not apply when late successional plant species are concerned. Not one of the five woody plants considered in this chapter is dominant towards the end of the successional sequence; it seems that late successional plants are never observed as being invasive. It is also generally observed that vegetation stages occurring at the end of successions are rarely invasible (but see Myers, 1983; Ewel, 1986). This may be explained, according to classical theory on succession (e.g. Odum, 1969; Whittaker, 1975) and some applications of the niche theory (see Bazzaz, 1986), by the increasingly fine adjustments between plant species as succession proceeds. It is suggested that one species which has not co-evolved with the others has little chance of success when it enters the plant community. Other explanations may be suggested and illustrated by the following results. Late successional plants often have large seeds (Salisbury, 1942; Grime & Jeffrey, 1965) which are heavily depredated; germination of their seeds needs moisture and often the presence of a humus layer and their seedlings grow well at low light densities (Canham & Marks, 1985). This combination of traits may limit colonisation ability and hinder establishment in open habitats different from those of the forest understorey. The rapid expansion after glaciation of tree species with similar life history traits is explained, however, by preferential colonisation of such species along rivers (Huntley & Birks, 1983). Forest dynamics concern isolated patches in a more or less homogeneous matrix; establishment is discontinuous in space and time (Pickett & White, 1985). Accordingly, the probability of a seed of a plant species not present widely in the patchy system reaching these areas by dispersal is very low.

Conclusion

Though perhaps different at an evolutionary level (Brown & Marshall, 1981; Gray, 1986) invasion processes are functionally equivalent to the usual processes occurring during succession (Johnstone, 1986). Dispersal, seed dormancy and germination and seedling survival are all significant in plant succession but are not so easily observed. The presence of a given plant species is only clearly perceived once these phases have been achieved. Then 'it will usually be impossible . . . to discover what were the direct causes of density, pattern and composition of a plant population' (Harper, 1977). Therefore, in many cases, a large part of the succession process remains insufficiently clear. On the other hand, invasions provide good examples for the study of dispersal and establishment because of the larger size of the spatial patterns involved and the often rapid increase in population size. Invasions are propitious in underlining the importance of disturbance to plant establishment and in casting dispersal in a different light. Investigations on both invasion and succession as complementary phenomena may thus provide the key to better understanding of plant dynamics.

References

Acherar, M. (1981). La colonisation des friches par le Pin d'Alep (*Pinus halepensis* Miller) dans les basses garrigues du Montpelliérais. Thèse 3ème cycle, Université des Sciences et Techniques du Languedoc, Montpellier.

Acherar, M., Lepart, J. & Debussche, M. (1984). La colonisation des friches par le Pin d'Alep (*Pinus halepensis* Miller) en Languedoc méditerranéen. *Acta Oecologica Plantarum*, **5**, 179–89.

Bannister, R. H. (1965). Variation in the breeding system of *Pinus radiata*. In *The Genetics of Colonizing Species*, ed. H. G. Baker & G. L. Stebbins, pp. 353–72. London: Academic Press.

Barry, J. P. (1952). Essai de cartographie parcellaire de la commune de Boissières (Gard). In *Atlas de la Cartographie Parcellaire*. Montpellier: Ecole Nationale Superiéure Agronomique.

Barry, J. P. (1960) Contribution à l'étude de la végétation de la région de Nîmes. *Année Biologique, 3ème série*, **36**, 311–540.

Bazzaz, F. A. (1986). Life history of colonizing plants: some demographic, genetic and physiological features. In *Ecology of Biological Invasions of North America and Hawaii*, ed. H. A. Mooney & J. A. Drake, pp. 96–110. Berlin: Springer-Verlag.

Boudy, F. (1950). *Economie Forestière Nord-Africaine. II. Monographie et Traitement du Pin d'Alep*, pp. 639–90. Paris: Larose.

Bourlière, F. (1956). Conséquences biologiques dues à la présence de la myxomatose. *La Terre et la Vie*, **10**, 123–34.

Brown, A. H. D. & Marshall, D. R. (1981). Evolutionary changes accompanying colonization in plants. In *Evolution Today*, ed. G. C. E. Scudder & J. L. Reveal, pp. 351–63. Pittsburgh: Hunt Institute for Botanical Documentation, Carnegie-Mellon University.

Camps, G. (1982). *La Préhistoire. A la Recherche du Paradis Perdu*. Paris: Librairie Académique Perrin.

Canham, C. D. & Marks, P. L. (1985). The response of woody plants to disturbance: patterns of establishment and growth. In *The Ecology of Natural Disturbance and Patch Dynamics*, ed. S. T. A. Pickett & P. S. White, pp. 197–218. London: Academic Press.

Clair-Maczulajtis, D. (1984). Quelques aspects de la biologie de l'*Ailanthus altissima* (Mill.). Thèse docteur-ès-sciences, Université de Paris VII.

Clements, F. E. (1928). *Plant Succession and Indicators. A Definitive Edition of Plant Succession and Plant Indicators*. New York: Hafner.

Connell, J. H. & Slatyer, R. O. (1977). Mechanisms of succession in natural communities and their role in community stability and organization. *American Naturalist*, **111**, 1119–44.

Crawley, M. J. (1987). What makes a community invasible? In *Colonization, Succession and Stability*, ed. A. J. Gray, M. J. Crawley & P. J. Edwards, pp. 429–53. Oxford: Blackwell Scientific Publications.

Darwin, C. (1859). *On the Origin of Species*. London: Murray.

Davis, M. B. (1981). Quaternary history and the stability of forest communities. In *Forest Succession. Concepts and Application*, ed. D. C. West, H. H. Shugart & D. B. Botkin, pp. 132–53. Berlin: Springer-Verlag.

Debussche, M. & Isenmann, P. (1985). Le régime alimentaire de la Grive musicienne (*Turdus philomelos*) en automne et en hiver dans les Garrigues de Montpellier (France méditerranéenne) et ses relations avec l'ornithochorie. *Revue d'Ecologie (Terre et Vie)*, **40**, 379–88.

Debussche, M., Escarre, J. & Lepart, J. (1980). Changes in Mediterranean shrub com-munities with *Cytisus purgans* and *Genista scorpius*. *Vegetatio*, **40**, 73–82.

Debussche, M., Escarre, J. & Lepart, J. (1982). Ornithochory and plant succession in mediterranean abandoned orchards. *Vegetatio*, **48**, 255–66.

Debussche, M., Lepart, J. & Molina, J. (1985). Le dissémination des plantes à fruits charnus par les oiseaux: rôle de la structure de la végétation et impact sur la succession en région méditerranéenne. *Acta Oecologica Generalis*, **6**, 65–80.

Debussche, M., Rambal, S. & Lepart, J. (1987). Les changements de l'occupation des terres en région méditerranéenne: évaluation des conséquences hydrologiques. *Acta Oecologica Applicata*, **8**, 317–32.

di Castri, F., Hansen, A. J. & Debussche, M. (ed.) (1990). *Biological Invasions in Europe and the Mediterranean Basin*. Dordrecht: Kluwer.

Dupuis, A. (1862). Notice sur la culture de l'Ailante glanduleux. *Bulletin de la Societé d'Acclimatation*, **10**, 877–89.

Ehrlich, P. R. (1986). Which animal will invade? In *Ecology of Biological Invasions of North America and Hawaii*, ed. H. A. Mooney & J. A. Drake, pp. 79–95. Berlin: Springer-Verlag.

Elton, C. S. (1958). *The Ecology of Invasions by Animals and Plants*. London: Methuen.

Eriksson, O., Inghe, O., Jerling, L., Tapper, P., Telenius, A. & Torstensson, P. (1983). A note on non-adaptation hypotheses in plant ecology. *Oikos*, **41**, 155–6.

Escarre, J. (1979). Etude des successions post-culturales dans les Hautes garrigues du Montpelliérais. Thèse 3ème cycle, Université des Sciences et Techniques du Languedoc, Montpellier.

Escarre, J., Houssard, C., Debussche, M. & Lepart, J. (1983). Evolution de la végétation et du sol après l'abandon cultural en région méditerranéenne; étude de succession dans les garrigues du Montpelliérais (France). *Acta Oecologica Plantarum*, **4**, 221–39.

Ewel, J. J. (1986). Invasibility: lessons from south Florida. In *Ecology of Biological Invasions of North America and Hawaii*, ed. H. A. Mooney & J. A. Drake, pp. 214–30. Berlin: Springer-Verlag.

Firbas, F. (1949). *Spät und Nacheiszeitliche Waldegeschichte Mitteleuropas Nordlich der Alpen*, vol. 1. Jena: Gustav Fisher.

Giban, J., Barthélémy, J. &Aubry, J. (1956). L'épizootie de myxomatose en France chez le lapin de garenne (*Oryctolagus cuniculus* L.). *La Terre et La Vie*, **10**, 167–78.

Gray, A. J. (1986). Do invading species have definable genetic characteristics? *Philosophical Transactions of the Royal Society of London*, **B314**, 655–74.

Grime, J. P. & Jeffrey, D. W. (1965). Seedling establishment in vertical gradients of sunlight. *Journal of Ecology*, **53**, 621–42.

Groves, R. H. & Burdon, J. J. (ed.) (1986). *Ecology of Biological Invasions: An Australian Perspective*. Cambridge: Cambridge University Press.

Harper, J. L. (1965). Establishment, aggression and cohabitation in weedy species. In *The Genetics of Colonizing Species*, ed. H. G. Baker & G. L. Stebbins, pp. 245–65. London: Academic Press.

Harper, J. L. (1977). *Population Biology of Plants*. London: Academic Press.

Héricourt, A. F. d' (1861). L'ailante glanduleux. *Annales Forestières et Métallurgiques*, **7**, 11–18.

Heurteaux, P. (1969). Recherches sur les rapports des eaux souterraines avec les eaux de surface, les sols halomorphes et la végétation en Camargue. Thèse Docteur-ès-Sciences, Université des Sciences et Techniques du Languedoc, Montpellier.

Huntley, B. & Birks, H. J. B. (1983). *An Atlas of Past and Present Pollen for Europe 0–13 000 Years Ago*. Cambridge: Cambridge University Press.

Johnstone, I. M. (1986). Plant invasion windows: a time-based classification of invasion potential. *Biological Reviews*, **61**, 369–94.

Kornberg, H. & Williamson, M. H. (ed.) (1986). *Quantitative Aspects of the Ecology of Biological Invasions*. London: The Royal Society.

Le Roy Ladurie, E. (1966). *Les Paysans du Languedoc*. Paris: Mouton.

Loisel, R. (1976). La végétation de l'étage méditerranéen dans le sud-est continental français. Thèse Docteur-ès-Sciences, Université de Marseille.

Mack, R. N. (1985). Invading plants: their potential contribution to population ecology. In *Studies on Plant Demography. John L. Harper Festschrift*, ed. J. White, pp. 127–42. London: Academic Press.

Mack, R. N. (1986). Alien plant invasion into the intermountain west. In *Ecology of Biological Invasions of North America and Hawaii*, ed. H. A. Mooney & J. A. Drake, pp. 191–213. Berlin: Springer-Verlag.

Macdonald, I. A. W., Kruger, F. J. & Ferrar, A. A. (ed.) (1986). *The Ecology and Management of Biological Invasions in Southern Africa*. Cape Town: Oxford University Press.

Molinier, R. & Tallon, G. (1965). Vers la forêt en Camargue. *La Terre et la Vie*, **1/2**, 135–85.

Mooney, H. A. & Drake, J. A. (ed.) (1986). *Ecology of Biological Invasions of North America and Hawaii*. Berlin: Springer-Verlag.

Mooney, H. A., Hamburg, S. P. & Drake, J. A. (1986). The invasions of plants and animals in California. In *Ecology of Biological Invasions of North America and Hawaii*, ed. H. A. Mooney & J. A. Drake, pp. 250–74. Berlin: Springer-Verlag.

Morel, A. (1956). Influence de l'épidémie de myxomatose sur la flore française. *La Terre et la Vie*, **10**, 226–38.

Myers, R. L. (1983). Site susceptibility to invasion by the exotic tree *Melaleuca quinquinervia* in southern Florida. *Journal of Applied Ecology*, **20**, 645–58.

Nahal, I. (1962). Le pin d'Alep (*Pinus halepensis* Mill.). Etude taxonomique, phytogéographique, écologique et sylvicole. *Annales de l'Ecole Nationale des Eaux & Forêts, Nancy*, **19**, 485–685.

Odum, E. P. (1969). The strategy of ecosystem development. *Science*, **164**, 262–70.

Oliver, C. D. & Stephens, E. P. (1977). Reconstruction of a mixed-species forest in central New England. *Ecology*, **58**, 562–72.

Pickett, S. T. A. & White, P. S. (1985). *The Ecology of Natural Disturbance and Patch Dynamics*. London: Academic Press.

Pons, A. (1984). La paléoécologie face aux variations spatiales du bioclimat méditerranéen. *Bulletin de la Societé botanique de France*, **131**, *Actualités botaniques (2–3–4)*, 77–83.

Pons, A., Toni, C. & Triat, H. (1979). Edification de la Camargue et histoire holocène de sa végétation. *La Terre et la Vie, Supplément*, **2**, 13–30.

Reille, M., Triat-Laval, H. & Vernet, J. L. (1980). Les témoignages des structures actuelles de végétation méditerranéenne durant le passé contemporain de l'action de l'homme. In *Colloque de la Fondation Emberger: La Mise en Place, l'Evolution et la Caractérisation de la Flore et de la Végétation Circumméditerranéenne*. Montpellier: Naturalia Monspeliensia, no. hors-série, 79–87.

Ridley, H. N. (1930). *The Dispersal of Plants Throughout the World*. Ashford, England: Lovell Reeve & Co.

Salisbury, E. J. (1942). *The Reproductive Capacity of Plants*. London: Bell.

Sébastian, C. (1956). Etude du genre *Phillyrea*. *Travaux de l'Institut Scientifique Chérifien, série botanique*, **6**, 1–102.

Strasberg, D. (1987). Ecologie de l'invasion de *Phillyrea angustifolia* en Camargue (Tour du Valat). *Mémoire de fin d'études, Ecole Nationale d'Ingénieur des Travaux Agricoles, Dijon*.

Struever, S. (1971). *Prehistoric Agriculture*. New York: The Natural History Press.

Tallon, G. (1931). Etude de l'association à *Phillyrea angustifolia* et *Jasminium fruticans*. *Bulletin de la Societé Nationale d'Acclimatation*, **55**, 29–32.

Tallon, G. (1954*a*). Influence des digues sur les conditions biologiques et l'évolution de la Camargue. *La Terre et la Vie*, **8**, 49–53.

Tallon, G. (1954*b*). Transformation de la Camargue par la riziculture. Evolution du Vaccarès. *La Terre et La Vie*, **8**, 65–79.

Tallon, G. (1955). Nouvelles observations au bois des Rièges. *La Terre et la Vie*, **9**, 225–32.

Thébaud, C. (1986). Perturbations naturelles et espèces végétales adventices: impact des crues sur les communautés végétales riveraines de l'Hérault, fleuve méditerranéen. *Mémoire de fin d'études Ecole Nationale d'Ingénieur des Travaux Agricoles, Dijon.*

Thébaud, C. & Debussche, M. (1990). Rapid invasion of *Fraxinus ornus* L. along the Hérault River system in southern France; the importance of seed dispersal by water. *Journal of Biogeography*, **17**, in press.

Thomas, A. S. (1960). Changes in vegetation since the advent of myxomatosis. *Journal of Ecology*, **48**, 287–306.

Thomas, A. S. (1963). Further changes in vegetation since the advent of myxomatosis. *Journal of Ecology*, **51**, 151–86.

Tilman, D. (1982). *Resource Competition and Community Structure*. Princeton, New Jersey: Princeton University Press.

Trabaud, L., Michels, C. & Grosman, J. (1985). Recovery of burnt *Pinus halepensis* Mill. forests. II. Pine reconstitution after wildfire. *Forest Ecology & Management*, **13**, 167–79.

Triat-Laval, H. (1978). Contribution pollenanalytique à l'histoire tardi- et post-glaciaire de la végétation de la Basse Vallée du Rhône. Thèse Docteur-ès-Sciences, Université d'Aix-Marseille III.

Tutin, T. G., Heywood, V. H., Burges, N. A., Moore, D. M., Valentine, D. H., Walters, S. M. & Webb, D. A. (ed.) (1972). *Flora Europaea*, vol. 3. Cambridge: Cambridge University Press.

Venable, D. L. & Lawlor, L. (1980). Delayed germination and dispersal in desert annuals: escape in space and time. *Oecologia*, **46**, 272–82.

Venable, D. L. & Levin, D. A. (1983). Morphological dispersal structures in relation to growth habit in the Compositae. *Plant Systematics & Evolution*, **143**, 1–16.

Vernet, J. L. (1980). Premières données sur l'histoire de la végétation post-glaciaire de la Provence central d'après l'analyse anthracologique. *Comptes Rendus de l'Académie des Sciences*, **291**, 853–5.

Villiers, T. A. & Wareing, P. F. (1960). Interaction of growth inhibitor and natural germination simulator in the dormancy of *Fraxinus excelsior* L. *Nature*, **185**, 112.

Watt, A. S. (1947). Pattern and process in the plant community. *Journal of Ecology*, **35**, 1–22.

Whittaker, R. H. (1975). *Communities and Ecosystems*, 2nd edn. New York: Macmillan.

14

Is fire an agent favouring plant invasions?

L. TRABAUD

By its frequent recurrence in the past and the present, fire is an important ecological force influencing the dynamics of plant communities in all regions of mediterranean climate. Fire more or less regularly interrupts the natural cycle of vegetation succession and its advent allows some stands to be rejuvenated. By disturbing the natural vegetation does fire permit introduced species to establish in burned areas? By disturbing vegetation (i.e. by opening gaps for potentially invasive species), does fire create areas where non-indigenous species may be able to establish? What are the types of invasive species: do they come from outside the communities but belong to local floras, or do they come from afar, from other countries? Do such species persist or are they only transitory? In short, is fire a factor in plant invasion or are local species so strongly adapted to fire that they do not permit any changes in the floristic composition of burned communities? Plant species display such a wide range of reproductive means that they are able to colonise a large variety of environments in such a manner that at least one species always has the potential to invade a burned area.

This research theme is an exciting and promising one for the understanding of vegetation dynamics and the past, present and future floristic composition and stucture of communities. Yet, unfortunately, only a few studies deal with plant invasion after fire in mediterranean-type ecosystems.

The Mediterranean Basin

In southern France, the dynamics of vegetation after fire have been studied in the garrigue shrublands on calcareous soils of Bas-Languedoc (Trabaud, 1970, 1980*a*, 1983; Trabaud & Lepart, 1980, 1981). After fire the vegetation of the communities returns quickly to its initial state. Most often the species which are present 12 years after the previous fire are the first ones to re-appear immediately after the fire and they become more and more numerous with time until the next fire. Thus, if fire opens gaps in the vegetation allowing alien species to invade burned areas, the quick and severe competition exerted by species present before fire prevents alien species from persisting.

Among species which try to establish in burned areas, some are foreign plants (coming from outside Europe), which were introduced a long time ago to the Bas-Languedoc region of the Mediterranean Basin, and which can be considered as invasive in the region without any effect of fire. Others are species which do not belong to the regional flora and come from more northern countries. With fire, their quantitative importance is low and they never persist over a long period.

This author (Trabaud, 1990) considered which of all these species were the most frequently recorded during the first 4 years after fire. The species *Erigeron canadense*, originally from North America, has been invasive in France for many years. It was observed on 26 per cent of the author's plots. *Sonchus oleraceus*, a Euro-Siberian plant which has invaded the region, was observed on 58 per cent of the plots. *Phleum boehmeri*, another plant of Euro-Siberian origin, and considered rare in the Mediterranean Basin, was observed on 14 per cent of the plots. *Epilobium tetragonum* ssp. *adnatum*, also of Euro-Siberian origin and rare in the Mediterranean, was observed on 2 per cent of the plots. These four species are not really invasive, but fire allows them to remain, or maintain sites which are not their own but on which they can reproduce and persist.

In a siliceous area of the Alberes in France, the author (Trabaud, 1980*a*) studied the recovery after fire of a maquis shrubland of *Erica* species and of a forest of *Quercus ilex*. A large number of exogenous species which did not occur in mature unburnt communities were observed invading the burnt areas. Among these latter species some (called xenophytes) do not belong to the original flora of the Mediterranean Basin. As for the burned communities in the calcareous area of Bas-Languedoc, the invasives were either species coming from outside Europe but introduced a long time ago (e.g. *Erigeron naudini*, a species from North America and observed on 60 per cent of the burnt plots) or else were species belonging to northern European floras, such as *Cardamine hirsuta* (circumboreal, encountered on 100 per cent of the burnt plots), *Sonchus oleraceus* (Euro-Siberian, recorded on 60 per cent of the burnt plots) and *Holcus lanatus* and *Barbarea vulgaris* (circumboreal, recorded on 40 per cent of the burnt plots).

In the same siliceous area, Prodon *et al.* (1984) studied six community types from grass swards to forests of *Quercus ilex* or *Q. suber* in plots of various ages from 1 to 4 years after fire. They too found that reappearance of species was immediately after fire. The species invading during the first years after fire included numerous annuals and, particularly in shrublands and forests, these annuals were species alien to the communities but endemic to the area; they were mainly species of the families Papilionaceae (12) and Compositae (2). Despite these changes, Prodon *et al.* concluded that the main species re-establishing after fire were present before.

At a similar site at Cape de Creus in north-western Spain, Franquesa (1987) studied matorral shrublands and reached a similar conclusion. Rapid colonisation was basically from resprouts of prolific species, with the vegetation composition stabilising rapidly. Invasive species (most often annuals and exogenous) appeared and disappeared quickly and were present only in the first few years. Franquesa stressed the importance of one invasive species, viz. *Galactites tomentosa*, which came from outside the unburned communities as air-borne propagules but belonged to the local flora. This species remained only for 2 years after fire, then disappeared totally.

In calcareous areas of southern France, the author (Trabaud, 1974; 1977; 1980*a*, *b*) followed an experimental approach using prescribed fires to better understand the effects of fire on plants and vegetation. A garrigue was burned at different frequencies and at different seasons. In spite of the repeated fire frequency there was no noticeable change in the flora of the *Quercus coccifera* garrigue (Trabaud, 1980*a*; Trabaud & Lepart, 1981). The floristic composition of the burned plots was on the whole very stable. After burning it was identical to that which existed at the beginning of the experiment. The dominant and characteristic species were present at all times. This relative stability comes from the differences in response of the species to fire. The perennial species which possess means of vegetative regeneration (resprouts) were present before the fires and always fully occupied the plots. On the other hand, invasive species appeared only in the first or second year after burning as individuals coming from seeds. They were rapidly eliminated in later years. However, the frequency and season of fire may lead to some minor changes in floristic composition. The influence of the season of fire was much more important than fire frequency; with autumn fires many alien short-lived species appeared, thereby increasing the floristic richness for a short time.

Species which had a tendency to invade the experimental plots can be classified into three groups:

1. Annuals or biennials alien to the community and invading the burnt plots. *Sonchus asper* is the only species of this group observed more than 40 times; it appeared with the help of fire and then quickly disappeared as soon as this influence diminished. Some other species acted in the same way but they were not encountered as frequently, viz. *Althaea hirsuta*, *Bromus madritensis*, *Carduus pycnocephalus*, *Cirsium acarna*, *Crepis pulchra*, *C. vesicaria* ssp. *taraxacifolia*, *Erigeron canadense*, *Geranium robertianum* ssp. *purpureum*, *Lactuca scariola*,

Papaver rhoeas, *Picris hieracioides*, *Rapistrum rugosum*, *Scleropoa rigida*, *Senecio vulgaris*, *Sherardia arvensis* and *Tragopogon australis*. All these species were not present in the plots at the beginning of the experiment and they disappeared with continued protection from fire.

2. Species not present at the beginning of the experiment but which establish. These truly invasive species, e.g. *Euphorbia nicaeensis*, *Hippocrepis comosa*, *Avena bromoides*, established on the plots owing to the conditions created by the fires. They did not exist before the fires, but once in place they persisted; they are therefore promoted by fire.

3. Species present at the beginning of the experiment but with increasing presence, e.g. *Sanguisorba minor*, *Arrhenatherum elatius*, *Galium asperum*, *Bupleurum rigidum* and *Sedum nicaeense*. All these species reacted positively to fire. Their presence value was higher at the end of the observation period.

In this field experiment, again, except for a few truly invasive species, fire was accompanied by the entry of a number of annual species, which quickly disappeared, because of competition from endigenous species and the waning influence of the fire.

The author (Trabaud, 1970, 1980*a*, *b*) and Boulet (1985) collected soil samples in order to know the germinative capacity of propagules buried in the soil. Seedlings appearing in the samples of species not present in the original communities represented only 40 per cent of the total, the most frequent species being *Erigeron canadense*, *E. naudini* and *Sonchus oleraceus*.

In the plant communities of the Mediterranean Basin fire does not seem to induce any important change to the current standing vegetation. After fire, communities revert towards a structure and a floristic composition identical to those of the pre-fire vegetation. Most plants regenerating after fire come from survival organs (rhizomes, stumps, bulbs, seeds, etc.) already present in the soil before the flames pass, or released (as seeds) immediately after fire from dead but standing plants or from plants located close to the fire. There are no species not present in the original stands which have been able to invade and persist in sites repeatedly burnt for millennia. All vegetatively regenerating species resprout during the first few months or years following fire whatever the season of burning. Germination of seed-reproducing plants occurs most commonly during the first 2 years after fire.

With regard to the floristic composition of communities after fire, new species which establish after fire are rare. There is no evidence of true 'invasion'. Species considered to be invasive because of their presence after fire and the absence of their living plants in the vegetation before fire do not play an important role.

The plant community returns to the initial state by an autochthonous process, i.e. the species existing before the fire reoccupy the burned sites – as opposed to an allochthonous process which would be characterised by a succession of stages. The dynamics of recolonisation after fire follow Egler's (1954) model of 'initial floristic composition', i.e. all the species are present on the site immediately after fire, even though changes in relative dominance of individuals may occur.

Species exist either in the soil or in the surrounding environments which could invade burned communities. However, because of competition from endogenous species these alien species do not appear or only appear in small numbers and for a short time.

In the Mediterranean Basin so-called 'pyrophytes', such as species of *Cistus* and *Pinus*, have long been considered by many authors to either invade communities or be favoured by fire (Braun-Blanquet, 1936; Kuhnholtz-Lordat, 1938; Kornas, 1958; Gaussen, 1970; Le Houérou, 1974). My observations and results (Trabaud, 1980*a*; Trabaud & Lepart, 1980; Trabaud *et al.*, 1985*a*, *b*), however, show that these types of plants do not invade burnt communities. Under natural conditions after wildfire there is no change in the floristic composition of communities and there is no invasion by *Cistus* or *Pinus* species provided they were not present before fire. New plants may come from seeds buried in the soil.

From results of experimental fires (Trabaud, 1980*a*; Trabaud & Lepart, 1981), repeated burning once every two years led to a disappearance of *Cistus*, particularly of *Cistus monspeliensis*. In the spring following a fire *C. monspeliensis* populations are composed only of seedlings. Two years after burning only a few individuals flowered. Full bloom was reached only during the third year. As this species of *Cistus* reproduces only sexually, the amount of seed buried in the soil tends to decrease, even to disappear, thereby producing a progressive decrease in the number of individuals present in the plots burnt every second year. Gerber (1898, as cited by Barry, 1960) stated that the same phenomenon held for *C. albidus*; when wildfires occurred in consecutive years the species disappeared. *Cistus* appears to be a species able to colonise bare areas rather than one that it is favoured by fire, for it is present in great numbers on many bare areas and in old fields. *Cistus* behaves as a light-demanding pioneer species which develops in the absence of competitors. As most of the above results are from calcareous sites, it could be different perhaps on sites with acid siliceous soils.

The same conclusions seem to apply to *Pinus halepensis* (Trabaud, 1980*a*). *Pinus halepensis* enlarges its area after fire only on the edges of sites in which it is (or was) present owing to its seeds being wind-dispersed. In the absence of fire *P. halepensis* easily and rapidly occupies abandoned or disturbed areas; it too is a light-demanding plant able to colonise old fields (Acherar *et al.*, 1984). *Cistus* and *Pinus* species are not fire-favoured species but rather opportunists able to occupy bare areas in the absence of aggressive competitors. Only in this way are populations maintained by fire.

Invasion in other countries with a mediterranean climate

In mature stands of Californian chaparral annual forbs and grasses are sparse (less than 1 per cent cover). Endemic forbs are abundant in the first year after fire (Sampson, 1944). During the second year, annual grasses and forbs become dominant (106 species), and then decline gradually until the fifth year after fire. By this time, herbaceous species are sparse and their density is equivalent to that in mature stands (Hanes, 1971). Chaparral seedlings and resprouts, on the other hand, cover all burned areas by the end of five years.

Forbs are numerous after fire in chaparral and occasional fires are essential for their survival (Sweeney, 1956). Common annual forbs belonging to the endemic flora include *Antirrhinum cornutum*, *Emmenanthe penduliflora*, *Lotus americanus*, *Navarettia mellita*, *Phacelia brachyloba*, *P. divaricata*, *P. suaveolens*, *P. grandiflora*, *Papaver californicum* and *Scutellaria tuberosa*. Weedy species, such as *Centaurea melitensis* (from the Mediterranean Basin) and *Senecio vulgaris* (from Eurasia), may invade burned areas from nearby roadsides (Sampson, 1944; Sweeney, 1956; Biswell, 1974). Annual grasses may invade from areas adjacent to the fire (Biswell, 1974; Wright & Bailey, 1982). A few annual grasses usually appear on burned areas in the first year after a fire and then may increase rapidly during the second and third years. Species in this group include *Aira caryophyllea*, *Gastridium ventricosum*, *Vulpia myuros* (= *Festuca megalura*), *Bromus rubens* and *B. rigidum*, all native to Europe.

As shrubs increase in stature, seed germination is inhibited (Christensen & Muller, 1974); thus annual species are gradually eliminated. Seeds of introduced annual species do not remain dormant for a long time and must re-invade burned areas from adjacent unburned stands (e.g. roadsides, old fields, etc.).

The same tendency to temporarily invade the plant community is also observed in the sclerophyll forests of southern Australia, where a luxuriant growth of herbs occurs only during the first few years after fire. Species such as *Senecio minimus*, *Cirsium vulgare* and *Urtica incisa* often became prolific (Gilbert, 1959). Other species are important colonisers in burned areas in different regions: for instance, in Victoria, *Senecio linearifolius*, *S. quadridentatus*, *S. vagus*, *S. velliodes*, *Solanum aviculare* and *Cardamine dictyosperma*; in south-west Western Australia, *Senecio ramosissimus*, *Helichrysum ramosum* and the introduced *Conyza* (*Erigeron*) *canadense* occur (Ashton, 1981); and in New South Wales, the introduced species *Phytolacca octandra* and *Solanum armatum* often thrive after fires of high intensity (Floyd, 1976).

During the first two years after a summer fire near Canberra, Purdie & Slatyer

(1976) and Purdie (1977*a*) observed the following introduced species to occur: *Aira cupaniana*, *Briza minor*, *Bromus mollis*, *Lolium rigidum*, *Vulpia bromoides* (= *V.uniglumis*), *Cerastium semidecandrum*, *Oxalis corniculata*, *Solanum nigrum*, *Conyza floribunda* (= *Erigeron naudini*), *Cirsium vulgare* (= *C. lanceolatum*), *Hypochoeris glabra*, *H. radicata*, *Lactuca serriola* and *Sonchus asper*.

These so-called 'fireweed' species are subsequently replaced by regenerating tree and shrub species of the natural communities (Gill, 1975; Purdie & Slatyer, 1976; Purdie, 1977*a*, *b*; Ashton, 1981; Christensen *et al.*, 1981; Specht, 1981). Species alien to the communities never increase. After the initial flush of these 'fireweeds' the course of pyric succession following a single fire in sclerophyll forests depends for many decades on the floristic composition of the initial vegetation (Ashton, 1981). Purdie (1977*b*) concluded that the 'floristic composition of the studied communities remained unchanged after burning'.

On the other hand, in some other mediterranean-climate countries, it seems that fire strongly favours invading species. In a burned woodland of *Acacia caven* in Central Chile, Altieri & Rodriguez (1974) recorded, one year after a wildfire, 44 200 plantlets per m^2 of *Vulpia*, *Koeleria* and *Bromus* species, all annual species native to Europe.

The region having the worst problems with invasive introduced species seems to be South Africa, however. In Cape fynbos, *Hakea* and *Acacia* species from Australia and *Pinus* species from Europe create a problem for conservation of native plant species (Taylor, 1977; Hall, 1979; Richardson *et al.*, 1987).

The Australian hakeas, originally introduced as hedge plants, are well adapted with their dense spiny-leaved growth to spreading and forming thicket-like infestations after fire. *Hakea sericea*, particularly, is by far the most widespread invasive plant in this group and now infests nearly all mountain fynbos areas (Hall, 1979). In the genus *Hakea* the seed is held in woody follicles on the plant until there is a fire. Shortly after the fire the follicles open, thereby releasing large numbers of viable winged seeds that may be blown considerable distances. This abundant release of seed leads to the establishment of quick-growing thickets that rapidly over-top the fynbos.

Of the various acacias, *Acacia cyclops* is a particular problem. Large seed stores have accumulated in the soil, 250 million seeds per hectare have been recorded (Taylor, 1977) in the top 10 cm of soil for *A. cyclops*. Unlike that of most fynbos species, *Acacia* seed is exceptionally long-lived. Thus frequent fires, instead of diminishing the invasion, greatly increase spread of the invasive shrubs, especially as *Acacia* seedling establishment from the uppermost layers of the soil is extremely efficient and seedlings soon outgrow their indigenous competitors (Roux & Middlemiss, 1963; Taylor, 1977; Hall, 1979). The germination of seeds of *A. cyclops* has been studied by Jones (1963) in South Africa, and Christensen (1978) in Western Australia. Jones found an increase in germination by heating the seed. In contrast, Christensen's data showed no increase in germination with heating. A major difference between the habitats of *A. cyclops* in South Africa and Australia seems to be the frequency of fire. Taylor (1977) noted that the South African vegetation in which it occurs was burned frequently. In Western Australia, Christensen & Kimber (1975) noted that the habitat was 'rarely subject to fire'.

Pines, particularly *Pinus halepensis* and *P. pinaster*, are a threat to fynbos because they invade after fires (D. Richardson, personal communication).

It seems, nevertheless, that plants of the fynbos or the forests resist fire very well and the autosuccessional inhibition model applies here too (references already quoted, but see also Frost, 1984; Kruger, 1984; Kruger & Bigalke, 1984). Introduced invasive woody plants may dominate endemic vegetation and perpetuate themselves. Fire is not the only factor favouring the invasion of these introduced species (for in the absence of fire introduced species also propagate). Sometimes prescribed fires may even be used to eliminate invasive species by using different types of management rotations.

Conclusion

According to the results presented in this chapter, there are two groups of countries with regard to the favouring of plant invasions by fire: one, comprising the Mediterranean Basin, California and Australia, and the other, Chile and South Africa. In the former group, invasive plants are usually only fugitive species (mostly annuals) which disappear soon after the fire, the natural perennial vegetation re-establishes and invasive species do not massively colonise the burned areas; in the latter group, the invasive plants establish massively and remain for a longer time, even in the absence of fire.

This grouping may be apparent only because, according to the authors, the original Chilean matorral and South African fynbos together with the species constituting them, re-establish after fire. What are the reasons for such a different behaviour of communities which allow or do not allow invasive plants to establish for a significant period of time? In the Mediterranean Basin, California and Australia, fire is an ancient factor which razes the communities fairly frequently. Individual species and the vegetation are adapted to fire by a variety of life forms which occupy all niches in the ecosystems. The same trend occurs in South Africa, although the invasion of fynbos by woody arborescent species (such as species of *Acacia*, *Hakea* and *Pinus*) might be because of the present lack of trees in this vegetation type (Moll *et al.*, 1980). Invasive trees or tall shrubs may fill a niche otherwise empty of competitors and because of this absence of competitors invasive species may thrive and develop in large numbers. In Chile, frequent fire is a more recent factor than in the other mediterranean-type countries: fire probably occurred in the distant past but not very frequently. The native Chilean vegetation is less fire-prone and less flammable than that of the other mediterranean-type countries. In addition, strong human pressure and a history of grazing of all the communities help to explain the fact that invasive annuals have greater cover values and a wider distribution.

The above two different explanations could be the reasons why invasive species are either more abundant or more persistent for a longer time. In fact, in all the mediterranean-climate countries, burned areas can act as reservoirs of non-indigenous species, as in other disturbed areas, thereby allowing these species to persist in sites where they otherwise would not exist, and from which they may become potentially invasive even in the absence of fire.

References

Acherar, M., Lepart, J. & Debussche, M. (1984). La colonisation des friches par le Pin d'Alep (*Pinus halepensis* Mill.) en Languedoc méditerranéen. *Oecologia Plantarum*, **4**, 179–89.

Altieri, M. A. & Rodriguez, J. A. (1974). Accion ecologica del fuego en el matorral natural mediterraneo de Chile, en Rinconada de Maipu. Thèse, Facultad de Agricultura, Universidad de Chile, Santiago.

Ashton, D. H. (1981). Fire in tall open forests (wet sclerophyll forests). In *Fire and the Australian Biota*, ed. A. M. Gill, R. H. Groves & I. R. Noble, pp. 339–66. Canberra: Australian Academy of Science.

Barry, J. P. (1960). Contribution à l'étude de la végétation de la région de Nîmes. *Année Biologique*, **36**, 311–550.

Biswell, H. H. (1974). Effects of fire on chaparral. In *Fire and Ecosystems*, ed. T. T. Kozlowski & C. E. Ahlgren, pp. 321–64. New York: Academic Press.

Boulet, C. (1985). Bilan floristique d'une garrigue de chêne kermès soumise à deux types de perturbations controlées. Thèse 3ème cycle, Ecologie Végétale, Faculté des Sciences et Techniques, Marseille.

Braun-Blanquet, J. (1936). La forêt d'yeuse languedocienne (Quercion ilicis). Monographie phytosociologique. *Memoires de la Société Etudes de Sciences Naturelles de Nîmes*, **5**, 1–147.

Christensen, N. L. & Muller, C. H. (1974). Effects of fire on factors controlling plant growth in *Adenostoma* chaparral. *Ecological Monographs*, **45**, 29–55.

Christensen, P. (1978). The concept of fauna priority areas. In *Proceedings of the Victorian Forests Commission & Monash University Fire Ecology Conference*, vol. 3, pp. 66–73.

Christensen, P. E. & Kimber, P. C. (1975). Effect of prescribed burning on the flora and fauna of south-west Australian forests. *Proceedings of the Ecological Society of Australia*, **9**, 85–106.

Christensen, P., Recher, H. & Hoare, J. (1981). Responses of open forests (dry sclerophyll forests) to fire regimes. In *Fire and the Australian Biota*, ed. A. M. Gill,

R. H. Groves & I. R. Noble, pp. 367–93. Canberra: Australian Academy of Science.

Egler, F. E. (1954). Vegetation science concepts. I. Initial floristic composition, a factor in old field vegetation development. *Vegetatio*, **4**, 412–17.

Floyd, A. G. (1976). Effect of burning on regeneration from seeds in wet sclerophyll forest. *Australian Forestry*, **39**, 210–20.

Franquesa, T. (1987). Regeneracio de los brolles silicicoles de la peninsula del Cap de Creus. *Quaderns Ecologia Aplicada*, **10**, 113–29.

Frost, P. G. H. (1984). The response and survival of organisms in fire-prone environments. In *Ecological Effects of Fire in South African Ecosystems*, ed. P. de V. Booysen & N. M. Tainton, pp. 273–309. Berlin: Springer-Verlag.

Gaussen, H. (1970). *Notice Explicative de la Carte de la Végétation de la Région Méditerranéenne*. Paris: UNESCO/FAO.

Gilbert, J. M. (1959). Forest succession in the Florentine valley, Tasmania. *Proceedings of the Royal Society of Tasmania*, **93**, 129–51.

Gill, A. M. (1975). Fire and the Australian flora: a review. *Australian Forestry*, **38**, 4–25.

Hall, A. V. (1979). Invasive weeds. In *Fynbos Ecology: A Preliminary Synthesis*, ed. J. Day, W. R. Siegfried, G. N. Louw & M. L. Jarman, pp. 133–47. Pretoria: South African National Sciences Progress Report No. 40.

Hanes, T. L. (1971). Succession after fire in the chaparral of southern California. *Ecological Monographs*, **41**, 27–52,

Jones, R. M. (1963). Studies in the autecology of the Australian acacias in South Africa. IV. Preliminary studies of the germination of seed of *Acacia cyclops* and *A. cyanophylla*. *South African Journal of Science*, **59**, 296–8.

Kornas, J. (1958). Succession régressive de la végétation de garrigue sur calcaires compacts dans la montagne de la Gardiole près de Montpellier. *Acta Societatis Botanicorum Poloniae*, **27**, 563–96.

Kruger, F. J. (1984). Effects of fire on vegetation structure and dynamics. In *Ecological Effects of Fire in South African Ecosystems*, ed. P. de V. Booysen & N. M. Tainton, pp. 220–43. Berlin: Springer-Verlag.

Kruger, F. J. & Bigalke, R. C. (1984). Fire in fynbos. In *Ecological Effects of Fire in South African Ecosystems*, ed. P. de V. Booysen & N. M. Tainton, pp. 69–114. Berlin: Springer-Verlag.

Kuhnholtz-Lordat, G. (1938). *La Terre Incendiée. Essai d'Agronomie Comparée*. Nimes: La Maison Carrée.

Le Houérou, H. N. (1974). Fire and vegetation in the Mediterranean Basin. *Proceedings of the Annual Tall Timbers Fire Ecology Conference*, **13**, 237–77.

Moll, E. J., McKenzie, B. & McLachlan, D. (1980). A possible explanation for the lack of trees in the fynbos, Cape Province, South Africa. *Biological Conservation*, **17**, 221–8.

Prodon, R., Fons, R. & Peter, A. M. (1984). L'impact du feu sur la végétation, les oiseaux et les micromammifères dans diverses formations méditerranéennes des Pyrénées orientales: premiers résultats. *Revue d'Ecologie Terre et Vie*, **39**, 129–58.

Purdie, R. W. (1977*a*). Early stages of regeneration after burning in dry sclerophyll vegetation. I. Regeneration of the understorey by vegetative means. *Australian Journal of Botany*, **25**, 21–34.

Purdie, R. W. (1977*b*). Early stages of regeneration after burning in dry sclerophyll vegetation. II. Regeneration by seed germination. *Australian Journal of Botany*, **25**, 35–46.

Purdie, R. W. & Slatyer, R. O. (1976). Vegetation succession after fire in sclerophyll woodland communities in south-eastern Australia. *Australian Journal of Ecology*, **1**, 223–36.

Richardson, D. M., van Wilgen, B. W. & Mitchell, D. T. (1987). Aspects of the reproductive ecology of four Australian *Hakea* species (Proteaceae) in South Africa. *Oecologia*, **71**, 345–54.

Roux, E. R. & Middlemiss, E. (1963). Studies in the autecology of the Australian acacias in South Africa. I. The occurrence and distribution of *Acacia cyanophylla* and *A. cyclops* in the Cape Province. *South African Journal of Science*, **59**, 286–94.

Sampson, A. W. (1944). Plant succession on burned chaparral lands in northern California. *California Agricultural Experimental Station Bulletin* No. 685.

Specht, R. L. (1981). Responses to fire of heathlands and related shrublands. In *Fire and the Australian Biota*, ed. A. M. Gill, R. H. Groves & I. R. Noble, pp. 395–415. Canberra: Australian Academy of Science.

Sweeney, J. R. (1956). Responses of vegetation to fire. A study of the herbaceous vegetation following chaparral fires. *University of California Publications in Botany*, **28**, 143–250.

Taylor, H. C. (1977). Aspects of the ecology of the Cape of Good Hope Nature Reserve in relation to fire and conservation. In *Proceedings of a Symposium on the Environmental Consequences of Fire and Fuel Management in Mediterranean Ecosystems*, ed. H. A. Mooney & C. E. Conrad, pp. 483–7. Washington, DC: USDA Forest Service General Technical Report WO-3.

Trabaud, L. (1970). Quelques valeurs et observations sur la phytodynamique des surfaces incendiées dans le Bas-Languedoc. *Naturalia Monspeliensia*, **21**, 231–42.

Trabaud, L. (1974). Experimental study of the effects of prescribed burning on a *Quercus coccifera* L. garrigue. *Proceedings of the Annual Tall Timbers Fire Ecology Conference*, **13**, 97–129.

Trabaud, L. (1977). Comparison between the effect of prescribed fires and wild fires on the global quantitative development of the kermes scrub oak (*Quercus coccifera* L.) garrigues. In *Proceedings of a Symposium on the Environmental Consequences of Fire and Fuel Management in Mediterrenean Ecosystems*, ed. H. A. Mooney & C. E. conrad, pp. 271–82. Washington, DC: USDA Forest Service General Technical Report WO-3.

Trabaud, L. (1980*a*). Impact biologique et écologique des feux de végétation sur l'organisation, la structure et l'évolution de la végétation des garrigues du Bas-Languedoc. Thèse Doctorat Etat Sciences, Université des Sciences et Techniques du Languedoc, Montpellier.

Trabaud, L. (1980*b*). Influence du feu sur les semences enfouies dans les couches superficielles du sol d'une garrigue de Chêne kermès. *Naturalia Monspeliensia*, **39**, 1–12.

Trabaud, L. (1983). Evolution après incendie de la structure de quelques phytocénoses méditerranéennes du Bas-Languedoc (Sud de la France). *Annales des Sciences Forestières*, **40**, 177–95.

Trabaud, L. (1990). Fire as an agent of plant invasion. A case study in the French Mediterranean vegetation. In *Biological Invasions in Europe and the Mediterranean Basin*, ed. F. di Castri, A. J. Hansen & M. Debussche, pp. 417–37. Dordrecht: Kluwer.

Trabaud, L. & Lepart, J. (1980). Diversity and stability in garrigue ecosystems after fire. *Vegetatio*, **43**, 49–57.

Trabaud, L. & Lepart, J. (1981). Changes in floristic composition of a *Quercus coccifera* garrigue in relation to different fire regimes. *Vegetatio*, **46**, 105–16.

Trabaud, L., Grosman, J. & Walter, T. (1985*a*). Recovery of burnt *Pinus halepensis* Mill. forests. I. Understorey and litter phytomass development after wildfire. *Forest Ecology & Management*, **12**, 269–77.

Trabaud, L., Michels, C. & Grosman, J. (1985*b*). Recovery of burnt *Pinus halepensis* Mill. forests. II. Pine reconstitution after wildfire. *Forest Ecology & Management*, **13**, 167–79.

Wright, M. A. & Bailey, A. W. (1982). *Fire Ecology. United States and Southern Canada*. New York: Wiley.

15

Plant invasion and soil seed banks: control by water and nutrients

R. L. SPECHT & H. T. CLIFFORD

It appears that few, if any, introduced species became established in undisturbed native plant communities prior to European settlement of Australia (Specht, 1981*a*). The invasion of introduced vascular plants into the disturbed mediterranean-type ecosystems of southern Australia has been almost continuous since European occupation of the continent. Some five to six introduced plant species have become naturalised every year since 1880 (Ewart, 1930; Specht, 1981*a*), a rate of invasion which can be projected back almost to the beginning of the nineteenth century (Kloot, 1985, and this volume).

Invasive plants flourished in the savanna communities (grasslands, savanna woodlands and savanna open-forests) on the more fertile soils, often replacing the indigenous ground stratum. In contrast, fewer invasions occurred in the sclerophyll communities (heathlands, mallee scrubs, woodlands and open-forests) that occur extensively in south-western Australia and occupy almost half of the mediterranean-type landscapes of south-eastern Australia. The soils associated with the sclerophyll communities are very low in plant nutrients.

Over the last 200 years, and especially over the last 50 years, urban and agricultural developments have expanded rapidly in the mediterranean-climate region of Australia. Over the latter 50-year period the discovery and amelioration of nutrient deficiencies of major (phosphorus, nitrogen, potassium and sulphur) and minor (copper, zinc, manganese and molybdenum) elements in these sclerophyll ecosystems has led to rapid agricultural development on these soils, with concomitant invasion of introduced plants.

Disturbance, which results in 'gaps' in the perennial native plant communities, seems to be the key to the successful invasion by introduced species (Specht, 1990). This chapter explores the dynamics of the soil seed bank following disturbance. We define the soil seed bank as the number of seeds stored per unit area of surface soil, as estimated by extraction of seed from soil by washing or by germination of the maximum number of seedlings in controlled environments.

Gaps in native plant communities: climatic control

The regions of the world experiencing a mediterranean-type climate, with summer drought alternating with a humid period of the year, show a gradient in severity of the dry summer season from perhumid to semi-arid (Emberger, 1955; Specht, 1987). The annual evaporative power of the atmosphere, as measured by pan evaporation, increases along this gradient.

As most of the overstorey vegetation of the five mediterranean-type regions is evergreen, monthly values of the water-balance budget can be used to establish the evaporative coefficient (k) of a particular plant community (Specht, 1972*a*).

$$\mathrm{Ea/Eo} = k\,(\mathrm{P} - \mathrm{R} - \mathrm{D} + \mathrm{Sext}) \qquad (1)$$

where Ea = actual evapotranspiration (cm per month)
Eo = pan evaporation (cm per month)
P = monthly precipitation (cm)
R = monthly runoff (cm)
D = monthly drainage losses (cm)
Sext = extractable soil water store (cm) at beginning of month
k = evaporative coefficient.

The evaporative coefficient (k) expresses the integration of resistances of canopy structure (external resistances) and of leaves (internal resistances) in the plant communities to the flow of available water from the soil (W = P − R − D + Sext) to the atmosphere (Ea) for each month of the year. In most mediterranean-climate regions available water is not in excess supply, and relative monthly evapotranspiration (Ea/Eo) is related linearly to available water (W), oscillating from a high value in winter and spring to a low value in summer, but rarely, if ever, reaching zero.

In any one locality, canopy structure and foliage resistances have evolved to equilibrate the community with the annual evaporative power (Eo) of the atmosphere. Variations in available water (W) from site to site within the same evaporative (Eo) region will result in different levels of productivity, *but essentially the same foliage projective cover*, of the overstorey of the plant communities. In effect, the evaporative coefficient (k) is a macro-climatic constant, unaffected by variations in the water relations of micro-habitats.

Variation in annual evaporation (Eo) from site to site induces variation in the evaporative coefficient (k) and with it, the foliage cover of the mature plant community (Table 15.1). Foliage projective covers (FPC) of both the overstorey and understorey strata of mature plant communities appear to be closely related to the

Table 15.1. *Foliage structure of mature evergreen plant communities in different mediterranean-climate zones of Australia*

Climatic zone	Evaporative coefficient (k)	Foliage projective cover (%)	
		Overstorey	Understorey
Perhumid	0.100–0.070	90–61	48–44
Humid	0.070–0.055	61–47	44–42
Subhumid	0.055–0.045	47–37	42–41
Semi-arid	0.045–0.035	37–28	41–40

evaporative coefficient (k) in Equation (1) (Specht, 1983):

Overstorey stratum of mature plant communities

$$\text{FPC}\ (\%) = 960\,k - 6.0 \quad (2)$$

$(n = 44, r^2 = 0.94)$

Understorey stratum of mature plant communities

$$\text{FPC}\ (\%) = 110\,k + 36.5 \quad (3)$$

$(n = 25, r^2 = 0.94)$

A few years after a perturbation such as fire, the sum of overstorey and understorey FPCs of an evergreen plant community reaches a constant value related to the evaporative coefficient in Equation (1) (Specht & Morgan, 1981; Specht, 1983):

$$\text{Total FPC}\ (\%) = 1070\,k + 30.5 \quad (4)$$

$(n = 29, r^2 = 0.93)$

At first, understorey FPC develops rapidly in the post-fire succession but, with time, the slowly regenerating overstorey FPC replaces the understorey FPC. Eventually, at maturity, constant values of overstorey and understorey FPCs (Equations (2) and (3)) result in equilibrium with the evaporative coefficient (k).

The foliage within the plant community thus develops into distinct ecophysiological strata (Specht & Specht, 1989; Specht *et al.*, 1990) with horizontal coverage in equilibrium with the evaporative power of the atmosphere (on an annual cycle). Conversely, the spaces between the horizontal foliage cover of both overstorey and understorey strata are predictable, thereby providing 'gaps' for seed invasion.

Gaps in the overstorey stratum: seed production by the understorey

A decrease in foliage projective cover of the overstorey stratum enables more of the incident solar radiation to penetrate through the canopy onto the understorey.

Concomitant with this increase in solar radiation, photosynthesis of the understorey increases, resulting in increased foliage cover, biomass production, flowering and seed production (Figure 15.1).

Gaps in native communities: soil seed banks

The foliage projective covers of overstorey and understorey strata of mature communities (and, conversely, the 'gaps' so created) are predictable and related to the evaporative coefficient (Equations (2), (3) and (4), Table 15.1).

The soil seed bank (the maximum number of seeds germinating) has been determined on random samples of soil (2.5 cm deep) collected from a series of plant communities in Victoria (Carroll & Ashton, 1965; Clifford & Mott, 1986; Clifford, unpublished). These values have been plotted (Figure 15.2) against the overstorey FPC estimated for each site (Clifford & Specht, unpublished). If the seed bank (N = number of seedlings per metre square) is converted to natural logarithms and plotted against overstorey FPC (%), a linear relationship is obtained:

$$\ln N = 10.21 - 0.050\,\mathrm{FPC} \quad (5)$$
$$(n = 12, r^2 = 0.84, p = 0.05)$$

Observations in subtropical Queensland indicate that although the seed bank

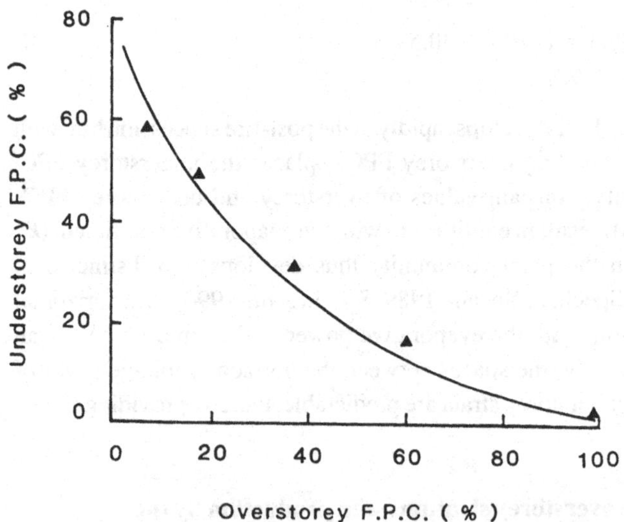

Figure 15.1*a* (above), *b* (opposite). Foliage projective cover (%) and number of inflorescences (/m²) in pastures under a tree density planting experiment at Samford, south-eastern Queensland (R. Grundy, personal communication).

is smaller than that recorded in temperate south-eastern Australia (Clifford & Mott, 1986), it shows a decrease with increasing overstorey FPC similar to that shown in Equation (5) (Clifford & Specht, unpublished).

$$\ln N = 8.38 - 0.024\,\text{FPC} \qquad (6)$$
$$(n = 21, r = 0.53, p = 0.02\text{–}0.01)$$

Grazed pastures: soil seed banks

No definitive studies have been conducted on seed banks under heavily grazed pastures in the mediterranean-climate region of Australia. The high values shown in Figure 15.3 probably result from the transformation of the savanna understorey from a perennial native grassland to a seasonal understorey of introduced annual grasses and herbs.

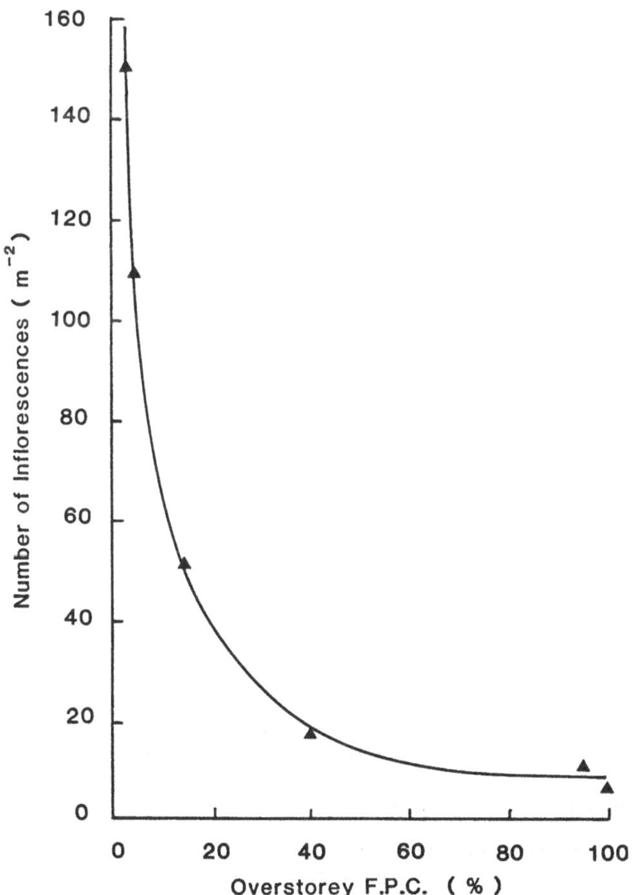

Disturbance in native plant communities: soil seed banks

The community 'gap' can be enlarged by manipulating either the overstorey, the understorey or both. There are no examples from the mediterranean-climate area of southern Australia. In subtropical Queensland, however, when overstorey trees were removed about 20 years ago from an open-forest of *Eucalyptus*, the soil seed bank increased by a factor of six (Table 15.2, from Grundy, 1986) to values that may be predicted by the relationship shown in Equation (6). Tree removal increased the overstorey 'gap', thereby allowing a large number of easily-dispersible seeds of introduced species to accumulate in the soil.

In the same area of subtropical Queensland a ground stratum of native perennial grasses (predominantly *Themeda triandra*) has been mown regularly over the last 20 years. Gradually, most of the native savanna species have been

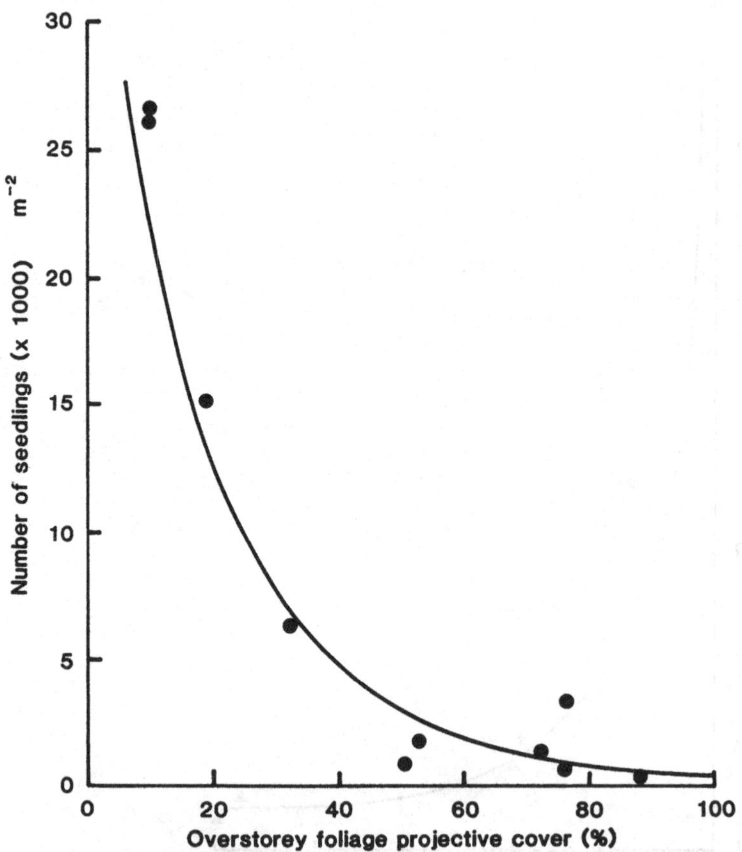

Figure 15.2. Relationship between soil seed bank (/m²) and overstorey foliage projective cover (%) in south-eastern Australian plant communities.

replaced by creeping grasses (especially *Digitaria didactyla*). The overstorey of eucalypt trees has been retained throughout the area. Repeated mowing has stimulated seed production in the understorey by a factor of 3.5 to 4 (Table 15.3; McIvor, 1987; R. M. Jones, personal communication).

Germination of seeds in the seed bank

Provided that the exposed soil is not compacted, an increase in the size of the community 'gap' following disturbance of the overstorey (± midstratum shrubs) will lead to an increased storage of seed in the soil seed bank. The magnitude of the seed bank appears to be predictably related to the size of the community 'gap'.

Often the seed bank (including occasional introduced species) of undisturbed communities fails to germinate (and/or establish) until a fire or other perturbation occurs (see contributions in Gill *et al.*, 1981). In contrast, the lack of competition from established plants in disturbed areas enables germination and establishment (especially of introduced species) to occur.

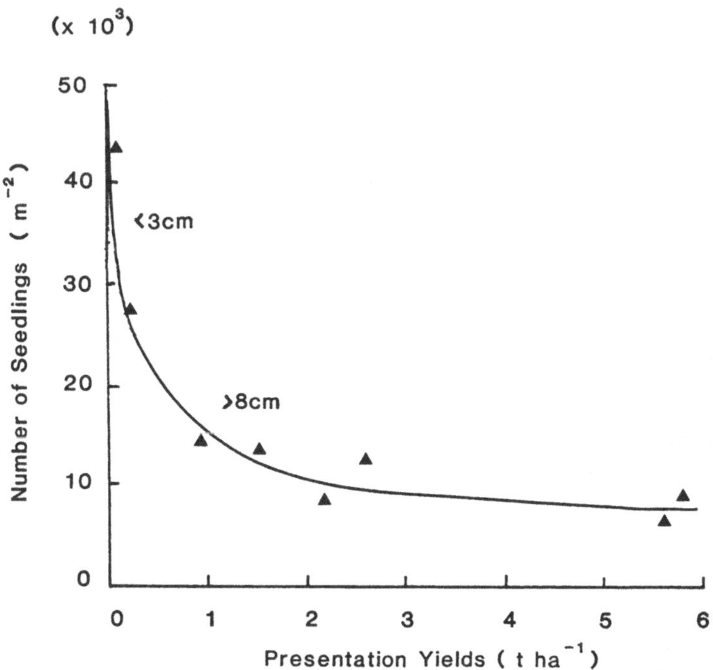

Figure 15.3. Relationship between soil seed bank (/m2) and presentation yield (t/ha) after ten years of a grazing intensity trial on pastures at Samford, south-eastern Queensland (R. M. Jones, personal communication).

Table 15.2. *Soil seed bank in mature eucalypt open-forest (with savanna ground stratum) and under power lines (gap) in Brisbane Forest Park (Grundy, 1986). Introduced herbs predominantly* Ageratum houstonianum

	Foliage projective cover (%)		Seed bank (/m²)		
Site	Overstorey	Understorey	Total seedlings /m²	Total herbs (%)	Introduced herbs (%)
North-facing					
Open-forest	78	55	660	46	39
Gap	3	87	4234	52	51
South-facing					
Open-forest	71	55	1272	40	34
Gap	12	96	6051	69	67

Table 15.3. *Foliage projective cover and soil seed bank in mature eucalypt open-forest with savanna ground stratum in Brisbane Forest Park: (1) undisturbed (apart from fire every 3 years); and (2) mown regularly over the last 20 years (but unburnt)*

	Ground stratum	
Index	Unmown but burnt	Mown but unburnt
Foliage projective cover (%)		
Overstorey	55	52
Understorey	58	20 (in 1966)[a] 54 (in 1986)[b]
Seed bank (/m²)		
Total seedlings	2530	8970
Native species		
Themeda triandra	24	0
Gnaphalium sphaericum	97	0
Introduced species		
Dactyloctenium sp.	30	1017
Digitaria ciliaris	9	948
Setaria sp.	0	78
Ageratum houstonianum	57	48
Anagallis arvensis	24	15

[a] Bases of *Themeda triandra* remaining after mowing.
[b] Lawn grass, *Digitaria didactyla*, at peak growing season.
Source: Day, 1986.

Table 15.4. *Mean number (and standard deviation) of* Hypochoeris radicata *and* Vulpia *spp. recorded in 20 random quadrats (20 cm square) on phosphate-fertilised and unfertilised plots at Dark Island heath*

	Number of plants (/m²)	
Introduced species	Phosphate-fertilised	Unfertilised
Hypochoeris radicata	475 + 350	0
Vulpia spp.	950 + 500	0

Source: Specht, 1963

Table 15.5. *Total number of* Hypochoeris radicata *and* Vulpia *spp. which germinated in ten samples of soil (200 cm² × 2 cm deep) collected between bushes at Dark Island heath in January 1987 and germinated in a controlled environment with and without fertiliser as a complete nutrient solution*

	Number of seedlings (/m²)	
Introduced species	Fertilised	Unfertilised
Hypochoeris radicata	160	120
Vulpia spp.	60	130
Total	220	250

Germination of seeds in the seed bank is stimulated by fire (Specht, 1981*b*). Infertile soils become much more fertile after fire because of the so-called 'ash-bed effect' (Humphreys & Craig, 1981). This increased fertility, together with reduced competition from established plants, soil micro-organisms and soil fauna, appears to favour germination and establishment.

The addition of phosphatic fertiliser to the nutrient-poor sands of Dark Island heath, South Australia, resulted within a few months in germination and establishment of introduced grasses and composites in the gaps between heath plants (Table 15.4, from Specht, 1963). Results of laboratory trials revealed that seed of these invasive species germinated equally well with and without the addition of phosphatic fertiliser (Specht, 1963). In the field, these tiny young seedlings may have germinated (as shown in Table 15.5) and died on the unfertilised plots before observation was possible. In particular, seeds of the composite *Hypochoeris radicata* are viable for only a year (L. Cameron, personal communication), the soil seed bank being recharged annually.

It is possible that seed may be dormant and that fire or some other perturbation is necessary to break dormancy. Keeley (1984) demonstrated that seed dormancy of two species common following fire in the Californian chaparral could be broken when the ash of stems of chaparral shrubs or even when one of the wood components (lignin) was mixed with the seed. Some water-soluble compound(s), released by heating wood structural components, may provide the clue. The question of whether such a compound(s) can be released by extended periods of solar heating, as experienced in the surface soil of enlarged 'gaps', then arises.

Seedling establishment

For successful establishment of seedlings from the soil seed bank, competition from established plants, soil micro-organisms and soil fauna, must be minimal. Improved water relations in the surface soil in the few months after fire (Specht, 1957) and increased nutrient level in the top few centimetres of soil following fire (Specht *et al.*, 1958) increase the chances of seedling survival.

As much of the Australian flora of both sclerophyll and savanna communities has evolved on soils low to very low in plant nutrients, their seedlings show

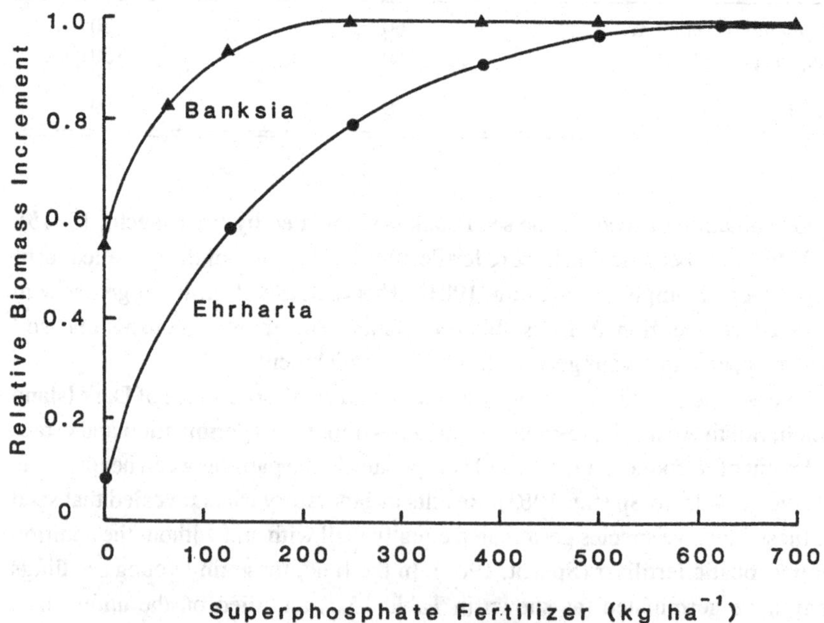

Figure 15.4. Biomass increment to added superphosphate fertiliser of *Banksia ornata* (a long-lived native shrub) and an invasive introduced grass (*Ehrharta calycina*) in the mediterranean-climate zone of southern Australia.

minimal response to soil nutrient improvement, either produced naturally after fire or following nutrient enrichment by some other means. Introduced invasive plants, on the other hand, show strong responses to nutrients (Figure 15.4). Dominant native plants in overstorey and understorey strata of mediterranean-climate ecosystems in Australia produce new shoot growth in late spring or summer when soil water is becoming limiting (Specht & Rayson, 1957; Specht *et al.*, 1981). Unfortunately, invasive plants grow vigorously in winter and spring (Figure 15.5), thereby utilising water and nutrient resources before the dominant native plants begin their annual growth cycle. Such strong competition for resources is ultimately detrimental to many of the native species (Specht, 1963, 1975, 1990).

Soil disturbance and fertiliser addition favoured the establishment and growth of both introduced and native seedlings in the cereal zone of Western Australia (Hobbs & Atkins, 1988).

In certain soils (such as in northern Israel (Rabinovitch-Vin, 1983) and in the Sierra Morena of Spain (Marañón, 1985)), high rates of soil nitrification, induced by disturbance, lead to the establishment of numerous (about 100 species per 0.1 ha), short-lived herbaceous species (Specht *et al.*, 1990).

Conclusions

The spatial distribution of the foliage of evergreen plant communities has evolved to achieve an evaporative balance with the environment, expressed by the evaporative coefficient (k = relative evapotranspiration (Ea/Eo)/available

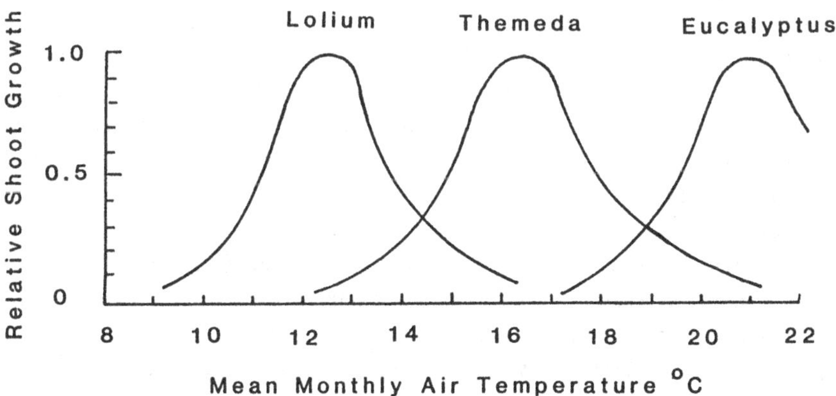

Figure 15.5. Seasonal shoot growth of invasive introduced species (*Lolium* sp.)as compared with native dominant species (*Eucalyptus* spp., *Banksia* spp.) and a native grass (*Themeda triandra*) in the mediterranean-climate zone of southern Australia.

water (W)). As the plant community regenerates after disturbance, the sum of the foliage projective covers of overstorey and understorey strata soon reaches a constant value in equilibrium with the evaporative coefficient. At maturity a balance is reached with constant values for foliage projective covers of overstorey and understorey strata. Thus, the spaces ('gaps') between the horizontal foliage cover of both overstorey and understorey are predictable.

This predictable community 'gap' has been shown to support a predictable soil seed bank of firstly, seed produced by resident understorey species, and secondly, of dispersible seed entering the plant community in turbulent down-draughts induced by the 'gaps'. Increased soil temperatures experienced in sub-tropical areas appear to reduce the magnitude of the seed bank (Equation (6)), compared with that recorded in the warm-temperate mediterranean region (Equation (5)). Disturbance of the overstorey will increase the size of the community 'gap', thereby enabling a larger seed bank to develop. In the humid/subhumid zone, the dispersible seeds of introduced species occupy the enlarged community 'gap'; in the semi-arid zone, where invaders are rare, native plants can expand.

In some plant communities in semi-arid areas, shrubby thickets may develop which will reduce the community 'gap' and hence the size of the soil seed bank (Hodgkinson *et al.*, 1980). Compacted surface soils are sites with a limited capacity for seed storage (Hodgkinson *et al.*, 1980).

Overgrazing or mowing of the savanna understorey (± overstorey trees) in some non-mediterranean areas favours creeping grasses and herbs with a high level of inflorescence production, thereby increasing the soil seed bank by a factor of 3.5 to 4 times.

Germination of seeds in the soil seed bank, including those of invasive species, appears to be favoured in 'gaps' enlarged by disturbance. Reduced competition for water would favour germination in 'gaps'. Extended periods of solar heating experienced in the enlarged 'gaps' appears to favour soil nitrification and thus the growth of pioneer and weedy plant species with high nitrate reductase activity in their leaves (Stewart *et al.*, 1990).

Reduced competition for both water and nutrients in enlarged 'gaps' favours survival of seedlings germinating from the soil seed bank. In the mediterranean-climate region of Australia, nutrient increment in the surface centimetres of soil following fire or from other forms of nutrient enrichment favours introduced species. The latter, with their vigorous shoot growth response in winter–spring, exploit resources before seedlings of dominant native species, with their spring–summer shoot growth rhythm, begin to grow.

After intense cultivation over the last 150 years, little native vegetation remains in the rural areas of the mediterranean-climate region of Australia.

Today, the pastures are composed almost entirely of introduced winter–spring annuals (Kloot, 1985, and this volume). The community 'gaps' in these seasonal pastures have maximised the soil seed bank (Equations (5) & (6)), thereby enabling the invasion of many prickly thistles (*Carduus tenuiflorus, Carthamus lanatus, Centaurea calcitrapa, Onopordum acanthium*) and of *Asphodelus fistulosus, Homeria collina, Oxalis pes-caprae*, etc. Some degree of control of these invasive plants can be achieved by increasing the deep-rooted perennial component of the pasture (Tiver & Crocker, 1951) and thereby reducing the size of the soil seed bank.

References

Carroll, E. J. & Ashton, D. H. (1965). Seed storage in soils of several Victorian plant communities. *Victorian Naturalist*, **82**, 102–10.

Clifford, H. T. & Mott, J. J. (1986). Functioning of tropical plant communities. V. Regenerative processes. In *Tropical Plant Communities: Their Resilience, Functioning and Management in Northern Australia*, ed. H. T. Clifford & R. L. Specht, pp. 68–77. St Lucia, Queensland: Botany Department, University of Queensland.

Day, K. A. (1986). An assessment of strategy theory based on seed populations stored in topsoils. B.Sc. Hons. thesis, Botany Department, University of Queensland.

Emberger, L. (1954). Une classification biogéographique des climats. *Recueil des Travaux des Laboratoires de Botanique, Géologie et Zoologie de la Faculté des Sciences de l'Université de Montpellier, Série Botanique*, **7**, 3–43.

Ewart, A. J. (1930). *Flora of Victoria*. Melbourne: Government Printer.

Gill, A. M., Groves, R. H. & Noble, I. R. (ed.) (1981). *Fire and the Australian Biota*. Canberra: Australian Academy of Science.

Grundy, R. (1986). The effect of disturbance (clearing) and aspect on vegetation in Brisbane Forest Park. B.Sc. Hons thesis, Botany Department, University of Queensland.

Hobbs, R. J. & Atkins, L. (1988). Effect of disturbance and nutrient addition on native and introduced annuals in plant communities in the Western Australian wheatbelt. *Australian Journal of Ecology*, **13**, 171–80.

Hodgkinson, K. C., Harrington, G. N. & Miles, G. E. (1980). Composition, spatial and temporal variability of the soil seed pool in a *Eucalyptus populnea* shrub woodland in central New South Wales. *Australian Journal of Ecology*, **5**, 23–9.

Humphreys, F. R. & Craig, F. G. (1981). Effects of fire on soil chemical, structural and hydrological properties. In *Fire and the Australian Biota*, ed. A. M. Gill, R. H. Groves & I. R. Noble, pp. 177–200. Canberra: Australian Academy of Science.

Keeley, S. C. (1984). Stimulation of post-fire herb germination in the California chaparral by burned shrub stems and heated wood components. In *Proceedings of the 4th International Conference on Mediterranean Ecosystems*, ed. B. Dell, pp. 79–80. Nedlands, Western Australia: Botany Department, University of Western Australia.

Kloot, P. M. (1985). Studies in the alien flora of the cereal rotation areas of South Australia. Ph.D. thesis, University of Adelaide.

Marañón, Arana T. (1985). Diversidad floristica y heterogeneidad ambietal en una dehesa de Sierra Morena. *Anales de Edafologia y Agrobiologia. II. Biologia Vegetal*, pp. 1183–97.

McIvor, J. G. (1987). Changes in germinable seed levels in soil beneath pastures near Townsville, north Queensland. *Australian Journal of Experimental Agriculture*, **27**, 283–9.

Rabinovitch-Vin, A. (1983). Influence of nutrients on the composition and distribution of plant communities in mediterranean-type ecosystems of Israel. *In Mediterranean-type Ecosystems. The Role of Nutrients*, ed. F. J. Kruger, D. T. Mitchell and J. U. M. Jarvis, pp. 74–85. Berlin: Springer-Verlag.

Specht, R. L. (1957). Dark Island heath (Ninety-Mile Plain, South Australia). V. The water relationships in heath vegetation and pastures on the Makin sand. *Australian Journal of Botany*, **5**, 151–72.

Specht, R. L. (1963). Dark Island heath (Ninety-Mile Plain, South Australia). VII. The effect of fertilizers on composition and growth, 1950–1960. *Australian Journal of Botany*, **11**, 67–94.

Specht, R. L. (1972*a*). Water use by perennial evergreen plant communities in Australia and Papua New Guinea. *Australian Journal of Botany*, **20**, 273–99.

Specht, R. L. (1972*b*). *The Vegetation of South Australia*, 2nd edn. Adelaide: Government Printer.

Specht, R. L. (1975). A heritage inverted, our flora endangered. *Search*, **6**, 472–7.

Specht, R. L. (1981*a*). Major vegetation formations in Australia. In *Ecological Biogeography of Australia*, vol. 1, ed. A. Keast, pp. 163–297. The Hague: Junk.

Specht, R. L. (1981*b*). Responses to fire of heathlands and related shrublands. In *Fire and the Australian Biota*, ed. A. M. Gill, R. H. Groves & I. R. Noble, pp. 395–415. Canberra: Australian Academy of Science.

Specht, R. L. (1983). Foliage projective covers of overstorey and understorey strata of mature vegetation in Australia. *Australian Journal of Ecology*, **8**, 433–9.

Specht, R. L. (1987). The effect of summer drought on vegetation structure in the mediterranean climate region of Australia. In *Plant Response to Stress – Functional Analysis in Mediterranean Ecosystems*, ed. J. D. Tenhunen, F. M. Catarino, O. L. Lange & W. C. Oechel, pp. 625–39. Berlin: Springer-Verlag.

Specht, R. L. (1990). Changes in the eucalypt forests of Australia as a result of human disturbance. In *The Earth in Transition. Patterns and Processes of Biotic Impoverishment*, ed. G. W. Woodwell, pp. 178–98. New York: Cambridge University Press.

Specht, R. L. & Morgan, D. G. (1981). The balance between the foliage projective covers of overstorey and understorey strata in Australian vegetation. *Australian Journal of Ecology*, **6**, 193–202

Specht, R. L. & Rayson, P. (1957). Dark Island heath (Ninety-Mile Plain, South Australia). I. Definition of the ecosystem. *Australian Journal of Botany*, **5**, 52–85.

Specht, R. L. & Specht, A. (1989). Canopy structure in *Eucalyptus*-dominated communities in Australia along climatic gradients. *Oecologia Plantarum*, **10**, 191–213.

Specht, R. L., Rayson, P. & Jackman, M. E. (1958). Dark Island heath (Ninety-Mile Plain, South Australia). VI. Pyric succession: changes in composition, coverage, dry weight, and mineral nutrient status. *Australian Journal of Botany*, **6**, 59–88.

Specht, R. L., Rogers, R. W. & Hopkins, A. J. M. (1981). Seasonal growth and flowering rhythms: Australian heathlands. In *Ecosystems of the World*, vol. 9B, *Heathlands and Related Shrublands. Analytical Studies*, ed. R. L. Specht, pp. 5–13. Amsterdam: Elsevier.

Specht, R. L. Grundy, R. I. & Specht, A. (1990). Species richness of plant communities – relationship with community growth and structure. *Israel Journal of Botany*, **39**, 465–80.

Specht, R. L., Clifford, H. T., Arianoutsou, M., Bird, L. H., Bolton, M. P., Forster, P. I., Grundy, R. I., Hegarty, E. E. & Specht, A. (1991). Structure, floristics and species richness of plant communities of southeast Queensland. *Proceedings of the Royal Society of Queensland*, **101**, in press.

Stewart, G. R., Gracia, C. A., Hegarty, E. E. & Specht, R. L. (1990). Nitrate reductase activity and chlorophyll content in sun leaves of subtropical Australian closed-forest communities. *Oecologia (Berlin)*, **82**, 544–51.

Tiver, N. S. & Crocker, R. L. (1951). The grasslands of south-east South Australia in relation to climate, soils and developmental history. *Journal of the British Grassland Society*, **6**, 29–80.

16

Invasion by annual brome grasses: a case study challenging the homoclime approach to invasions

J. ROY, M. L. NAVAS & L. SONIÉ

The convergence of the biotas of the five regions of the world with a mediterranean-type climate was actively studied over a decade ago (di Castri & Mooney, 1973; Mooney, 1977; Cody & Mooney, 1978). Already at that time, the large interchange of plant species between the five regions was mentioned (di Castri & Mooney, 1973) but not much documented. The present volume aims to fill that gap. Such an enterprise gains support from the assumption that the success of invasions strongly depends upon the similarity of the source and reception areas in terms of climate, life forms and structure of the biota (Baker, 1986) and that, as a consequence, the five mediterranean-type regions constitute a rather specific network within which invasions are better analysed, understood and amenable to predictions than they would be using larger geographical units composed of several climatic types.

In this chapter, in addition to documenting the invasion patterns of a genus containing ecologically and economically important species, we want to challenge the above assumption. In fact, because of life form, phenology, plasticity or ecotypic differentiation, many species have distributions encompassing regions of several climatic types. For such species, only extremes in climate may prevent invasion. Moreover, climatic similarity is a condition neither sufficient nor necessary for an invasion to occur (as shown by the literature on biological control of insects and weeds, e.g. Harris, 1984). More interestingly, results of recent studies (Forcella & Wood, 1984; Forcella *et al.*, 1986) suggest a good relation between the invasive potential of a species and the width of its native distribution. Although these studies deal with the geographical and not the climatic width of species distributions, they suggest that an analysis restricted to regions within a single type of climate may pass over critical points.

Annual species of bromes seem to be an appropriate group of species in which to investigate the soundness of the homoclime approach to invasion because, firstly, they are annuals and, for most of them, Eurasian in origin, like the majority of the invaders of the mediterranean-type regions (Fox, 1990); and secondly, they allow comparisons to be made, both within the same genus and within the same life form, of the invasive capabilities of species with different native climatic distributions.

Materials and methods

The world distribution of annual brome grasses was derived from a survey of about 300 floras in the libraries of scientific institutions at Montpellier and Paris (France), Stanford and Berkeley (California), and Rabat (Morocco). The data for the mediterranean regions are from Munz & Keck (1959) and Munz (1974) for California, Matthei (1986) for Chile, H. P. Linder (personal communication) for South Africa, Gardner (1952) and Jessop & Toelken (1987) for Australia. For Eurasia and northern Africa, we used county and national floras. The list of all floras used is available from the authors.

The following information was recorded: synonymy, life form (annual or perennial), distribution and habitat. Measures of abundance were not considered because of the different terminologies used by different authors. Nomenclature follows that of Tutin *et al.* (1980) for the European species, of Hitchcock (1971) for the American and Australian species, of Rechinger (1970) for the Middle Eastern ones and of Andrew (1956) for *Bromus adoensis*.

For our analysis, we considered only the species described in at least two floras of each region. As a result, eleven species were eliminated. For several countries of Africa and South America, we found only lists of species without indications of life form. When the life form could not be found from other floras, these species (35 in all) were then eliminated. The number of insufficiently defined, and then discarded, species is higher than the number of taxonomically well-defined species. It stresses the need for additional taxonomic studies and may slightly restrict the validity of our conclusions. *Bromus willdenowii* was not included in our analysis because it has been widely introduced by humans as a forage species and because it is sometimes reported to be perennial.

World maps were drawn at a scale of 1/110000000 and maps of North America and Eurasia were drawn at a scale of 1/20000000. Distributions were established from data recorded in the more recent floras, especially for the invaded regions. Limits were established including all the sites where the species was recorded.

Maps of the nine major climatic zones and their intermediates, as defined by Walter *et al.* (1975), were superimposed on the distribution maps of the species and the climates covered by each species recorded.

Native distribution of annual bromes

The phylogeny of the grass family is discussed in relation to continental drift by Clayton (1975). Grasses arose in the rainforests of tropical Gondwanaland. It is thought that tribes had differentiated by the end of the Cretaceous before Gondwanaland broke apart. The Pooideae, the sub-family of the grasses to which *Bromus* belongs, is the single predominant sub-family which migrated to the temperate climates of the northern landmasses of Europe, Asia and North America. The decreasing number of genera from Eurasia to North America and to South America suggest that Eurasia was the centre of migration and that only some of the evolving species were present in North America when it broke away in the Eocene. Migration from North America to South America occurred along the American cordilleras in the Pliocene. The distribution of the annual bromes follows this general pattern: 32 species are native to Eurasia, 2 to North America and 1 to South America. *Bromus arenarius*, a native of Australia, is the only species which does not fit into this palaeoevolutionary scheme. It may have reached Australia early in historic times.

Fox (1990) showed that annual plants reach their maximum diversity in areas of mediterranean-type climates. Whilst the proportion of annuals in the world flora is about 13 per cent, it reaches 30 per cent and 27 per cent in the floras of the Mediterranean Basin and of the California floristic province respectively. In the mediterranean regions of South Africa and Australia, it is close to the world figure. The number of native annual brome species found in the Eurasian and northern African climatic zones as defined by Walter *et al.* (1975) is shown in Figure 16.1. Among the 35 native species, a higher number of species (24) indeed occurs in the regions of mediterranean climate but the annual brome species are numerous in most climates. They are absent only from the humid-equatorial climatic regions. This extensive climatic distribution is obtained because of species having a large amplitude (e.g. *B. hordeaceus* covers climatic zones 34 to 9 with the exception only of zones 47 and 78) rather than because of species being specifically adapted to certain climates (with the exception of *B. adoensis* found only in zones 2, 3 and 23).

Present distribution in mediterranean regions

Annual bromes are ruderal species and their present distribution has been shaped by human movements both within Eurasia (Smith, 1986) and between continents. Intercontinental invasions became significant subsequent to the European conquest of the sea around AD 900 (Crosby, 1986). Our records show that invasions by brome grasses really got under way, however, during the second half of the nineteenth century (Table 16.1).

In Australia, the first list of introduced species dates from 1802 (see Maiden, 1916). The first record of *Bromus* sp. (possibly *B. hordeaceus*), however, was made in 1847 by McEwin (Kloot, 1983, 1984). In South Africa, the first herbarium specimen of a *Bromus* species dates from 1832, now deposited at the Pretoria National Herbarium. In California, the first specimens (six species) were collected in 1862 (Watson, 1880). The early rate of spread, as estimated from the number of specimens collected and kept in herbaria, is shown for the most common species (Figure 16.2). The data for California, expressed as a percentage of the number of specimens of *B. carinatus* (native), suggest that the increase in specimen number at the turn of the century is not only a consequence of an increase in botanical activity, but also corresponds to the spread of the species in the invaded countries.

The current result of species fluxes between the five mediterranean regions of the world is synthesised in Figure 16.3. A list of the species is given in Table 16.2. As already stated, Europe is the main source of invaders, but the number of

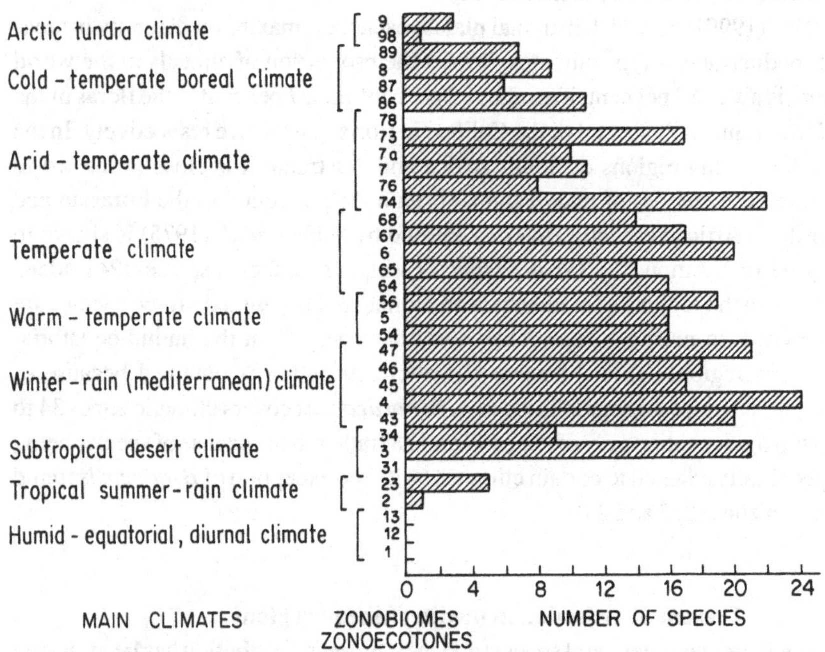

Figure 16.1. Number of native annual *Bromus* species per climatic zone along a latitudinal gradient from the equator to the North Pole in Africa and Eurasia. The zonobiomes, as defined by Walter *et al.* (1975) and represented by a single value, correspond typically to the main climates, whilst the zonoecotones correspond to a transition from a main climate to another, represented respectively by a first and a second value.

Table 16.1. *Date of the first collection of specimens of* Bromus *species in South Africa (National Herbarium, Pretoria), California (University of California, Berkeley, and Rancho Santa Ana Herbaria) and Australia (National Herbarium of New South Wales)*

	South Africa	California	Australia
Discovery by Europeans	1497	1542	1616
First significant settlement	1652	1769	1788
Nationality of settlers	Dutch	Spanish	English
Bromus species			
B. alopecuros			1916
B. arenarius		1906	
B. commutatus	1917	1881	
B. diandrus	1897	1881	1880
B. hordeaceus	1913	1882	1885
B. japonicus	1841	1905	
B. madritensis		1887	1898
B. racemosus		1896	1901
B. rubens		1880	1902
B. secalinus		1890	1899
B. squarrosus			1966
B. sterilis			1907
B. tectorum		1899	1910
B. trinii		1881	

invading species differs greatly between the invaded continents. The low number found in the mediterranean region of Chile is surprising in view of its early settlement by Mediterranean colonists.

Fox (1990) emphasised the heterogeneity of the mediterranean regions: the Mediterranean Basin, Chile and California have in common a Pleistocene assemblage of species and a settlement and trade by Mediterranean colonists, whilst South Africa and Australia share a Gondwanan assemblage of species (with fewer annuals), oligotrophic soils and a pattern of settlement from northern Europe. The qualitative analysis of the species exchanges between mediterranean regions was done using a factorial correspondence analysis on a matrix of the five mediterranean regions × presence or absence of the 29 species (cf. Table 16.2). The first three axes comprise 88 per cent of the variance. The projection of the continents and species on axes 1 and 2 is shown in Figure 16.4. Axis 1 detaches the Mediterranean Basin and the other continents, whilst axis 2 detaches the American mediterranean regions on the one hand and Australia plus South Africa on the other. The solid lines group continents and the more

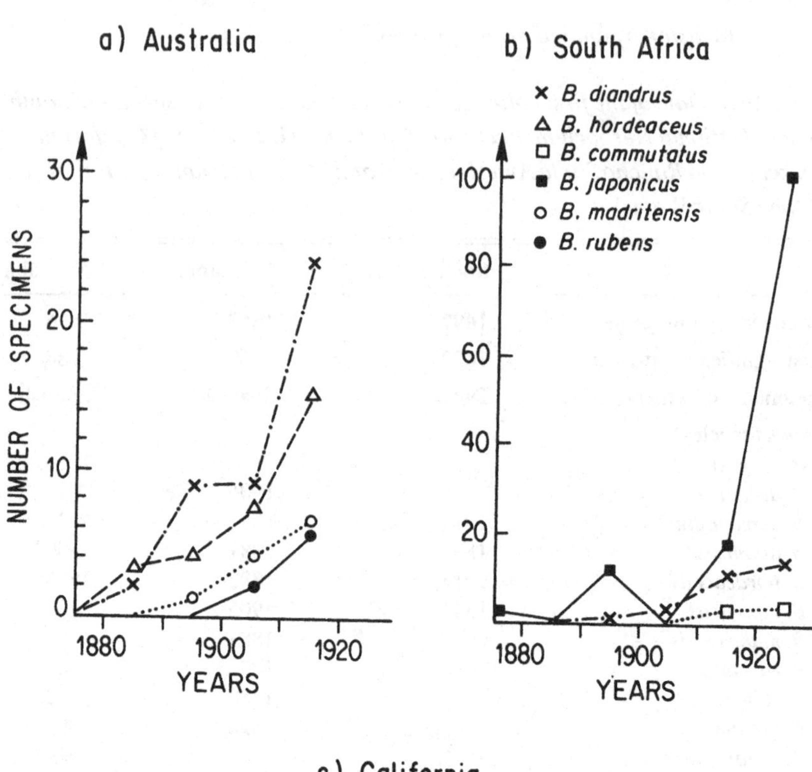
a) Australia
b) South Africa
× B. diandrus
△ B. hordeaceus
□ B. commutatus
■ B. japonicus
○ B. madritensis
● B. rubens
NUMBER OF SPECIMENS
YEARS
1880
1900
1920
0
10
20
30
20
40
60
80
100

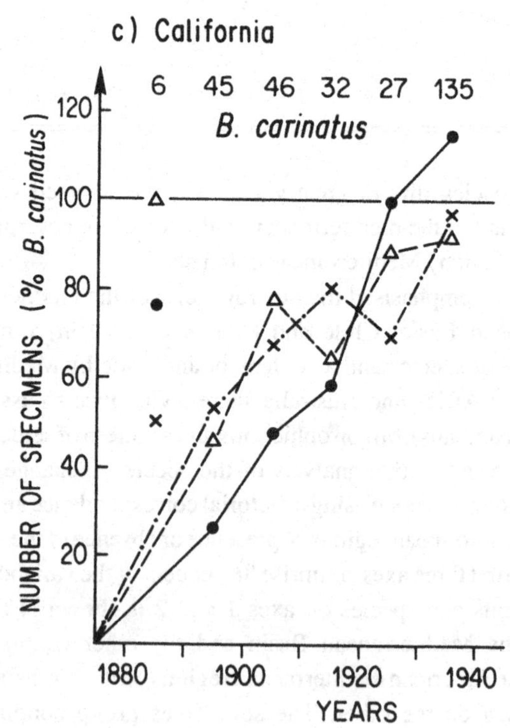
c) California
6
45
46
32
27
135
B. carinatus
NUMBER OF SPECIMENS (% B. carinatus)
YEARS
1880
1900
1920
1940
20
40
60
80
100
120

closely linked species. The Mediterranean Basin, Chile and California have characteristic species whilst Australia is linked to species which are also close to the Mediterranean Basin and Chile. South Africa does not have characteristic species and appears closer to Australia than to the other continents. The distribution of annual brome species does not obviously support the grouping of mediterranean-climate regions of Fox (1990).

Challenging the homoclime approach to invasions

Of a total of 45 successful invasions of mediterranean regions (Figure 16.3), only one (*B. briziformis* from Europe) involves species not found in mediterranean-climate regions on their native continents. This result seems to support the homoclime hypothesis; in fact, it does not because the proportion of species whose native distributions do not include areas of mediterranean climate is low (9 out of 36) and seven of these nine species are not invasive at all, not even in regions climatically similar to their native environment.

The low predictive capacity of the homoclime approach can be seen by comparing the number of invasive species with the number of potentially invasive species (i.e. species native to the other continents). Their ratio gives a measure of invasion status: 1 means the continent received all possible invaders and a ratio of 0 means the continent is free of invasive species. Summed over all continents, only 37 per cent of the invasions able to be predicted on the basis of climatic similarity actually occurred (Table 16.3). The analysis also shows that the continents differ in their invasion status – it ranges from 0.58 for California to 0.18 for Chile. California and Chile differ statistically from the mean estimate of invasion status (the log likelihood ratio test of Sokal & Rohlf, 1981). Differences probably lie in the properties of the recipient lands (their invasibility) or in the history of human activities between the continents. No correlation was found between invasion status and geographic distance to the source, although invasion status did appear to be proportional to a measure of the economic activity of the invaded

Figure 16.2 (opposite). Early rate of spread of invading annual brome grasses in (a) Australia, (b) South Africa, and (c) California, as estimated by the number of specimens found in the herbaria at Sydney, Pretoria and Berkeley plus Claremont respectively. Data for the decade 1880–1890 have been shown but because of the low numbers of specimens of both native and introduced species, they are indicative only. For California, data are expressed as a percentage of *B. carinatus*, a native species, to correct for the variation with time of collectors' activities. Data for Australia were compiled by J. Dalby and S. Jacobs. Computer lists for South Africa (PRECIS) were made available by G. E. Gibbs Russell. The curators of the University of California at Berkeley and Rancho Santa Ana herbaria opened their collections to the first author.

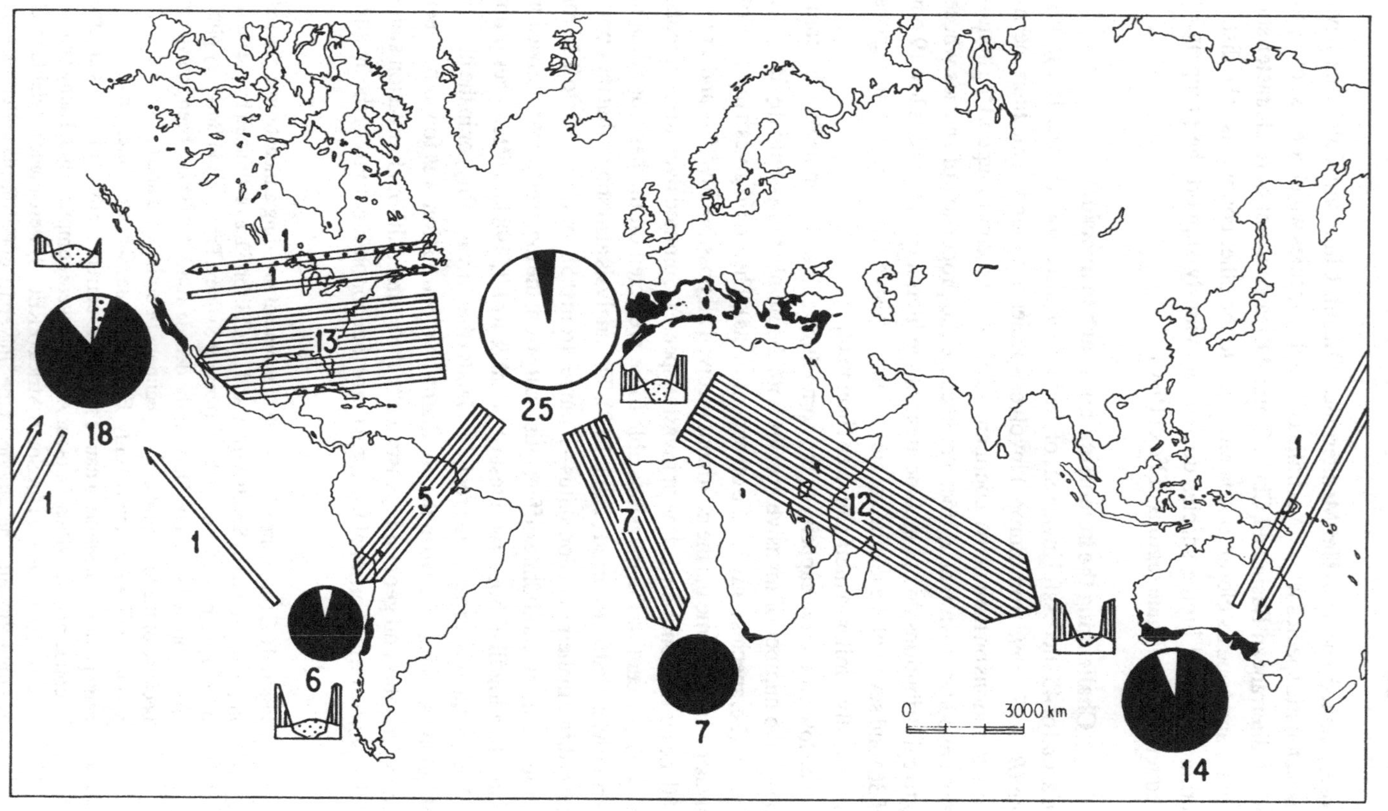

1
1
13
25
18
1
1
5
6
7
7
12
0
3000 km
1
14

Table 16.2. *List of the* Bromus *species found in the different regions of mediterranean climate*

Species	Symbols	Continents
B. alopecuros	Al	EU, AU
B. arenarius	Ae	AU, NA
B. arizonicus	Ai	NA
B. arvensis	Av	EU, NA
B. brachystachys	Ba	EU
B. briziformis	Bi	NA
B. carinatus	Ca	NA, EU, AU
B. commutatus	Co	EU, AU, NA, AF
B. danthoniae	Da	EU
B. diandrus	Di	EU, AU, NA, AF
B. fasciculatus	Fa	EU
B. grossus	Gr	EU
B. haussknechtii	Ha	EU
B. hordeaceus	Ho	EU, AU, SA, NA, AF
B. intermedius	In	EU
B. japonicus	Ja	EU, AU, NA, AF
B. lanceolatus	La	EU, AU
B. lepidus	Le	EU
B. madritensis	Ma	EU, AU, SA, NA, AF
B. racemosus	Ra	EU, AU, NA
B. rigidus	Ri	EU, AU, SA, NA, AF
B. rubens	Ru	EU, AU, NA
B. scoparius	Sc	EU, SA, NA
B. secalinus	Se	EU, NA
B. sericeus	Sr	EU
B. squarrosus	Sq	EU
B. sterilis	St	EU, AU, SA, NA
B. tectorum	Te	EU, AU, NA, AF
B. trinii	Tr	SA, NA

The species names are followed by the continents occupied by these species (AU, Australia; NA, North America; SA, South America; EU, Eurasia; AF, South Africa). The native continent is underlined. (*B. briziformis* is native to Eurasia but is not found in the Mediterranean region of Eurasia.)

Figure 16.3 (opposite). Exchanges of *Bromus* species between the five regions of mediterranean climate. Circle areas and arrow widths are proportional to the number of species. White sectors of the circles represent species native to the Mediterranean Basin, dark sectors represent species introduced from the Mediterranean. The dotted sector represents non-Mediterranean Basin species found in the mediterranean-climate region of North America. The map of the mediterranean-climate regions (coloured in black) is from di Castri (1981). The climatic diagrams are those of Long Beach, California; Valparaiso, Chile; Rabat, Morocco; Dasseneiland, South Africa; and Cape Naturaliste, Western Australia.

region. As for the qualitative analysis, the quantitative analysis of invasion by annual bromes does not show similarities between the Mediterranean Basin, Chile and California on the one hand and Australia plus South Africa on the other, as advanced by Fox (1990).

We will now examine some characteristics of the native distribution of the species. We asked if a mediterranean region marginal to the native distribution was associated with a low capacity to invade the other mediterranean countries. We considered that the zonobiomes and zonoecotones as defined by Walter (see Figure 16.1) constitute a grid (9 × 9), with each climatic zone being one point. For example, zonoecotone 56 corresponds to the point x = 5, y = 6 and zonobiome 4 to the point x = 4, y = 4. The climatic distribution of each species covers part of this grid and can be characterised by its gravity centre. We calculated the gravity centre of the native distributions of the invasive and non-invasive Eurasian bromes. The means, followed by the standard errors, are x = 5.21 (0.17)

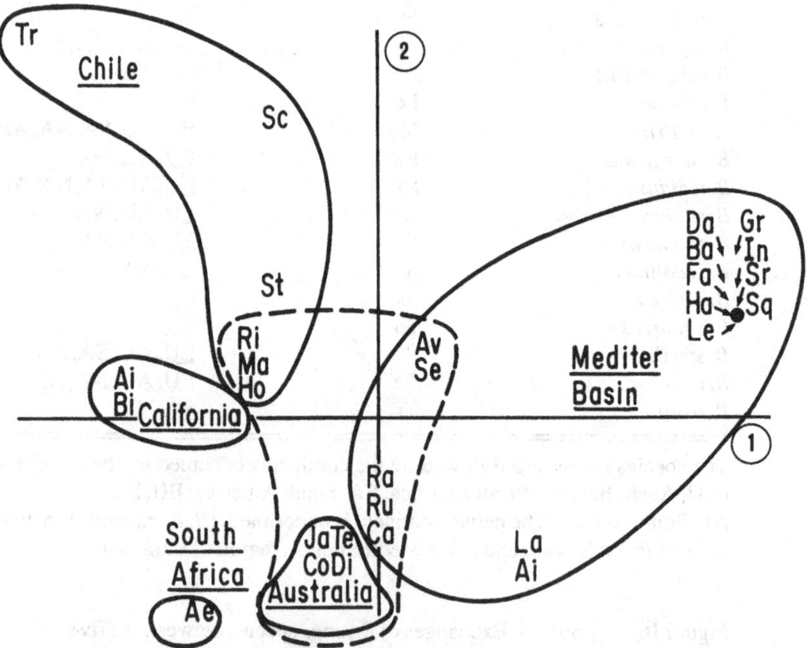

Figure 16.4. Factorial correspondence analysis of the distribution of annual brome grasses in the five regions of mediterranean climate (data from Table 16.2). Species abbreviations are given in Table 16.2. Full and broken lines encompass variables with respectively high and medium links. These links were derived from the analysis of the position of the variables on axis 1, 2 and 3; 88 per cent of the variance is accounted for.

Table 16.3. *Invasion status of the five regions of mediterranean climate as measured by the ratio of actual vs. potential invasion by annual species of* Bromus

	Mediterranean Basin	California	Australia	South Africa	Chile	Total
Number of actual invaders	1	15	13	7	5	41
Number of potential invaders	4	26	28	29	28	115
Invasion status	0.25	0.58	0.46	0.24	0.18	0.37
G-test	—	$.05>P>.02$	$.20>P>.10$	$.20>P>.10$	$.05>P>.02$	

The proportion of actual to potential invaders of each region was tested against the average proportion using the log likelihood ratio test (G-test).

and y = 5.20 (0.18) for the invasive species and x = 4.76 (0.23) and y = 4.47 (0.28) for the non-invasive species. Only the ordinates significantly differ ('t' test, $P < 0.05$). On average, the Eurasian species which invaded the mediterranean regions of the other continents have a native climatic distribution which is less centred around a mediterranean climate (x = 4, y = 4), compared to the non-invasive species. It is illustrated by the world distributions of *B. hordeaceus* and *B. japonicus* whose centres of distribution are central Europe, even though these two species have invaded respectively four and three regions with a mediterranean climate. This situation contrasts with the typically Mediterranean species *B. lanceolatus* and *B. fasciculatus* which have invaded one and zero mediterranean regions respectively (Figures 16.5 & 16.6).

Comparison of Figures 16.5 and 16.6 (distribution of successful and unsuccessful invaders respectively) suggests, however, that a relation exists between the size of the native distribution and the invasive potential of a species, as found by Forcella & Wood (1984) and Forcella *et al.* (1986) for the genera *Carduus, Centaurea, Onopordum* and *Echium.*

Instead of relating the invasive potential of a species to the width of its native distribution as measured by the number of countries it covers, as the above authors did, we related it rather to the climatic width of the native distribution as measured by the number of climatic zones covered (as defined by Walter *et al.*, 1975). The results show a positive relation between the number of occupied mediterranean-climate regions and the number of climatic zones included in the native country (Figure 16.7). If this relation were the only determinant of an

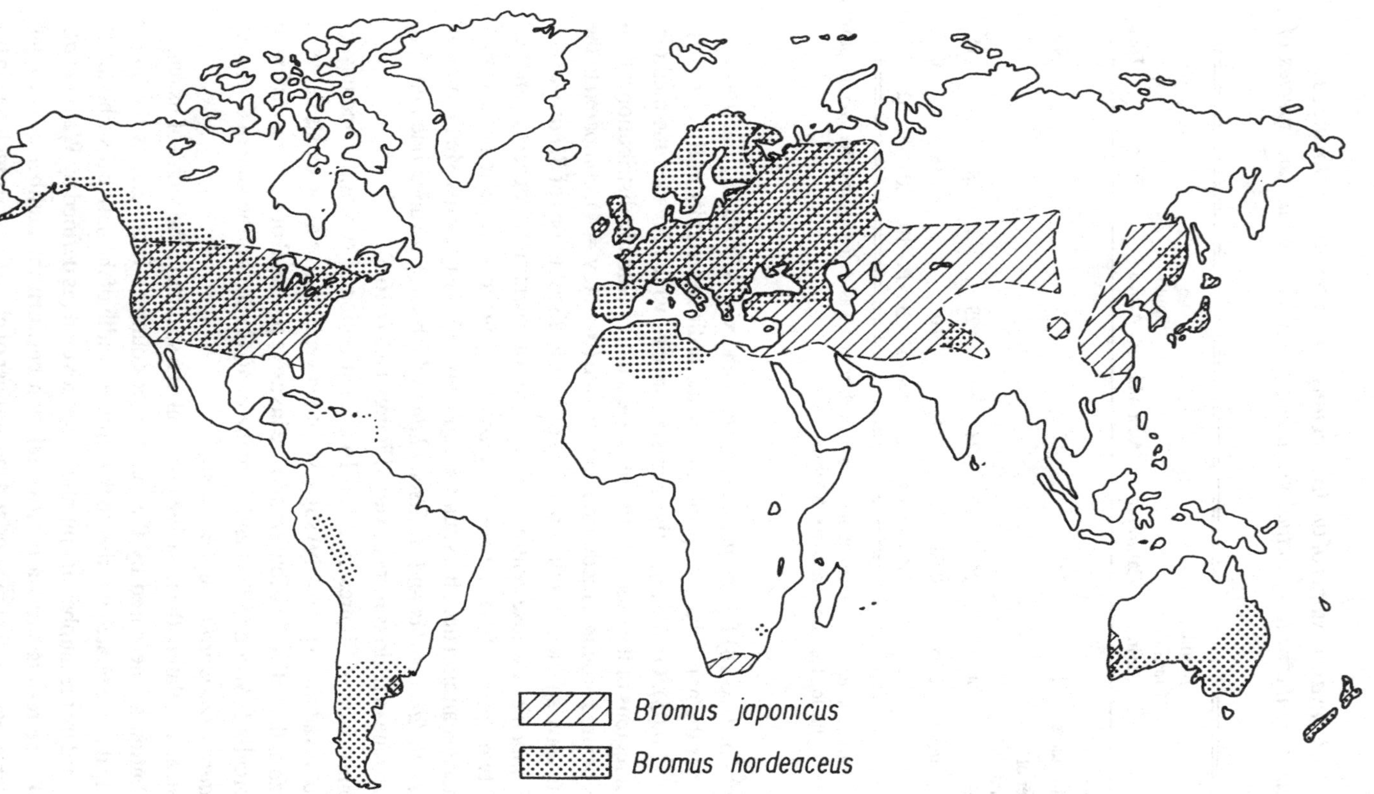

Figure 16.5. World distribution of *Bromus hordeaceus* and *B. japonicus*.

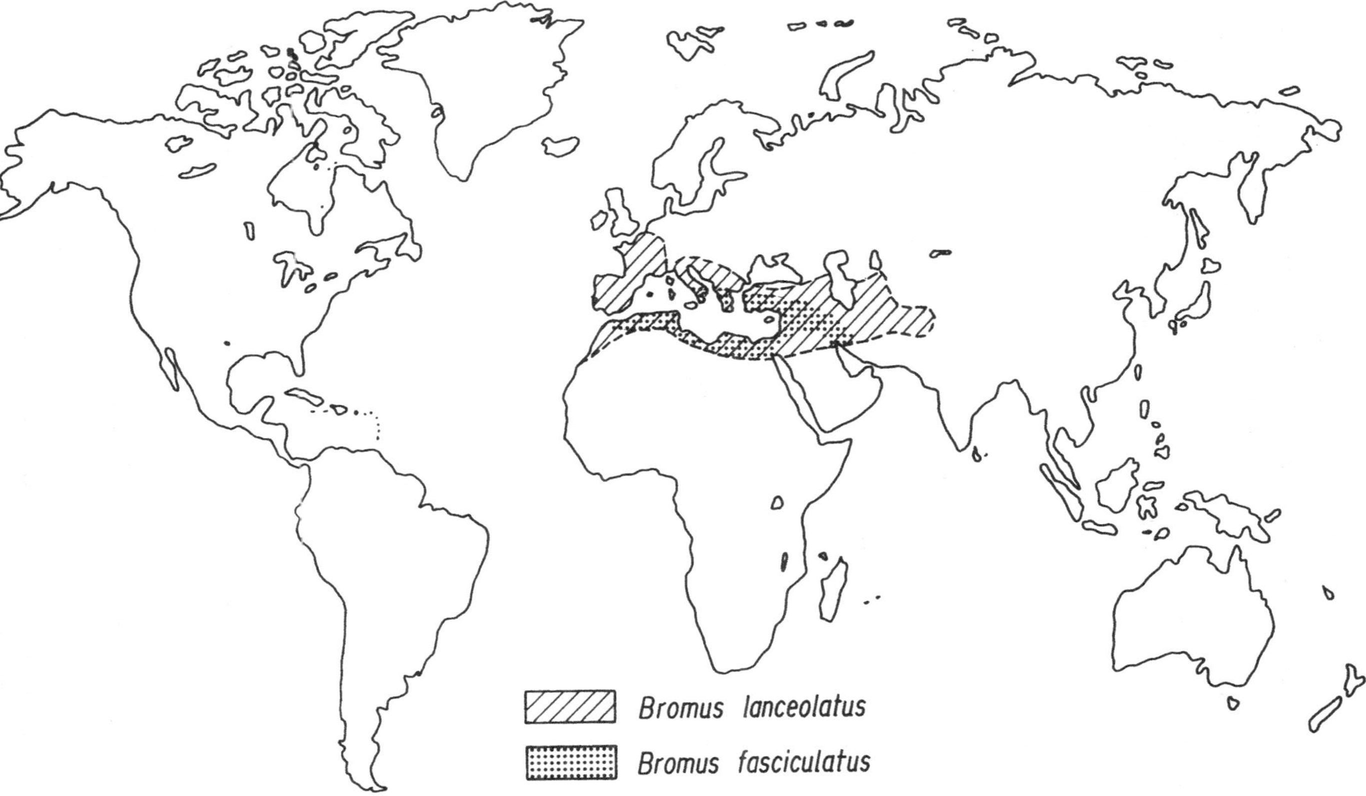

Figure 16.6. World distribution of *Bromus lanceolatus* and *B. fasciculatus*.

invasion, many species would probably invade other regions and move horizontally on the graph to join the lower line. Although determinants other than native distribution play a role in the invasion process, Figure 16.7 allows us to determine which species are the more likely to invade new territories; it is the species on the upper left of the figure, especially *B. squarrosus*, *B. arvensis* and *B. secalinus*. *Bromus squarrosus*, native to Eurasia, is in fact found also in North

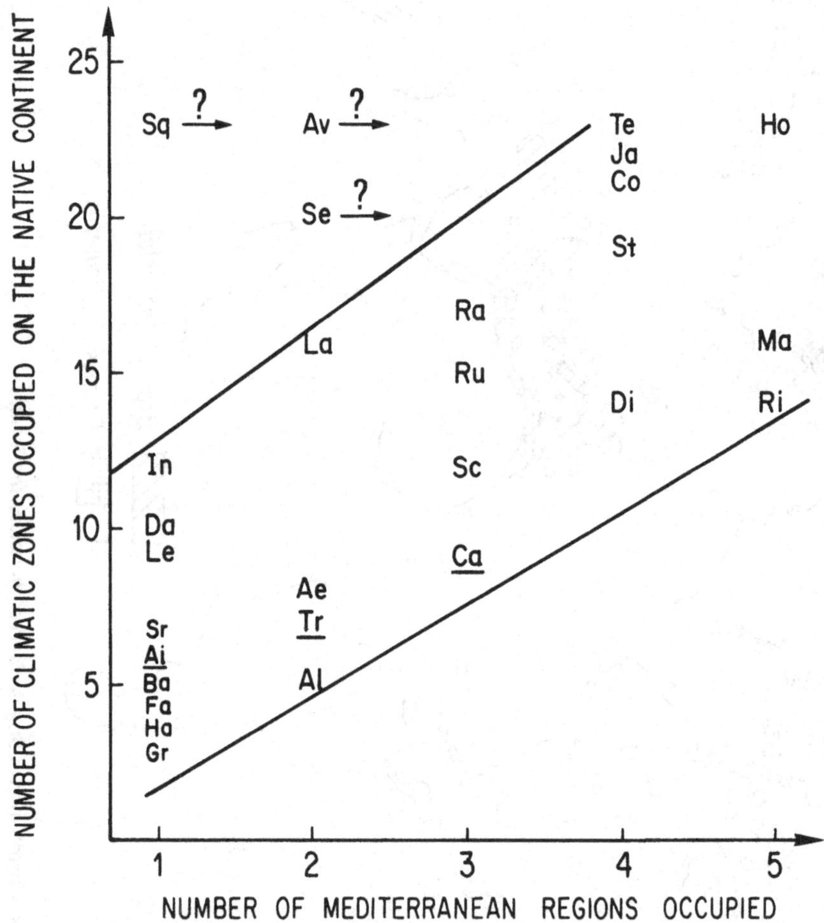

Figure 16.7. Relationship between the number of mediterranean regions occupied by each annual *Bromus* species and the number of climatic zones covered by their distribution in the native continent. Arrows indicate a probable extension of the invasion by (1) *B. squarrosus*, (2) *B. arvensis* and (3) *B. secalinus*. See Table 16.2 for key to species. The underlined species are not native to Eurasia.

America but not in California and one specimen was collected in Chile in 1986 (Matthei, 1986). *Bromus arvensis*, present in Mediterranean Europe and North America, is also found in non-mediterranean South America and southern Africa and in New Zealand. *Bromus secalinus*, present in Mediterranean Europe and North America, is also found in non-mediterranean areas of South America and Australia. The three next potential invaders are connected by the upper line on Figure 16.7. They are, from left to right, *B. intermedius*, *B. lanceolatus* and *B. tectorum* (the latter species being found locally in Chile but only in a warm-temperate climatic area).

Forcella & Wood (1984) and Forcella *et al.* (1986) concluded that the positive relation they obtained between width of native distribution and invading capacity arose from the fact that the propagules of widespread species have a higher probability of transport to other countries. Although this explanation is not to be rejected totally, we are inclined to agree with Noble (1989) that, with the considerable increase in intercontinental exchanges since the beginning of this century, invasion by a species depends more on the interaction between its biological properties and those of the recipient region than on the probability of it reaching that region. There are examples of *Bromus* species which have been introduced early with no subsequent proof of their existence in the receiving continent, as, for instance, in the case of *B. alopecuros* in North America (Hitchcock, 1971) and in South Africa (Linder, personal communication) and in the cases of *B. lanceolatus* and *B. racemosus* in Chile (Matthei, 1986). We suggest that the same biological traits that enable some species to spread across their native continents also make them able to invade new continents. Using a climatic criterion to define the width of the native distribution (instead of a geographical one) allows us to refine this relation: it is the species which spread across climatic barriers in their native continents which invade other continents. Plasticity, ecotypic differentiation, as well as an appropriate phenology, are among the traits allowing a species to occupy different climates. We are currently investigating physiological and genetic traits of widely distributed and endemic species of annual brome grasses to further test this hypothesis.

Conclusion

Annual brome grasses reach their highest specific diversity in areas of mediterranean climate, but many species are found under most climatic conditions except for humid-equatorial ones.

Most *Bromus* species are native to Eurasia. Invasions from the Mediterranean Basin to other regions of mediterranean-type climate date back to the second half of the nineteenth century. The invaded regions have received between 5 (Chile) and 13 (California) of the 24 species native to the Mediterranean Basin.

Of the invasions that would be possible if the only condition of success was the climatic similarity between the source and the recipient continents, only 37 per cent can be demonstrated. We found that the climatic width of the native distribution is positively correlated with the number of invaded mediterranean regions. We suggest that the few species with a wide native climatic distribution, but found currently only in one or two regions (viz. *B. squarrosus*, *B. arvensis* and *B. secalinus*), are serious potential invaders.

Our results advocate a global biogeographic approach to invasions instead of an analysis within regions of similar climate. They suggest that the same biogeographical properties that allowed some species to spread within their native continents (and across climatic barriers) also allowed them to invade other continents.

Acknowledgements. We thank very much the persons who have helped us obtain information on the distribution of brome grasses: J. Dalby, M. Fox, R. Groves, J. M. d'Herbes, S. Jacobs, J. P. Jessop, Jiang Shu, H. P. Linder, G. Montenegro, G. Russell and the staff of the herbaria at the University of California at Berkeley, University of California at Los Angeles and Rancho Santa Ana. We also thank the graphic section of CEPE, CNRS for their work. Part of this study was done while M. L. Navas and J. Roy were based at the Stanford University Laboratory of Prof. H. A. Mooney and supported by the French Ministry of Agriculture and a CNRS-NSF exchange program respectively. This study was also supported by a grant PIREN-Environnement to J. Roy.

References

Andrew, F. W. (1956). *The Flowering Plants of Sudan*, vol. 3. Khartoum: Sudan Government.

Baker, H. G. (1986). Patterns of plant invasion in North America. In *Ecology of Biological Invasions of North America and Hawaii*, ed. H. A. Mooney & J. A. Drake, pp. 44-57. New York: Springer-Verlag.

Clayton, W. D. (1975). Chorology of the genera of Gramineae. *Kew Bulletin*, **30**, 111–32.

Cody, M. L. & Mooney, H. A. (1978). Convergence versus non-convergence in mediterranean-climate ecosystems. *Annual Review of Ecology & Systematics*, **9**, 265–321.

Crosby, A. W. (1986). *Ecological Imperialism. The Biological Expansion of Europe, 900–1900*. Cambridge: Cambridge University Press.

di Castri, F. (1981). Mediterranean-type shrublands of the world. In *Ecosystems of the World*, vol. II, *Mediterranean-type Shrublands*, ed. F. di Castri;, D. W. Goodall & R. L. Specht, pp. 1–52. Amsterdam: Elsevier.

di Castri, F. & Mooney, H. A. (ed.) (1973). *Mediterranean-type Ecosystems: Origin and Structure*. New York: Springer-Verlag.

Forcella, F. & Wood, J. T. (1984). Colonization potentials of alien weeds are related to their native distributions: implications for plant quarantine. *Journal of the Australian Institute of Agricultural Science*, **50**, 35–40.

Forcella, F., Wood, J. T. & Dillon, S. P. (1986). Characteristics distinguishing invasive weeds within *Echium* (Bugloss). *Weed Research*, **26**, 351–64.

Fox, M. (1990). Mediterranean weeds: exchanges of invasive plants between the five mediterranean regions of the world. In *Biological Invasions in Europe and the Mediterranean Basin*, ed. F. di Castri, A. J. Hansen & M. Debussche, pp. 179–200. Dordecht: Kluwer.

Gardner, C. A. (1952). *Flora of Western Australia*, vol. 1, part 1, *Gramineae*. Perth: Government Printer.

Harris, P. (1984). Current approaches to biological control of weeds. In *Biological Control Programmes against Insects and Weeds in Canada 1969–1980*, ed. J. S. Kelleher & M. A. Hulme, pp. 95–103. Farnham Royal, U.K.: Commonwealth Agricultural Bureaux.

Hitchcock, A. S. (1971). *Manual of the Grasses of the United States*, 2nd edn, rev. A. Chase. Washington, D.C.: US Government Printing Office.

Jessop, J. P. & Toelken, H. R. (ed.) (1987). *Flora of South Australia*, part IV. Adelaide: South Australian Government Printing Division.

Kloot, P. M. (1983). Early records of the alien flora naturalised in southern Australia. *Journal of the Adelaide Botanic Garden*, **6**, 93–131.

Kloot, P. M. (1984). The introduced elements of the flora of southern Australia. *Journal of Biogeography*, **11**, 63–78.

Maiden, J. (1916). Weeds at Sydney in 1802–4. *Agricultural Gazette of New South Wales*, **27**, 40.

Matthei, O. (1986). El genero *Bromus* L. (Poaceae) en Chile. *Gayana, Botanico*, **43**, 47–110.

Mooney, H. A. (ed.) (1977). *Convergent Evolution in Chile and California. Mediterranean Climate Ecosystems*. Stroudsburg: Dowden, Hutchinson & Ross.

Munz, P. A. (1974). *A Flora of Southern California*. Berkeley: University of California Press.

Munz, P. A. & Keck, D. D. (1959). *A California Flora*. Berkeley: University of California Press.

Noble, I. R. (1989). Attributes of invaders and the invading process: terrestrial vascular plants. In *The Ecology of Biological Invasions: A Global Perspective*, ed. J. A. Drake, H. A. Mooney, F. di Castri, R. H. Groves, F. J. Kruger, M. Rejmanek & M. Williamson, pp. 301–13. Chichester: Wiley.

Rechinger, K. H. (1970). *Flora Iranica*. Graz: Akademische Druck.

Smith, P. M. (1986). Native or introduced? Problems in the taxonomy and plant geography of some widely introduced annual brome grasses. *Proceedings of the Royal Society of Edinburgh*, **89B**, 273–81.

Sokal, R. R. & Rohlf, F. J. (1981). *Biometry*. New York: Freeman.

Tutin, T. G., Heywood, V. H., Burges, N. A., Moore, D. M., Valentine, D. H., Walters, S. M. & Webb, D. A. (ed.) (1980). *Flora Europaea*, vol. 5. Cambridge: Cambridge University Press.

Walter, H., Harnickell, E. & Mueller-Dombois, D. (1975). *Climate Diagram Maps of the Individual Continents and the Ecological Climate Regions of the Earth*. New York: Springer-Verlag.

Watson, S. (1880). *Geological Survey of California, Botany*, vol. II. Cambridge, Mass.: John Wilson.

Part IIIb Mammals

In general, mammals have not been as invasive in regions of mediterranean climate as have plants. But there are some notable exceptions of which the rabbit (*Oryctolagus cuniculus*) is the main example. The rabbit is native to Spain where it is not a serious pest. On introduction to Chile and to southern Australia it has become a major factor in the present ecology of these two large regions. Yet it has never multiplied in North America, to which it has been introduced a number of times. In South Africa, the rabbit is found in any numbers only on the offshore islands and has not as yet colonised the mainland.

Other mammals have become invasive in all regions of mediterranean climate, especially the rodents, some of which are not confined in their native European distributions to the Mediterranean Basin. The house mouse *Mus musculus* is strongly invasive in most regions, although it forms plagues periodically in the characteristically mediterranean-climate region of southern Australia in which cereals are cropped. The only other area where the house mouse may reach similar plague proportions is, however, in a region of China with a very different climatic regime. Obviously, any attempt to generalise about the invasiveness of mammals in the five regions of mediterranean climate will be difficult, as the following five chapters show.

17

Patterns of Pleistocene turnover, current distribution and speciation among Mediterranean mammals

G. CHEYLAN

There is no unequivocal definition of the Mediterranean Basin. More often, this area is understood in terms of its climatic pattern; namely, an area with a low annual precipitation, over two-thirds of which occurs during the mild winter months, whereas the summers are dry and often hot. Places sharing this climatic regime have distinctive semi-arid plant communities, known locally as 'monte bajo' in Spanish, 'garrigue' in French, 'macchia' in Italian and 'xerovumi' in Greek.

According to Quézel (1981), the distribution of these sclerophyllous scrub communities is mainly coastal and much more restricted geographically than the Mediterranean bioclimatic region as defined by Emberger *et al.* (1962). There is thus a considerable discrepancy between botanical and climatological maps, depending mainly upon geomorphology (Figure 17.1). In places where an elevated mountain range rises steeply from the sea, as in France, Italy and parts of the Balkans and Turkey, the extension of Mediterranean habitats is greatly reduced and both botanical and climatological limits more or less overlap. This is not the case, however, in North Africa, Iberia and the Middle East.

In North Africa, hereafter called Maghreb, the Mediterranean climate ends with the 150 mm isohyet, thereby including the pre-desertic mountains known as the Saharan Atlas (Emberger *et al.*, 1962). Unlike the map of Emberger *et al.*, Quézel's map does not expand the Mediterranean region farther south than the main Atlas range, but includes the Sous plain and the Draa valley (Atlantic Morocco), home of several tropical species of mammal, such as *Mastomys, Xerus* and three species of *Crocidura*.

In Spain and the Balkans, Mediterranean shrublands are absent from the northernmost portions of these peninsulas, although they experience a mediterranean-type climate.

Major discrepancies are found in Asia Minor and the Middle East. According

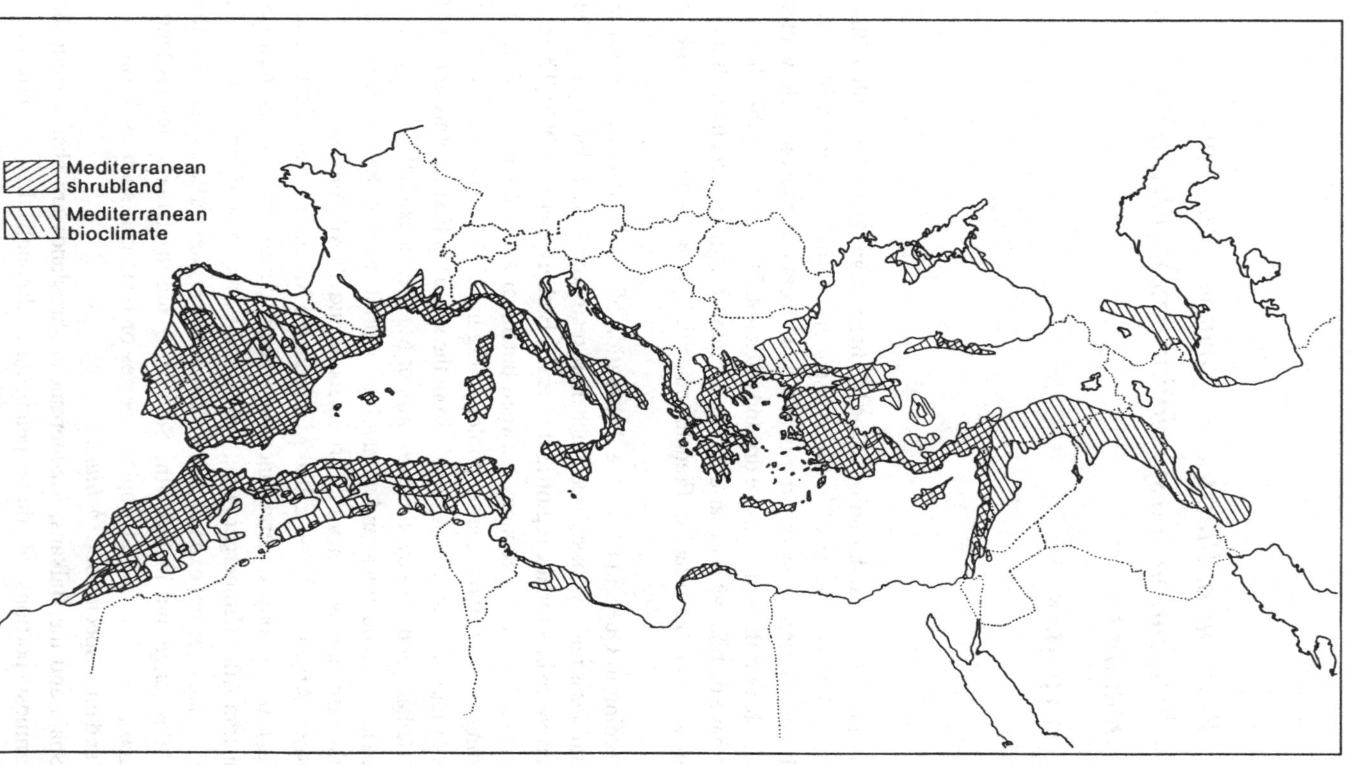

Figure 17.1. Boundaries of the Mediterranean Basin region.

to climatologists, the Mediterranean region stretches in these areas eastwards to Iran, terminating with the oak forests of the Zagros mountains. Likewise, from the gulf of Iskenderum (Turkey) to the Zagros range, the Mediterranean region overlaps the Turkey–Syria and the Iraq–Iran borders for more than 1500 km. All this area, lying between the Syrian desert and the cold steppes of the Anatolian plateau, is, however, excluded from the botanical definition of the term 'mediterranean'.

A number of smaller, isolated areas are excluded as well, namely: the Crimean peninsula in the northern Black Sea, the south-western coast of the Caspian Sea from Azerbaijan (USSR) to Tehran (Iran), and the Tripolitanian coast of Libya where the Mediterranean region merges into the Sahara. Both the Barka plateau (Cyrenaica, Libya) and the northern Turkish coast, near Sinope, are unquestionably Mediterranean.

Given these limitations, the Mediterranean region, from a botanical point of view, has an area of 410890 km^2. This estimate of Le Houérou (1981) doesn't take into account forested and cultivated lands, however, whilst no data are available for Albania and Bulgaria. Three countries make up 65 per cent of this total: Spain (160490 km^2), Turkey (55000 km^2) and Morocco (53000 km^2). These estimates can be compared with those of Blondel (1982), who gave much higher estimates based on bird distributions. From a total area of 2968000 km^2, Iberia–France and Italy spread over 756300 km^2, the Balkans over 270200 km^2, Asia Minor and the Middle East over 1271100 km^2 and North Africa over 580000 km^2.

A comparison of the maps of Blondel (1982) and Emberger *et al.* (1962) reveals good agreement over North African and Iberian-French-Italian boundaries, but they differ for the Balkans (where Blondel includes southern Yugoslavia and most of Bulgaria in the Mediterranean) as well as for Asia Minor, where Turkey as a whole (with the exception of the Black Sea coast), Azerbaijan and the western half of Iran are considered as Mediterranean.

In compiling the list of the Mediterranean mammals to be presented subsequently as Appendix 17.1, I have generally defined the region in the broadest sense, with the exception of the Middle East, where I have treated separately the faunas of coastal Turkey, Syria, Lebanon and Israel, and those along the Turkey–Syria and Iraq–Iran borders. I present the fauna of these regions under six subdivisions, corresponding to well-recognised geographical units:

1. The Iberian peninsula (and southern France)
2. Italy
3. The Balkans: the Yugoslavian and Albanian coasts, most of Greece, Thrace (European Turkey) and south-eastern Bulgaria

4. Asia Minor and the Near East: western Turkey, coastal Syria, Lebanon and Israel
5. Middle East: south-eastern Turkey, northern Syria and Iraq, Zagros range (Iran)
6. Maghreb: northern Morocco, Algeria and Tunisia.

Compilation of a list of Mediterranean mammals is not easy because many faunal lists are published in journals difficult to locate. The region still lacks a comprehensive coverage with good maps. The author thus apologises for those references he has overlooked.

Because of the lack of references or the difficulties in finding them, the following areas have not been included in Appendix 17.1: Tripolitana and Cyrenaica (Libya), Crimea (USSR), the Black Sea coast near Sinope (Turkey) and Azerbaijan plus the south-western coast of the Caspian Sea (USSR and Iran).

For the European part of this region, references used are Van Den Brink & Barruel (1967), Schilling *et al.* (1986) and Niethammer & Krapp (1978–1986), the latter unfortunately not yet completed (only Rodentia and part of Artiodactyla and Carnivora are available to date).

For the Near East, there is Harrison (1964–1972), supplemented by Attalah (1977–1978) who gives distributions in Israel, Lebanon, Jordan, Syria, Iraq and the eastern half of Turkey. For Turkey: Kumerloeve (1967, 1975), Niethammer & Krapp (1978–1986), Spitzengberger (1968, 1970) for shrews, and Osborn (1962, 1963, 1964) for bats, Arvicolidae, hares, squirrels, jerboas and dormice. For Iran Misonne (1958) was the source.

The basic reference for the Maghreb is Heim de Balsac (1936), supplemented by Rode (1947, 1948) for hedgehogs, shrews, bats and rodents, and Joleaud (1934) for Moroccan mammals. More recent references include Panouse (1951, 1957), Saint Girons & Petter (1965), Dupuy (1967), Bernard (1969), Aellen & Strinati (1969), Willan (1973), Anciaux de Faveaux (1976), Gaisler (1983), Aulagnier & Destre (1985) and Hutterer (1986).

Unless otherwise stated, the classification followed is by Honacki *et al.* (1982). An exception was the classification of shrews of the genus *Crocidura*, for which genetic studies have been made by Catalan (1984) for the western Mediterranean and by Catzeflis (1983), Catzeflis *et al.* (1985) and Vögel *et al.* (1986) for the eastern part of the Mediterranean Basin. Similarly, results of recent genetic studies of Spanish hares (Palacios, 1979; Bonhomme *et al.*, 1986) have shown that *Lepus granatensis* must be considered distinct from *L. europaeus*. Unfortunately, similar studies on North African and Sardinian hares, which are morphologically close to *L. granatensis*, are still lacking, as well as studies on *L. europaeus/L. capensis* populations which are doubtfully conspecific.

A number of other groups need to be investigated using similar methods, e.g. the genera *Gerbillus* and *Dipodillus*, of which not less than five species were described recently for North Africa where they are known only from the type localities. This unusually high number of species endemic to the same area will probably be much reduced as a result of genetic studies. Similarly, the taxonomy of the genus *Acomys* is not fully understood and the status of two endemic taxa (*A. minous* in Crete and *A. cilicicus* in Turkey) needs further study, as this group could be an aggregate of sibling species.

The list of Mediterranean mammals (Appendix 17.1) includes every species known in this area since classical times. This means that eight species are included which have been eradicated by humans over the past three millennia, namely the lion (*Panthera leo*), the cheetah (*Acinonyx jubatus*), the African ass (*Equus asinus*), the onager (*Equus onager*), the African elephant (*Loxodonta africana*), the Asian elephant (*Elephas maximus*), the Bubal hartebeest (*Alcephalus busephalus*) and the red gazelle (*Gazella rufina*). Other species are extinct over most of their former range or are on the verge of extinction in the Mediterranean as a whole, e.g. the brown bear (*Ursus arctos*), the serval (*Felis serval*), the panther (*Panthera pardus*), the caracal (*Felis caracal*), the hyena (*Hyaena hyaena*), the fallow deer (*Cervus dama*) and Cuvier's gazelle (*Gazella cuvieri*). The past distribution of the aurochs (*Bos taurus*), now extinct in the wild but the ancestor of domestic cattle, is unclear, as it is not always easy to tell this species from primitive strains identified from fossil bones. Some populations probably survived until the classical period, or even later, in parts of the eastern Mediterranean (Bodenheimer, 1960; Kumerloeve, 1975).

Species richness

Altogether, 197 species of mammals have been recorded so far for the Mediterranean region (Table 17.1, Appendix 17.1). Given the limitations expressed earlier (uncertainty about species ranges and questionable taxonomy), the total richness of this area probably exceeds 200 species.

Eleven species enter the Mediterranean only in its easternmost extension (i.e. the Middle East): *Triaenops persicus*, *Pipistrellus coromandra*, *Herpestes auropunctatus*, *Gazella subgutturosa*, *Mesocricetus brandti*, *Calomyscus bailwardi*, *Ellobius lutescens*, *Meriones persicus*, *M. vinogradovi*, *Tatera indica* and *Allactaga elater*.

Other species (bracketed in Appendix 17.1) are found in only very reduced portions of the Mediterranean; these species are:

Table 17.1. *Distribution of mammals on larger Mediterranean islands*

Species	Maj	Min	Ibi	Sar	Cors	Sic	Mal	Corf	Cre	Rho	Cyp
Hemiechinus auritus											+
Erinaceus europaeus				+	+	+	?				
E. concolor								+	+	+	
E. algirus	+	+	+				+				
Crocidura russula			+	+		+	+				
C. suaveolens		+			+			+	+		+
C. zimmermanni[a]									+		
Suncus etruscus				+	+	+	+	+			+
Talpa romana						?		+			
Oryctolagus cuniculus	+	+	+	+	+	+			+		
Lepus capensis	+			+	+			+	+	+	+
Hystrix cristata						+					
Eliomys quercinus	+	+		+	+	+					
Myoxus glis				+	+	+		+	+		
Muscardinus avellanarius						+		+			
Pitymys savii						+					
Arvicola terrestris						?					
Apodemus sylvaticus	+	+	+	+	+	+		+	+		
A. flavicollis								+			
A. mystacinus								+	+	+	
Rattus rattus	+	+	+	+	+	+	+	+	+	+	+
R. norvegicus	+	+	+	+	+	+	+	+	+		+
Mus musculus	+	+	+	+	+	+	+	+	+	+	+
M. spretus	+	+	+								
M. abbotti											+
Acomys cahirinus											+
A. minous[a]									+		
Vulpes vulpes				+	+	+		+			+
Canis aureus								+			
Meles meles									+	+	
Mustela nivalis	+	+		+	+	+	+	+	+		
Lutra lutra								+			
Martes foina			+					+	+	+	
M. martes	+	+		+	+	+					
Genetta genetta	+		+								
Felis silvestris	+	+	?	+	+	+			+	?	
Sus scrofa				+	+						
Cervus elaphus				+							
Ovis musimon[a]				+	+						+
Capra hircus									+		

Table 17.1 (*cont.*)

Species	Maj	Min	Ibi	Sar	Cors	Sic	Mal	Corf	Cre	Rho	Cyp
Surface area ($km^2 \times 100$)	36	7	5	241	87	257	2.5	6	83	14	93
Number of species	13	12	10	18	17	18	7	18	17	7	11

Maj, Majorca; Min, Minorca; Ibi, Ibiza; Sar, Sardinia; Cors, Corsica; Sic, Sicily; Mal, Malta (Cheylan, 1984); Corf, Corfu (Neithammer, 1962); Cre, Crete (Zimmerman *et al.*, 1942–1949); Rho, Rhodes (Niethammer, 1962); Cyp, Cyprus (Spitzenberger, 1978, 1979).
[a] Endemic species.

1. Tropical species penetrating into the Mediterranean by two favoured routes: the Atlantic Moroccan coast, from the lower Draa valley to the Sous plain, and Israel, sometimes heading north to Lebanon and the gulf of Iskenderum (Turkey)
2. Saharo-Sindian species belonging mainly to the genus *Gerbillus* and several bats, which reach the edges of the Mediterranean region in North Africa, Syria and Jordan
3. Boreal species, belonging mainly to the Arvicolidae and Muridae and especially numerous in Italy, the Balkans and Asia Minor.

When species whose distributions are mainly non-Mediterranean are excluded from the list, the total richness of the Mediterranean region is reduced to 155 species.

A comparison of the species richness of the five zoogeographical sub-regions referred to above (excluding the Middle East whose bat fauna is poorly known and its species richness thus probably underestimated) yields some interesting trends. Asia Minor is by far the richest area (117 species of which 23 are locally distributed), followed by the Maghreb (91 species) and the Balkans (82 species). Lastly, Iberia (74 species) and Italy (71 species) have depauperate faunas when compared with the third European peninsula, the Balkans.

Unexpectedly, there is no relationship between surface area and species richness of these sub-regions, as Asia Minor is only third in area but first in species richness, whereas Iberia, by far the most extensive region, ranks fourth in species richness (Table 17.2).

The similarity between these faunas was compared using the index 'I' of Jaccard (1908), in which

$$I = c / a + b - c \quad (1)$$

Table 17.2. *Species richness and area of five Mediterranean sub-regions*

	Iberia	Maghreb	Asia Minor	Italy	Balkans
Extent of shrubland (km^2)	193 290	91 900	63 790	33 200	26 680
Number of mammal species	61 + (13)	71 + (20)	94 + (23)	58 + (13)	67 + (15)

Areas from Le Houérou (1981) with steppes, cultivated and forested areas excluded. The number of species of uncertain status (? in Appendix 17.1) or with only local distributions is given in brackets.

where a and b are the number of items in list a and b respectively, and c is the number of items shared in common by both lists. The closest biogeographical relationships were found between Asia Minor and North Africa (I = 0.70); the latter region is distinct from Europe (I = 0.56), although very close geographically (the Strait of Gibraltar is only 14 km wide). The relationship between Europe and Asia Minor is somewhat intermediate (I = 0.59).

The close relationship between North Africa and Asia Minor is a consequence of the desert belt which links the southern margins of the two regions. Many species of the families and sub-families Rhinopomatidae, Hipposiderinae, Equidae, Antelopinae, Gerbillinae and Dipodidae are characteristic of these hyper-arid lands. Conversely, several boreal families found either in southern Europe and Asia Minor are lacking in North Africa, namely Soricidae, Talpidae, Spalacidae, Cricetidae and Arvicolidae.

Species invasion stages and extinction processes

Faunal exchange between Europe and Africa is a very uncommon event. Mammals of both continents evolved independently during most of the Quaternary (Arambourg, 1962), given that the terrestrial connections between the two continents were precluded by two barriers impassable to most of the wingless vertebrates.

To the west, the Strait of Gibraltar allowed terrestrial mammal migrations only during the Messinian (Miocene/Pliocene boundary, 7 to 5.5 m.y. ago). During this period, the tectonic exhaustion which locked this strait for a time led to the subsequent desiccation of the Mediterranean, an event which left impressive evaporite layers on the bottom of the sea (Hsu *et al.*, 1973). This event, known as the salinity crisis, allowed some mammalian exchanges between Spain and North Africa, as summarised by Jaeger *et al.* (1987). Seven species of rodent and one lagomorph migrated from Europe to North Africa, where some survived to the end of the Pliocene. Northward migrations allowed the colonisation of Spain by five genera of African rodents. The most noteworthy was the Gerbillinae, whose fate paralleled that of European invaders of North Africa (Jaeger *et al.*, 1987). There was consequently no faunal exchange by the western route during the Pliocene and the Pleistocene. Even during maximal sea regressions the Gibraltar channel and others, such as the Sicilian, were deep enough to impede any land connection.

The eastern route was only temporarily open to mammal migrations. The first clues of desertification of the Sahara are noticeable during the Miocene, and during most of the Pleistocene this area experienced periods of moist climate ('pluvials') and intervening hyper-arid episodes (Chaline, 1985). The rapid diversification of the genera *Gerbillus* and *Gazella* in North Africa during the Pleistocene illustrates the progressive aridification of the area during the Quaternary (Arambourg, 1957; Jaeger, 1975*a*).

None of the Pleistocene glaciation covered the Mediterranean. Indeed, during its southernmost extension (Mindel 1? m.y. to 0.65 m.y.), the ice sheet was situated 600 km north of the northernmost limit of present-day Mediterranean climate (West, 1977).

During glacial maxima, biomes were shifted southward from where they are located today by as much as 10° to 20° latitude. Mediterranean vegetation was almost completely wiped out of Europe during the Würm glacial maximum and survived only in a few scattered refugia in Andalucia, Sicily, Peloponnisos and Crete. Most of southern Europe was covered by temperate broadleaf forests, whereas a tundra-like vegetation developed in northern latitudes (Flint, 1971). Mediterranean vegetation was more widespread in Asia Minor along the southern slopes of the Anatolian plateau, whereas North Africa remained essentially unchanged (Flint, 1971).

During moist periods the Mediterranean vegetation spread across the desert, creating a forested steppe whose main components were *Quercus ilex*, *Q. suber*, *Olea europaea*, *Pistacia atlantica*, *P. lentiscus*, *Myrtus communis*, *Arbutus unedo*, *Ceratonia siliqua* and several species of palm. Some of these trees are still found in the Hoggar mountains of Algeria, where several endemic taxa evolved from widespread Mediterranean species, e.g. *Cupressus deprezianus*, *Myrtus nivellei* and *Olea laperrenei* (Gattefossé, 1933). Similarly, moist periods allowed northward expansion of Sudanian species, now restricted to a few spots along the northern shore of the desert, such as *Acacia gummifera*, *A. tortilis* and *A. seyal*. Consequently, the Maghreb has not always been as isolated as it is today and the development of grassland and savanna-like habitats allowed many mammals to migrate.

Table 17.3. *Changes in the fauna of North Africa during the Middle/Upper Pleistocene and the Holocene*

Species	1	2	3	4	5	6	7	8
Elephas iolensis	+	+						
Loxodonta atlantica			+					
L. africana						+	+	+[a]
Ceratotherium simum	+	+	+	+	+	+	+	
Rhinoceros mercki					+			
Equus mauritanicus	+	+	+	+	+			
E. melkiensis					+			
E. algiricus					+			
E. asinus							+	+[a]
Phacochoerus aethiopicus			+	+	+	+		
Sus scrofa				+	+	+	+	+
Hippopotamus amphibius	+	+	+	+	+	+	+	
Giraffa camelopardalis					+	+	+	
Giraffa sp.	+							
Gazella atlantica	+	+	+	+				
G. dorcas			+	+	+	+	+	+
G. rufina			+	+	+	+	+	+[a]
G. cuvieri			+	+	+	+	+	+
G. tingitana					+			
Oryx sp.	+							
O. dammah					+	+	+	+[a]
Hippotragus sp.		+	+					
Redunca maupasi	+	+	+	+				
R. sp.					+	+	+	
Alcephalus probubalis	+	+	+	+				
A. busephalus					+	+	+	+[a]
Connochaetes taurinus	+	+	+	+	+	+	+	
Taurotragus derbianus	+							
Bos primigenius	+	+	+	+	+	+	+	
Bubalus antiquus					+	+	+	
Rabaticeras arambourgi			+					
Ammotragus lervia					+	+	+	+
Cervus sp.		+?						
C. elaphus								+
Megaceroides algericus					+	+		
Crocuta crocuta	+	+	+	+	+	+	+	
Hyaena hyaena	+	+	+	+	+	+	+	+
Canis aureus	+	+	+	+	+	+	+	+
Vulpes atlantica		+						
V. vulpes					+	+	+	+
Mellivora capensis	+	+			+	+	+	+
Mustela nivalis						+?	+	+

Table 17.3 (*cont.*)

Species	1	2	3	4	5	6	7	8
Panthera pardus	+				+	+	+	+
P. leo	+				+	+	+	+[a]
Ursus arctos	+				+	+	+	+[a]
Macaca sylvanus	+	+	+	+	+	+	+	+

1, Amirian; 2, Anfatian; 3, Tensiftian (?200000–100000 BP); 4, Pre-Soltanian (about 95000 BP); 5, Soltanian (?80000–11000 BP); 6, Capsian (11000–5000 BP); 7, Neolithic (5000–3000 BP); 8, Recent.
[a] Extinct species within the Mediterranean region.
Source: Arambourg, 1970; Jaeger, 1975*b*.

Changes in the Pleistocene mammal fauna of North Africa

North African palaeontology is now known satisfactorily. Arambourg (1970) and Jaeger (1975*a*, *b*) have updated our knowledge of this rich fauna, documented originally by Romer (1928) and Pomel (1894). Most of this fauna is Ethiopian. Few Eurasian mammals managed to colonise North Africa during the Quaternary, following the eastern route which stretches along the Libyan coast. From the Lower Pleistocene to the Holocene, the main change has been the gradual extinction of archaic lineages during the Middle Pleistocene and the more violent eradication of numerous large species during the last 5000 years (Arambourg, 1962).

From the Amirian (Lower Middle Pleistocene) to the Pre-Soltanian (terminal Riss/Würm interglacial), the number of large mammals (Proboscidea, Perissodactyla and Artiodactyla), remained fairly constant, varying from 10 to 15 species (Table 17.3). This number quickly increased to 21 during the Soltanian, roughly contemporaneous with the Würm glacial in Europe, owing to the arrival of several Eurasian species, such as *Rhinoceros mercki*, *Bubalus antiquus*, *Ammotragus lervia* and *Megaceroides algericus*, as well as many Ethiopian species (e.g. *Giraffa camelopardalis*, *Oryx dammah*). By the Neolithic, this number had already dwindled to 16, every Eurasian mammal having vanished, save for *Sus scrofa*. This number was further reduced to ten during Roman times, then to five following European colonisation a century ago.

Unfortunately, the extermination of North African big mammals is not over; three species are presently endangered (two species of *Gazella* and the Barbary sheep) (Willan, 1973; Aulagnier & Thevenot, 1986). The wild boar, consumption of whose meat is prohibited by religious laws, is currently the only fairly abundant large species in North Africa.

The gradual deterioration of the Saharan vegetation and subsequent isolation of the Maghreb can explain many Holocene extinctions of hoofed mammals. Moreover, the patchy distributions of many species by the end of the last pluvial (Neolithic) may have rendered these populations more vulnerable than they were to the ever-increasing hunting pressure and competition from domesticated animals, thereby further prompting their extinction. This process of deterioration has not been compensated by the immigration of Eurasian species, since, unlike rodents, none appeared during the Holocene.

From the Amirian to the Neolithic, about one Eurasian mammal immigrated per geological epoch (Table 17.3). A major wave took place during the Soltanian (= Würm glacial in Europe). Four species invaded North Africa during this stage, together with two insufficiently known species of *Equus* related to the caballine and asine species (Bagtache *et al.*, 1984).

Ethiopian species experienced two major invasions: one during the Tensiftian (*Loxodonta*, *Phacochoerus*) and another during the Soltanian. Thus, the Soltanian has induced for both faunas the major migrations of big mammals. A pluvial period is known to have occurred in the Sahara during the early stages of this epoch, between 40000 to 30000 BP. Between that pluvial and the next, which occurred from 12000 to 4000 BP, the Sahara was mainly hyper-arid, just as it is today (Klein, 1984).

Remnants of these Eurasian invasions are found in Cyrenaica, where some migrants were trapped by the advance of the desert, and apparently never went further west. Three rodents are indicators of this eastern route, discussed earlier by Heim de Balsac (1936): namely *Microtus guentheri*, *Allactaga tetradactyla* and *Spalax ehrenbergi*.

The colonisation of North Africa by Ethiopian mammals should have occurred along a wider front, although some routes can be traced from the present distributions of some species. Out of 23 Ethiopian species, 16 are, or were, widely distributed in North Africa, and thus prevent any information being gained about the route they followed when they invaded this area. The other seven (to which the bat *Hipposideros caffer* could also be added) are three species of *Crocidura*, and *Mellivora*, *Xerus*, *Mastomys* and *Acomys*, all restricted to Morocco and/or some oases of southern-central Algeria. They thus clearly indicate a coastal migration, following the Atlantic shore of the Sahara. Similar examples can be found among birds (e.g. *Melierax metabates*, *Asio capensis*, *Numida meleagris*) and reptiles (e.g. *Naja haje*, *Bitis arietans* and formerly *Crocodylus niloticus*).

At the other end of the Sahara, species have travelled northwards following

Table 17.4. *Changes in the rodent fauna of North Africa during the Middle/Upper Pleistocene*

Species	Middle Pleistocene	Upper Pleistocene	Recent
Leggada sp.	+		
Acomys cahirinus			+
Apodemus sylvaticus			+
Arvicanthis sp.	+		
Lemniscomys barbarus		+	+
Mastomys erythroleucus			+
Mus sp.	+	+	+
Paromys sp.	+		
Paraethomys cf. *filfilae*	+	+	
Rattus			+
Dipodillus			+
Gerbillus sp.	+		
G. cf. *campestris*		+	+
Meriones cf. *shawi*	+	+	+
Pachyuromys			+
Psammomys			+
Ellobius barbarus	+		
E. cf. *fuscocapillus*	+		
Jaculus sp.	+	+	+
Eliomys quercinus		+	+
E. sp.	+		
Ctenodactylus		+	+
Atlantoxerus getulus	+	+	+
Xerus erythropus			+
Hystrix cristata	+	+	+

Source: Jaeger, 1975*a*, *b*.

the Nile valley; these species are likely to be found in Israel, sometimes reaching Lebanon and southern Turkey: e.g. species of *Mellivora*, *Genetta*, *Herpestes*, *Procavia*, *Alcephalus* and *Acomys*. Two bat genera (*Rousettus* and *Taphozous*) can be added to this list.

Unlike large mammals, several rodents entered North Africa during the last pluvial (Table 17.4). The newcomers are either Eurasian (e.g. *Apodemus*, *Rattus*), Ethiopian (e.g. *Mastomys*, *Xerus*) or desertic mammals (e.g. *Acomys*, *Pachyuromys*, *Psammomys*). Furthermore, the sudden radiation of the genera *Dipodillus* and *Gerbillus* increased the species richness of rodents in the Maghreb.

Consequently, rodent and hoofed mammal changes give a very different figure in the Maghreb. From ten genera of rodents recorded during the Soltanian, the present faunal assemblage consists of 17 genera. Nevertheless, several rodent genera presently have very reduced distributions, such as *Xerus* and *Mastomys*, and are prone to extinction in the near future. The time span since the last pluvial could have been too short to allow many extinctions, as is predictable for this saturated fauna (Diamond, 1972).

Changes in the Pleistocene mammal fauna of the Levant

Tchernov (1984) has accurately summarised our present knowledge about mammalian successions in Israel (and Syria, in part) (Tables 17.5 & 17.6). The fate of the megafauna paralleled those of North Africa: a gradual increment in species richness during the last interglacial and a rapid decrease during the Holocene (Table 17.5).

Starting with six species during the Mindel/Riss interglacial, the number of large mammal species (Perissodactyla and Artiodactyla) increased to 17 during the last glacial. Currently it is reduced to three. Interestingly, two of these three survivors are found in North Africa as well (namely *Gazella* and *Sus*). Two additional species have been eradicated by humans fairly recently in this region (*Dama* and *Capreolus*). Very few large Ethiopian species invaded the Levant during the Middle and Upper Pleistocene. There are only two (*Phacochoerus* and *Alcephalus*), in striking contrast to the situation in North Africa.

Unlike large mammals, the curves of species richness for rodents have very different shapes for North Africa and the Levant. Whilst rodent species increased steadily during the Holocene in North Africa (from 10 to 17 genera), the number of Levantine rodents remained fairly constant at around 15 genera from the last interglacial to the present (two unrecorded fossil genera have been added to the present faunal list: *Psammomys* and *Jaculus*). Ethiopian rodents are scarce and – save for *Acomys* – are extinct presently: *Arvicanthis*, *Mastomys* and *Cryptomys*. The climatic variations have had very different effects on rodent assemblages on both sides of the Sahara. Whereas a major increment in species richness has been recorded in North Africa during the last pluvial, allowing many Ethiopian and Eurasian rodents to invade the area, a major increase has been recorded in the Levant during the last interglacial (Riss/Würm) with no subsequent impoverishment.

Changes in the Pleistocene mammal fauna of Mediterranean Europe

Pleistocene faunal successions in Mediterranean Europe are still poorly understood. Most of the fossil deposits excavated so far are located at the northern margins of this area, notably in Spain, Italy and southern France, and have

Table 17.5. *Changes in the large mammal fauna (Perissodactyla, Artiodactyla and Carnivora) of Israel during the Pleistocene/Holocene*

Species	1	2	3	4	5	6	7
Dicerorhinus hemitoechus	+	+	+	+			
D. (= *Rhinoceros*) *mercki*		+					
Equus caballus		+	+	+	+	+	
E. hydruntinus		+	+	+	+	+	
E. hemionus		+	+	+	+		
Sus scrofa		+	+	+	+	+	+
Phacochoerus garrodae			+				
Camelus sp.	+						
C. dromedarius				+	+		
Metridiochoerus evronensis	+						
Megaloceros sp.	+						
Dama sp.	+	+	+	+	+	+	+[a]
D. mesopotamica			+	+	+	+	
Cervus elaphus			+	+	+	+	
Capreolus capreolus			+	+	+	+	+[a]
Bison priscus	+	+	+	+			
Bos primigenius			+	+	+	+	
Hemibos sp.		+	+	+	+	+	
Capra aegagrus (= *hircus*)			+	+	+	+	
C. ibex			+	+	+	+	+
Alcephalus busephalus			+	+	+	+	
Gazella gazella			+	+	+	+	+
Nyctereutes vinetorum			+	+			
Vulpes vulpes	+	+	+		+		
Canis aureus	+	+	+		+	+	+
C. lupaster			+	+	+		
C. lupus			+	+	+	+	+
Ursus arctos				+	+	+	+
Vormela peregusna			+	+	+	+	+
Meles meles				+	+	+	+
Martes foina				+	+	+	+
Felis silvestris			+	+	+	+	+
F. chaus				+	+	+	+
F. (= *Panthera*) *pardus*			+	+	+	+	+
F. (= *Panthera*) *leo*				+	+		
Herpestes ichneumon				+	+	+	+
Crocuta crocuta	+	+	+	+			
Hyaena hyaena	+	+	+	+	+		

1, Mindel/Riss (650000–500000 BP); 2, Riss (500000–150000 BP); 3, Riss/Würm (150000–70000 BP); 4, Mousterian (70000–40000 BP); 5, Late Würm (40000–7000 BP); 6, Neolithic (7000–5500 BP); 7, Present.

[a] Extinct species within the Mediterranean region.

Source: Tchernov, 1984.

Table 17.6. *Changes in species of rodents of the Pleistocene/Holocene of Israel*

Species	1	2	3	4	5	6	7
Sciurus anomalus			+	+	+	+	+
Myominus judaicus	+	+					
M. roachi			+	+	+	+	
Dryomys nitedula					+	+	+
Eliomys melanurus (= *quercinus*)					+	+	+
Allocricetus bursae	+	+					
A. jesreelicus			+				
A. magnus			+	+			
Cricetus cricetus	+	+					
Mesocricetus sp.		+					
M. auratus			+	+	+	+	
Cricetulus migratorius				+	+	+	+
Spalax ehrenbergi		+	+	+	+	+	+
'Meriones' obeidiensis	+	+					
Meriones sacramenti						+	+
M. tristami			+	+	+	+	+
Gerbillus sp.	+	+	+	+			
G. dasyurus					+	+	+
G. pyramidum						+	+
G. allenbyi						+	+
Ellobius fuscocapillus			+	+			
Microtus guentheri			+	+	+	+	+
Arvicanthus ectos	+	+	+				
Apodemus flavicollis		+		+	+		
A. sylvaticus	+	+	+	+	+	+	+
A. mystacinus	+	+	+	+	+	+	+
Mus 'musculus' (see text)	+	+	+	+	+	+	+
Rattus haasi			+				
R. rattus					+	+	+
Mastomys batei			+	+			
Acomys cahirinus					+	+	+
Cryptomys asiaticus		+					
Jordanomys haasi		+					
Hystrix indica			+	+	+	+	+

Recent species unknown as fossils are not listed. 1, Mindel/Riss; 2, Riss; 3, Riss/Würm; 4, Mousterian (Early Würm); 5, Late Würm; 6, Neolithic; 7, Recent.
Source: Tchernov, 1984.

yielded confusing assemblages of species. Unfortunately, excavations located in the presumed southern refugia are not numerous enough to allow an accurate reconstruction of faunal successions during the last stages of the Pleistocene.

The Villafranchian, or Lower Pleistocene, European fauna was still basically tropical. Its transition to the Middle Pleistocene or Mindel glacial (about 1 m.y.) traces the extinction of most of these genera: a cheetah *Acinonyx*, a hunting-dog *Lycaon*, a raccoon-dog *Nyctereutes*, the mastodonts, a tapir, *Tapirus*, several species of gazelles, antelopes and goat-antelopes: *Gazella*, *Gazellospira*, *Procamptoceras*, *Pliotragus* and *Gallogoral*, a relative of the African buffalo *Syncerus*, primitive sheep and oxen *Megalovis* and *Leptobos*, and a relative of colobus monkeys *Dolichopithecus* (Kurten, 1968).

Many genera known since the Villafranchian in Europe continued to thrive throughout the Pleistocene: *Elephas* (*Mammuthus*), *Equus*, *Rhinoceros* (*Dicerorhinus*), *Hippopotamus*, *Sus*, *Cervus*, *Alces*, *Megaceros*, *Dama*, *Panthera*, *Lynx*, *Canis*, *Vulpes*, *Cuon*, *Hyaena*, *Ursus* and *Macaca*, but by the end of the Würm glacial, which ended roughly 12000 years BP, most of these species had become extinct. Exceptions were four species scattered in three different refugia: the lion and the fallow deer in Greece, the porcupine in Italy and the Barbary ape in Spain, and seven ancient lineages still widespread in Europe: *Sus*, *Cervus*, *Alces*, *Ursus*, *Canis*, *Vulpes* and *Lynx*. Well after their extinction from continental Europe, some Villafranchian lineages survived on Mediterranean islands, where they underwent rapid speciation (Table 17.7). *Elephas*, *Hippopotamus*, *Cervus*, *Megaceros*, *Cuon* (*Cynotherium*), *Myotragus* and several species of otters are examples of the effectiveness of these insular refugia for Plio-Pleistocene mammals (Palombo, 1985).

Continental refugia as illustrated by the distributions of the lion, the fallow deer, the porcupine and the Barbary ape have allowed very little speciation among European mammals, save for the prominent radiation of the family Arvicolidae (Kurten, 1968; Chaline, 1974).

The onset of the ice age, particularly during the Gunz and Mindel glacials, witnessed the arrival of a new fauna originating mainly from temperate and boreal Asia: *Bison priscus*, *Bos primigenius*, *Ovibos moschatus*, *Rangifer tarandus*, *Capreolus capreolus* and *Ursus thibetanus*. Only two tropical Asian species made a quick appearance during interglacials: *Bubalus murrensis* (Mindel/Riss) and *Felis chaus* (Riss/Würm). Following these early newcomers, others invaded Europe starting with the Riss, and more commonly during the Würm; these are the much popularised mammoth *Elephas primigenius*, woolly rhinoceros *Rhinoceros tichorhinus* (= *Coleodonta antiquitatis*), *Equus hemionus*, *Saiga tatarica*, *Capra ibex*, *Hemitragus bonali* and *Rupicapra rupicapra*. These species usually did not enter the Mediterranean. They are totally unknown in

Table 17.7. *The Upper Pleistocene mammalian faunas of some large Mediterranean islands*

	Majorca	Corsica	Sardinia	Crete
Soricidae	*Episoriculus hidalgo*	*E. corsicanus*	*E. similis*	*Crocidura* sp.
Canidae	—	*Cynotherium (Cuon) sardous*	*C. sardous*	—
Mustelidae	—	*Cyrnaonyx majori*	*C. majori*	*Isolutra cretensis*
Bovidae	*Myotragus balearicus*	—	—	—
Cervidae	—	*Megaceros cazioti*	*M. cazioti*	*M.* sp.
Gliridae	*Eliomys (Hypnomys) morpheus*	—	—	Gliridae species
Muridae	—	*Rhagamys orthodon*	*R. orthodon*	*Mus minotaurus*
Arvicolidae	—	*Microtus (Tyrrhenicola) henseli*	*Microtus henseli*	—
Leporidae	—	*Prolagus sardus*	*P. sardus*	—
Upper Pleistocene findings of questionable stratigraphy				
	—	—	*Talpa tyrrhenaica* *Macaca majori* *Vulpes vulpes* *Megalenhydris barbaricina* *Nesolutra ichnusae* *Elephas lamarmorae* *Sus scrofa*	*Meles arcalus* *Martes foina* *Elephas creticus* *Hippopotamus creutzburgi* *Kritimys catreus*

Source: Alcover *et al.*, 1981; Palombo, 1985; Dermitzakis & de Vos, 1987; Vigne, 1990.

North Africa, where the most prominent Eurasian invaders for this period were two thermophilous species: *Elephas* (*Mammuthus*) *meridionalis* and *Rhinoceros* (*Dicerorhinus*) *mercki*. The fate of these immigrants paralleled that of the Villafranchian fauna. Most species vanished during the Würm or the early Holocene. The ibex and the chamois sought refuge in the mountains, whilst the boreal species retreated beyond the Arctic Circle.

The Holocene extinction rate of European megafauna was comparable to that of North Africa. Six species still roam freely on the southern regions of Europe (namely *Rupicapra*, two species of *Capra*, *Sus*, *Cervus* and *Capreolus*), whereas two more vanished a few centuries ago (*Saiga* and *Dama*) (note that *Dama*, *Bison* and *Ovis* have been reintroduced to Europe in recent times). This total of eight species compares with seven species which survived in North Africa up to the nineteenth and twentieth centuries. The species richness during the Würm glacial was very similar in Europe and in North Africa, amounting to 23 species in the former location, insular species of the Mediterranean excluded, and 21 in North Africa.

Island changes

During the past decade, island faunas have received increasing attention from European palaeontologists. Results of these studies have given new insights to the terminal Pleistocene/Holocene mammalian successions, revealing an intense recent immigration and widespread extinctions. The dramatic changes experienced on these islands have been synthesised by Vigne & Alcover (1985) for Majorca, Minorca, Corsica and Sardinia, and by Dermitzakis & de Vos (1987) for Crete. Alcover *et al.* (1981) and Palombo (1985) provided much information on the larger Mediterranean islands, whereas Vigne (1990) has further synthesised the Corso-Sardinian turnover.

Balaeric and Corsican–Sardinian palaeontology is best known. Sicily has a more confusing history, because of frequent connections with continental Italy, whereas Cyprus has been less accurately investigated than other islands.

Upper Pleistocene mammals of large Mediterranean islands are summarised in Table 17.7. In striking contrast to continental faunas, Mediterranean islands shared an impressive number of endemic species, belonging to the families Soricidae, Mustelidae, Canidae, Bovidae, Cervidae, Elephantidae, Hippopotamidae, Gliridae, Muridae, Cricetidae, Arvicolidae and Leporidae.

Save for some poorly known taxa, such as *Crocidura* in Crete, *Vulpes* and *Lutra* in Sicily, every Mediterranean island species and more than half the genera alike were endemic. This unusually high rate of speciation indicates a fairly long unbroken isolation. Sicily appears to be the sole island inhabited by several continental genera such as *Cervus*, *Bos*, *Lutra*, *Vulpes* and *Apodemus*. Another unusual feature of Sicily is its high species richness, which could be attributed either to repeated mainland connections, unsatisfactory stratigraphy, or both.

A comparison between Upper Pleistocene and present faunas emphasises the devastating effects of Holocene humans upon these islands. This devastation has been studied in the Balaerics, Corsica and Sardinia by Vigne & Alcover (1985). The first clues to human presence in Corsica and Sardinia date from the ninth millennium BP (Vigne, 1990), whereas in the Balaerics, humans probably appeared during the seventh millennium BP (Reumer & Sanders, 1984). In the course of 7000 years, the archaic fauna was completely extirpated from these islands, save for *Prolagus* which survived up to the eighteenth century AD on one Sardinian islet. The gradual extinction of Pleistocene mammals was compensated for by the arrival of many modern taxa inadvertently introduced by humans, such as rats and mice, or deliberately brought, such as game and domestic animals. By the time of the Roman conquest, the faunal turnover was complete, leaving a supersaturated fauna two to five times richer than during Pleistocene times (Cheylan, 1984).

Post-Pleistocene invasions

With the advent of a warmer climate, significant northward animal migrations should have occurred. Surprisingly, there seem to have been few changes in the course of the last 12000 years. The present Mediterranean mammals are basically the same as they were during most of the last glaciation, with the exception of the extinctions of prominent big mammals.

For Europe Kurten (1968) listed only five post-glacial immigrants, all of them Mediterranean colonisers. Two invaded Spain (*Herpestes ichneumon* and *Genetta genetta*) and three the Balkans (*Vormela peregusna*, *Canis aureus* and *Microtus guentheri*). Another possible newcomer is *Crocidura russula* whose appearance in France is not older than 5500 BP (Poitevin *et al.*, 1986). As pointed out by Poitevin, records for this species in the Levant are erroneous, and its presence in the fossil record of this region needs confirmation (Tchernov, 1984).

It is noteworthy that all the Holocene immigrants came from large southern refugia – the Maghreb and the Levant – where they survived glacial periods. So few post-glacial immigrants and such an unbalanced immigration pattern (4 carnivores and 1 rodent) are not easily explained. The Mediterranean Sea seems to have acted as an impassable barrier, impeding any significant recolonisation of Mediterranean habitats by southern mammals.

The introduction of the genet in Spain has been allegedly attributed to the Arabs by Hugues (1928). The same fate could have happened to the mongoose, whereas invaders of the Balkans could have reached Europe by travelling around the Black Sea.

Examination of the changes in the faunal assemblage in the Levant yields interesting results. The accurate list provided by Tchernov (1984) reveals eight post-glacial immigrants, of which five appeared during the period 11 000 to 7000 BP and three during the period 7000 to 5500 BP. Five of these species are desert rodents: *Acomys cahirinus*, *Gerbillus dasyurus*, *G. pyramidum*, *G. allenbyi* and *Meriones sacramenti*. The other three are a desert hedgehog *Hemiechus auritus*, a European species *Eliomys quercinus* (= *E. melanurus*) and an oriental invader *Rattus rattus*. This fauna clearly indicates a deterioration of Mediterranean habitats, colonised by a new set of species which were absent during the last glacial stage.

The last European immigrants are a highly successful group of rodents, namely rats and mice. *Mus 'musculus'* fossils are known from the Mindel/Riss interglacial (650000 to 500000 BP) in Hungary and Greece (Kurten, 1968). A recent revision of European *Mus* (Bonhomme *et al.*, 1984) has split the taxon previously known as *Mus musculus* into five different species or semi-species. Thus, the eastern European *Mus* fossils previously attributed to *M. musculus* can be either *M. abbotti*, *M. hortulanus* or *M. musculus*. In Israel, *Mus* fossils are common throughout the Upper Pleistocene (Tchernov, 1984). A re-examination of the extensive material formerly assigned to *M. musculus* revealed the continuous occurrence of *Mus abbotti* (= *M. spicilegus*) beginning with the Acheulian (120000 BP) (Auffray, 1988). *Mus musculus* is a much more recent invader, entering the Levant scene not before the Natufian, and perhaps even the Aurignacian epoch (22000 to 12000 BP) (Auffray, 1988). On Mediterranean islands, *M. musculus* appears during the third millennium BP (Vigne & Alcover, 1985), but in Sardinia it is quite possible that this species invaded the island during the Neolithic (Sanges & Alcover, 1980). The colonisation of continental Europe could have been delayed, since no house mice are known before Roman times.

Rattus rattus, similarly, is unknown from European deposits before 2000 to 2200 BP (Armitage *et al.*, 1984), although some island excavations have revealed its presence during the fifth millennium BP (Storch, 1970; Sanges & Alcover, 1980). In Egypt, the species has been known since 3500 BP (Armitage *et al.*, 1984).

The last invader of Europe is the Norway rat *Rattus norvegicus*, which appears somewhere around the Middle Ages followed by a major invasion at the end of the eighteenth century (Niethammer & Krapp, 1978–1986).

The oldest anthropogenic transportation of a mammal could be the introduction of the rabbit to North Africa. The Palaeolithic material attributed to this species is represented by two questionably old findings from Algeria and Morocco (Romer, 1928; Gobert & Gaufrey, 1932). The abundance of the species in Neolithic deposits (Romer, 1928; Hopwood & Hollyfield, 1954) suggests an early introduction from Iberia, where the species has been known since at least the Mindel.

During Roman times, humans began to play an increasing role in mammalian transportation, almost inevitably linked to hunting practices. During the last 2000 years, eight mammal species have been added to the Mediterranean fauna, of which five are game animals, one is used to track rabbits into their dens (polecat) and two escaped from captivity where they were bred for their fur (nutria and muskrat). The Romans brought the fallow deer *Cervus dama* and the red deer *C. elephus* into southern Europe and North Africa respectively. These early introductions were followed by many others – several ungulates by the end of the nineteenth century, of which only one, the muflon *Ovis musimon* from Corsica and Sardinia, successfully established healthy populations everywhere (mostly in France, but also in Spain, Italy, Yugoslavia and other European countries). The nutria *Myocastor coypus* and the muskrat *Ondatra zibethicus*, introduced in the 1940s from the Americas, are restricted to Provence in the Mediterranean. More recent introductions are the Barbary sheep *Ammotragus lervia* to Spain, the cottontail *Sylvilagus floridanus* to Italy and France, and the mongoose *Herpestes ichneumon* to Yugoslavia. It is not known how long the polecat *Mustela putorius* has been in Morocco, but most authors agree that it was an anthropogenic introduction.

Conclusions

The mammal fauna of the Mediterranean still awaits extensive treatment comparable to Voous' (1960) treatise on birds. Basic accounts, such as those of Heim de Balsac (1936) and Misonne (1958), are restricted to certain parts of the Mediterranean and need to be updated.

In this chapter palaeontological data have been used when available to assign probable origins to the better known species, according to the biogeographical units of Trouessart (1922) and Schmidt (1954):

1. Palaearctic region
   (a) Euro-Siberian sub-region (includes the Mediterranean)
   (b) Irano-Turanian sub-region (cold central Asian deserts)
   (c) Saharo-Sindian sub-region (hot desert belt stretching from Morocco to the Sind desert in India)
2. Ethiopian region (sub-Saharan Africa)
3. Oriental region (Asia south of the Himalayas).

From this classification, it appears that southern European mammals are basically Euro-Siberian, a category which accounts for between 70 and 80 per cent of the total. For Asia Minor, Euro-Siberian mammals still predominate although all other sources are well represented (about 10 to 20 per cent of the total). The most distinctive mammal assemblage is that of the Maghreb, where a large number of species are either Ethiopian or Saharo-Sindian (about 30 per cent), whilst Euro-Siberian species rank third.

Mediterranean mammals are thus representative of a wide range of geographical origins and habitats. Of particular interest is the fairly large number of browsers characteristic of open landscapes, such as asses, gazelles, hamsters, gerbils and jerboas.

The diverse origins of Mediterranean mammals have allowed various invasions in the course of the Middle and Upper Pleistocene: Eurasian and Ethiopian to North Africa, Irano-Turanian and Boreal to Europe. But these multiple invasions did not always lead to reciprocal extinctions as relaxation times seem to have been protracted in continental biotas. Consequently, every Mediterranean megafauna, either in southern Europe, North Africa or the Levant, had increased from 100 to 300 per cent by the beginning of the Würm, and decreased steadily only during the Holocene. This rapid disappearance during the post-glacial can be made to fit the prehistoric overkill theory of Martin (1984). Nevertheless, the changing environment induced by early herders and cultivators is probably as important a factor as increased hunting pressure in explaining these early extinctions.

The saturated glacial faunas may have impeded high speciation rates in Mediterranean refugia. It is noticeable that most Mediterranean Pleistocene endemics evolved from depauperate insular faunas during several hundred thousands of years of isolation. By comparison, the last glacial stage covered only 60 000 to 70 000 years, quite a short period of time for differentiation of most mammalian populations. Furthermore, none of the species which presumably originated in Mediterranean refugia was thermophilous, as expected from the allopatric speciation model of Mayr (1942, 1963). The Italian porcupine is the last archaic species of tropical descent in Europe, although this population is still undifferentiated from the African stock. All other relictual lineages which sought refuge on the southern peninsulas are now extinct in Europe and none evolved to the species level.

The best supported example of speciation in European refugia comes, surprisingly, from the Arvicolidae, a family whose origin can be traced back to Boreal Asia (Chaline, 1974). This is interesting, as Boreal species may have had continuous distributions during glacial stages, thereby allowing gene flow which presumably impeded allopatric speciation.

Conversely, islands and mountains have been as efficient in promoting speciation among Mediterranean biotas as Pleistocene refugia. This is best illustrated by the origin of 16 endemics in the Maghreb. Five species evolved from relictual populations isolated by physical barriers: one species of probable Eurasian descent (*Crocidura whitakeri*); one of Oriental stock (*Macaca sylvanus*); two species of Ethiopian stock (*Elephantulus rozeti* and *Atlantoxerus getulus*); and one of Irano-Turanian stock (*Meriones shawi*). But, strangely enough, the majority of this fauna of 11 species is Saharo-Sindian in origin and evolved in new habitats induced by the desertification of the Sahara, as for two species of *Gazella*, four of *Gerbillus*, three of *Dipodillus*, and one each of *Jaculus* and *Ctenodactylus*. Moreover, most of these species are fairly recent, first appearing during the Upper or terminal Middle Pleistocene (cf. the genera *Gazella* and *Gerbillus*; see Arambourg, 1957; Jaeger, 1975*a*).

Adaptive radiation in a patchy environment, which probably reduced gene flow among populations, thus accounts for most of the Pleistocene speciations in North Africa.

Acknowledgements. I am indebted to Jean-Denis Vigne and Jacques Michaux for their criticisms of an early draft of this chapter and for the useful references they drew to my attention.

Appendix

Appendix 17.1. *Mammal species of different regions of the Mediterranean Basin excluding islands*

Family & species	Maghreb	Iberia	Italy	Balkans	Asia Minor	Middle East
Erinaceidae						
Erinaceus europaeus		+	+			
E. concolor				+	+	+
E. algirus	+	+				
Hemiechinus auritus					+	+
Soricidae						
Sorex minutus		(+)	+	+	(+)	
S. araneus			+	(+)		
S. granarius		+				
S. samniticus			+			
Neomys fodiens		(+)	+	(+)		
N. anomalus		+	(+)	+	+	
Suncus etruscus	+	+	+	+	+	
Crocidura leucodon (1)			+	+	+	+
C. suaveolens (2)		+	+	+	+	
C. russula	+	+				
C. lusitanica (3)	(+)					

Appendix 17.1 (*cont.*)

Family & species	Maghreb	Iberia	Italy	Balkans	Asia Minor	Middle East
C. whitakeri (4)	+					
C. tarfayaensis (3)	(+)					
C. viaria (3)	(+)					
Talpidae						
Talpa caeca		+	+	+	+	+
T. romana		+	+			
T. europaea		+	(+)	(+)		
Galemys pyrenaicus		+				
Pteropodidae						
Rousettus aegyptiacus					+	
Rhinopomatidae						
Rhinopoma hardwickei	(+)				(+)	
R. microphyllum					(+)	
Emballonuridae						
Taphozous nudiventris					(+)	
Nycteridae						
Nycteris thebaica					(+)	
Rhinolophidae						
Hipposideros caffer	(+)					
Asellia tridens	(+)				(+)	
Triaenops persicus						(+)
Rhinolophus ferrumequinum	+	+	+	+	+	+
R. hipposideros	+	+	+	+	+	
R. euryale	+	+	+	+	+	
R. blasii	+		(+)	+	+	
R. mehelyi	+	+	(+)	+	+	
Vespertilionidae						
Myotis capaccinii	+	+	+	+	+	
M. mystacinus	+	+	+	+	+	
M. emarginatus	+	+	+	+	+	
M. nattereri	+	+	+		+	
M. bechsteini		(+)	+	(+)	+	
M. myotis		+	+	+	+	+
M. blythii (5)	+	+	+	+	+	
M. nathalinae		+				
M. daubentoni		+	(+)	(+)	+?	
Pipistrellus coromandra						+
P. pipistrellus	+	+	+	+	+	+
P. kuhlii	+	+	+	+	+	+
P. nathusii		(+)	+	+	+	

Appendix 17.1 (*cont.*)

Family & species	Maghreb	Iberia	Italy	Balkans	Asia Minor	Middle East
P. savii	+	+	+	+	+	
Nyctalus lasiopterus	+	(+)	(+)	+	+	
N. noctula	+	+	+	+	+	
N. leisleri	+	(+)	(+)	+	+	
Eptesicus bottae (6)					+	+
E. serotinus (7)	+	+	+	+	+	+
Vespertilio murinus			(+)	(+)		(+)
Otonycteris hemprichii	(+)				+	
Barbastella barbastellus	+	+	+		+	
Plecotus austriacus	+	+	+	+	+	
P. auritus		(+)	(+)	(+)	+?	
Miniopterus schreibersi	+	+	+	+	+	
Molossidae						
Tadarida teniotis	+	+	+	+	+	
Cercopithecidae						
Macaca sylvanus	+	(+)				
Canidae						
Canis aureus	+			+	+	+
C. lupus		+	+	+	+	+
Vulpes vulpes	+	+	+	+	+	+
Ursidae						
Ursus arctos (8)	#	#	#	+	+	+
Mustelidae						
Mustela nivalis	+	+	+	+	+	+
M. putorius (9)	+?	+	+			
Vormela peregusna				+	+	+
Martes foina		+	+	+	+	+
M. martes			+			+
Lutra lutra	+	+	+	+	+	+
Meles meles		+	+	+	+	+
Poecilictis libyca	+					
Mellivora capensis (3)	(+)				(+)	+
Viverridae						
Genetta genetta	+	+			(+)	
Herpestidae						
Herpestes ichneumon	+	+			+	
H. auropunctatus						(+)
Hyaenidae						
Hyaena hyaena	+				+	+

Appendix 17.1. (*cont.*)

Family & species	Maghreb	Iberia	Italy	Balkans	Asia Minor	Middle East
Felidae						
Felis silvestris (10)	+	+	+	+	+	+
F. serval (11)	+					
F. chaus					+	+
Lynx lynx			#	+	+	+
L. pardinus		+				
L. caracal	+				+	+
Panthera pardus	+				+	+
P. leo (12)	#			#	#	#
Acinonyx jubatus (13)	#				#	#
Equidae						
Equus asinus (14)	#					
E. onager (15)					#	
Elephantidae						
Loxodonta africana (16)	#					
Elephas maximus (17)						#
Procaviidae						
Procavia capensis (18)					(+)	
Suidae						
Sus scrofa	+	+	+	+	+	+
Cervidae						
Cervus elaphus (19)	+	+	+	+	+	+?
C. dama (20)				#	#	+
Capreolus capreolus (21)		+	+	+	+	+
Bovidae						
Capra pyrenaica		+				
C. ibex					+	
C. hircus (22)				+?	+	+
Ammotragus lervia	+					
Ovis aries					+	+
Alcephalus busephalus (23)	#				(#)	
Gazella cuvieri	+					
G. rufina (24)	#					
G. gazella (25)					(+)	
G. dorcas	+				+	#
G. subgutturosa						+
Sciuridae						
Sciurus vulgaris		+	+	+		
S. anomalus					+	+
Atlantoxerus getulus	+					

Appendix 17.1. (*cont.*)

Family & species	Maghreb	Iberia	Italy	Balkans	Asia Minor	Middle East
Xerus erythropus (26)	(+)					
Spermophilus citellus				(+)		
Gliridae						
Eliomys quercinus (27)	+	+	+	(+)	+	
Dryomys nitedula			(+)	+	+	+
Myomimus roachi				+	+	
Myoxis glis		(+)	+	+	+	
Muscardinus avellanarius			+	+	+	
Hystricidae						
Hystrix cristata	+		+			
H. indica					+	+
Castoridae						
Castor fiber		(+)			+?	+?
Spalacidae						
Spalax nehringi					+	
S. leucodon				+		
S. ehrenbergi					+	
Muridae						
Apodemus sylvaticus	+	+	+	+	+	+
A. mystacinus				+	+	
A. flavicollis			+	+	+	+
A. microps				+?	+	
A. agrarius				(+)		
Micromys minutus		(+)	(+)	(+)		
Rattus rattus	+	+	+	+	+	+
R. norvegicus	(+)	+	+	+	+	+
Mus spretus	+	+				
M. abbotti				+	+	+
M. domesticus	+	+	+	+	+	+
Acomys cahirinus (3)	(+)				+	+
A. cilicicus (28)					+	
Lemniscomys barbarus	+					
Mastomys erythroleucus (29)	(+)					
Nesokia indica					+	+
Cricetidae						
Mesocricetus auratus					+	
M. brandti						+
M. newtoni				+		
Cricetulus migratorius				+	+	+
Calomyscus bailwardi						+

Appendix 17.1. (*cont.*)

Family & species	Maghreb	Iberia	Italy	Balkans	Asia Minor	Middle East
Gerbillus hoogstraali (30)	+					
G. allenbyi (31)					+	
G. campestris	+					
G. henleyi	(+)				(+)	
G. hesperinus (32)	+					
G. jamesi (33)	+					
G. dasyurus					(+)	
G. pyramidum	(+)				(+)	
G. occiduus (34)	+					
Dipodillus zakarai (35)	+					
D. maghrebi (36)	+					
D. simoni	+					
Meriones shawi	+					
M. persicus						+
M. vinogradovi						+
M. tristami					+	+
M. libycus	(+)					+
M. crassus	(+)				(+)	+
Tatera indica						+
Psammomys obesus	(+)				(+)	
Pachyuromys duprasi	(+)					
Arvicolidae						
Clethrionomys glareolus		(+)	(+)	(+)		
Dinaromys bogdanovi				+		
Arvicola sapidus		+				
A. terrestris			+	+	+	+
Microtus cabrerae		+				
M. guentheri				+	+	
M. epiroticus				+	+?	
M. arvalis		(+)			+	+?
M. agrestis		(+)				
M. irani					+	+
M. nivalis		+	+	+	+	
Pitymys duodecimcostatus		+				
P. lusitanicus		+				
P. savii			+			
P. thomasi				+		
P. multiplex			(+)			
P. subterraneus				(+)		
P. majori					+	
Ellobius lutescens						+

Appendix 17.1. (*cont.*)

Family & species	Maghreb	Iberia	Italy	Balkans	Asia Minor	Middle East
Dipodidae						
Allactaga elater						+
A. euphratica					+	+
Jaculus orientalis	+				(+)	
J. jaculus	(+)				(+)	
Ctenodactylidae						
Ctenodactylus gundi	+					
Leporidae						
Lepus granatensis (37)		+				
L. capensis (38)	+	(+)	+	+	+	+
Oryctolagus cuniculus (39)	+	+				
Macroscelididae						
Elephantulus rozeti	+					

+, widely distributed species; (+), locally distributed species; #, extinct species within the Mediterranean region.

(1) Includes *C. lasia* (Honacki *et al.*, 1982).
(2) Includes *C. gueldenstaedti* (Catzeflis *et al.*, 1985).
(3) In North Africa, restricted to southern Morocco.
(4) Sometimes listed as conspecific with *C. suaveolens*.
(5) Most records of *M. myotis* in the Mediterranean are probably referable to the sibling species *M. blythii*.
(6) Includes *E. anatolicus* (Honacki *et al.*, 1982).
(7) Includes *E. isabellinus* (Aellen & Strinati, 1969).
(8) Last mentioned Maghreb early nineteenth century AD (Loche, 1867).
(9) In North Africa, restricted to northern Morocco where probably introduced by humans.
(10) Includes *F. libyca* (Honacki *et al.*, 1982).
(11) In North Africa, restricted to coastal Algeria, where probably extinct by now (H. Kowalski, personal communication).
(12) In Europe, present up to the first century AD in Thrace and Macedonia (Dorst, 1965). Last killed Turkey 1870, Syria 1896, and Iraq 1914 (Kumerloeve, 1975). In Iran, survived in the Zagros range up to the 1950s (Misonne, 1958). Last killed North Africa (Morocco) 1922 (Dorst, 1965).
(13) Known to have survived to 1930 in the North African high plateaux (Panouse, 1957). Last mentioned Turkey 1879 (Kumerloeve, 1975).
(14) A re-examination of some old records previously attributed to *Equus mauritanicus*, a fossil zebra from North Africa, has shown that these bones are really a fossil ass, *E. melkiensis*, from the Middle Palaeolithic (Bagtache *et al.*, 1984). Wild asses are common in wall paintings of the Saharan Neolithic and survived to Roman times in North Africa (Monod, 1933; Lhote, 1985).

(15) Extinct Syria early twentieth century.
(16) Known to have inhabited Morocco up to third century AD, possibly seventh century AD (Panouse, 1957).
(17) Extinct eighth century BC in Orontes valley (Turkey/Syria) (Kumerloeve, 1975).
(18) Northern limit of distribution in Lebanon.
(19) In North Africa, restricted to a patch along the Tunisian/Algerian border, where probably introduced during Roman times. This species is unknown from North African fossil deposits.
(20) Feral origin of European 'wild' populations are very questionable.
(21) Recently extinct in Israel and Lebanon (Harrison, 1964–1972).
(22) Aegean populations probably originated from a domesticated stock.
(23) Extinct in North Africa between 1930–1950 (Panouse, 1957); in Palestine around 1900 (Heim de Balsac, 1936).
(24) Formerly restricted to north-central Algeria (Chelif valley) where now extinct.
(25) In eastern Mediterranean, restricted to Israel and formerly Lebanon.
(26) Outside sub-Saharan Africa, restricted to the Sous plain, Atlantic Morocco.
(27) Includes *E. melanurus*; often treated as a distinct species.
(28) Only known occurrence near Silifke, Turkey.
(29) Only known occurrence outside sub-Saharan Africa at Essaouira, Atlantic Morocco.
(30) Known only from the Sous plain, Atlantic Morocco.
(31) Known only from the coastal dunes of Israel.
(32) Known only from coastal northern Morocco.
(33) Known only from Sousse region, Tunisia.
(34) Known only from Draa valley, southern Morocco.
(35) Known only from Kerkennah island, Tunisia.
(36) Known only from Fes region, Morocco.
(37) Listed as distinct from *L. europaeus* by Palacios (1979) and Bonhomme *et al.* (1986).
(38) Includes *L. europaeus*. Genetic relationships between these two taxa need investigation.
(39) Indigenous to Iberia only. Now introduced worldwide.

References

Aellen, V. & Strinati, P. (1969). Liste des Chiroptères de Tunisie. *Revue Suisse de Zoologie*, **76**, 421–31.

Alcover, J. A., Moya-Sola, S. & Pons-Moya, J. (1981). *Les Quimeres del Passat. Els Vertebrats Fossils del Plio-Quaternari de les Balears i Pitiuses*. Cuitat de Mallorca: Moll Publishers.

Anciaux de Faveaux, M. (1976). Distribution des Chiroptères en Algérie, avec notes écologiques et parasitologiques. *Bulletin Société Histoire Naturelle Afrique du Nord*, **67**, 69–80.

Arambourg, C. (1957). Observations sur les Gazelles fossiles du Pléistocène supérieur de l'Afrique du nord. *Bulletin Société Histoire Naturelle Afrique du Nord*, **48**, 49–81.

Arambourg, C. (1962). Les faunes mammalogiques du Pléistocène circumméditerranéen. *Quaternaria*, **6**, 97–109.

Arambourg, C. (1970). Les vertébrés du Pléistocène de l'Afrique du nord. *Archives Museum National Histoire Naturelle, Paris*, **10**, 1–126.

Armitage, P., West, B. & Steedman, K. (1984). New evidence of Black Rat in Roman London. *The London Archaeologist*, **4**, 375–83.

Attalah, S. I. (1977–1978). The mammals of the eastern Mediterranean region: their ecology, systematics and zoogeographical relationships. *Säugetierkundliche Mitteilungen*, **25**, 241–320; **26**, 1–50.

Auffray, J. C. (1988). Le commensalisme chez la souris domestique. Origine, écologie et rôle dans l'évolution chromosomique. Thèse, Université Sciences et Techniques de Languedoc, Montpellier.

Aulagnier, S. & Destre, R. (1985). Introduction à l'étude des Chiroptères du Tafilalt (sud-est marocain). *Mammalia*, **49**, 329–38.

Aulagnier, S. & Thevenot, M. (1986). Les Ongulés sauvages du Maroc: constat d'une régression alarmante. *Le Courrier de la Nature*, **104**, 16–25.

Bagtache, B., Hadjouis, D. & Eisenmann, V. (1984). Présence d'un *Equus* caballin (*E. algericus* n. sp.) et d'une autre espèce nouvelle d'*Equus* (*E. melkiensis* n. sp.) dans l'Altérien des allobroges, Algérie. *Comptes Rendus Académie Sciences Paris*, **298**, II (14), 609–12.

Bernard, J. (1969). Les mammifères de Tunisie et des régions voisines. I. Rongeurs. *Bulletin Faculté Agronomie Tunis*, **24/25**, 41–160.

Blondel, J. (1982). Caractérisation et mise en place des avifaunes dans le bassin méditerranéen. *Ecologia Mediterranea*, **8**, 253–72.

Bodenheimer, F. S. (1960). *Animal and Man in Bible Lands*. Leiden: Brill.

Bonhomme, F., Catalan, J., Britton-Davidian, J., Chapman, V. M., Moriwaki, D., Nevo, E. & Thaler, L. (1984). Biochemical diversity and evolution in the genus *Mus*. *Biochemical Genetics*, **22**, 275–303.

Bonhomme, F., Fernández, J., Palacios, F., Catalan, J. & Machordon, A. (1986). Caractérisation biochimique du complexe d'espèce du genre *Lepus* en Espagne. *Mammalia*, **50**, 495–506.

Catalan, J. (1984). Application de méthodes génétiques à la systématique des musaraignes (Soricidés) d'Europe méridionale. *Mémoires Ecole Pratique Hautes Etudes Montpellier*, **15**, 1–96.

Catzeflis, F. (1983). Analyse cytologique et biochimique des Crocidures de l'île de Chypre (Mammalia, Insectivora). *Revue Suisse Zoologie*, **90**, 407–15.

Catzeflis, F., Maddalena, T., Hellwing, S. & Vogel, P. (1985). Unexpected findings on the taxonomic status of east Mediterranean *Crocidura russula* auct. (Mammalia, Insectivora). *Zeitschrift fur Säugetierkunde*, **50**, 185–201.

Chaline, J. (1974). Palingenèse et phylogenèse chez les Campagnols (Arvicolidae, Rodentia). *Comptes Rendus Académie Sciences Paris*, **D278**, 437–40.

Chaline, J. (1985). *Histoire de l'Homme et des Climats au Quaternaire*. Paris: Doin.

Cheylan, G. (1984). Les mammifères des îles de Provence et de Méditerranée occidentale: une exemple de peuplement insulaire non equilibré? *Revue Ecologie*, **39**, 37–54.

Dermitzakis, M. D. & de Vos, J. (1987). Faunal succession and the evolution of mammals in Crete during the Pleistocene. *Neues Jahrbuch Geologie Palaontologie Abhaudlungen*, **173**, 377–408.

Diamond, J. (1972). Biogeographic kinetics: Estimation of relaxation times for avifaunas of south-west Pacific islands. *Proceedings of the National Academy of Sciences, USA*, **69**, 3199–203.

Dorst, J. (1965). *Avant que Nature Meure*. Neuchâtel: Delachaux & Niestlé.

Dupuy, A. R. (1967). Distribution actuelle des espèces menacées d'Algérie. *Bulletin Société Sciences Naturelles Physiques Maroc*, **47**, 355–85.

Emberger, L., Gaussen, H. & De Philippis, A. (1962). *Carte Bioclimatique de la Region Méditerranéenne*. Paris: UNESCO-FAO.

Flint, R. F. (1971). *Glacial and Quaternary Geology*. New York: Wiley.

Gaisler, J. (1983). Nouvelles données sur les Chiroptères du nord Algérien. *Mammalia*, **47**, 359–70.

Gattefossé, J. (1933). Evolution biologique du Sahara occidental. *La Terre et la Vie*, **3**, 323–35.

Gobert, E. & Gaufrey, R. (1932). Deux gisements extrèmes d'Ibero-Maurusien. *L'Anthropologiste*, **42**, 449–90.

Graf, J. D., Hausser, J., Farina, A. & Vogel, P. (1979). Confirmation du statut spécifique de *Sorex samniticus* Altobello 1926. *Bonner Zoologische Beiträge*, **30**, 14–21.

Harrison, D. L. (1964–1972). *The Mammals of Arabia*, 3 vols. London: Benn.

Hausser, J., Graf, J. D. & Meylan, A. (1975). Données nouvelles sur les *Sorex* d'Espagne et des Pyrénées. *Bulletin Société Vaudoise Sciences Naturelles*, **38**, 241–52.

Heim de Balsac, H. (1936). Biogéographie des oiseaux et mammifères de l'Afrique du nord. *Bulletin Biologique France Belgique Supplément*, **21**, 1–446.

Honacki, J. H., Kinman, K. E. & Koeppl, J. W. (1982). *Mammal Species of the World*. Lawrence, Kansas: Allen Press & the Association of Systematics Collections.

Hopwood, A. T. & Hollyfield, J. P. (1954). An annotated bibliography of the fossil mammals of Africa. *Fossil Mammals of Africa*, **8**, 1–194.

Hsu, K. J., Ryan, W. B. F. & Cita, M. B. (1973). Late Miocene desiccation of the Mediterranean Sea. *Nature*, **242**, 240–4.

Hugues, A. (1928). Note sur la Genette en France. *Revue Française Mammalogie*, **2**, 52–60.

Hutterer, R. (1986). The species of *Crocidura* (Soricidae) in Morocco. *Mammalia*, **50**, 521–34.

Jaccard, P. (1908). Nouvelles recherches sur la distribution florale. *Bulletin Société Vaudoise Sciences Naturelles*, **44**, 223–79.

Jaeger, J. J. (1975*a*). Les Rongeurs du Miocène à l'actuel en Afrique nord-occidentale. Thèse, Université Sciences et Techniques de Languedoc, Montpellier.

Jaeger, J. J. (1975*b*). The mammalian faunas and hominid fossils of the Middle Pleistocene of the Maghreb. In *After the Australopithecines*, ed. K. W. Butzer & G. L. L. Isaac, pp. 400–18. The Hague: Mouton.

Jaeger, J. J., Coiffait, B., Tong, H. & Denys, C. (1987). Rodent extinctions following Messinian faunal exchanges between western Europe and northern Africa. *Mémoires Société Géologique France*, **150**, 153–8.

Joleaud, L. (1934). Considérations générales sur la faune des mammifères du Maroc. *La Terre et La Vie*, **4**, 259–71.

Klein, R. G. (1984). Mammalian extinctions and stone age people in Africa. In *Quaternary Extinctions*, ed. P. S. Martin & R. G. Klein, pp. 553–73. Tucson: The University of Arizona Press.

Kumerloeve, H. (1967). Zur Verbreitung kleinasiatischer Raub und Huftiere sowie einiger Grossnager. *Säugetierkundliche Mitteilungen*, **15**, 337–409.

Kumerloeve, H. (1975). Die Säugetiere (Mammalia) der Turkei. *Veröffentlichungen Zoologischen Staatssammlung München*, **18**, 69–158.

Kurten, B. (1968). *Pleistocene Mammals of Europe*. London: Weidenfeld & Nicolson.

Le Houérou, H. N. (1981). Impact of man and his animals on Mediterranean vegetation. In *Ecosystems of the World*, vol. 11, *Mediterranean-type Shrublands*, ed. F. di Castri, D. W. Goodall & R. L. Specht, pp. 479–517. Amsterdam: Elsevier.

Lhote, H. (1985). Origine du cheval et de l'âne en Afrique du nord. *Mammalia*, **49**, 387–94.

Loche, V. (1867). *Histoire Naturelle des Mammifères de l'Algérie*. Paris: Arthus Bertrand.

Martin, P. S. (1984). Prehistoric overkill: the global model. In *Quaternary Extinctions*, ed. P. S. Martin & R. G. Klein, pp. 354–403. Tucson: The University of Arizona Press.

Mayr, E. (1942). *Systematics and the Origin of Species*. New York: Columbia University Press.

Mayr, E. (1963). *Animal Species and Evolution*. Cambridge, Mass.: Harvard University Press.

Misonne, X. (1958). Analyse zoogéographique des mammifères de l'Iran. *Mémoires Institut Royal Sciences Naturelles Belgique*, **59**, 1–157.

Monod, T. (1933). Anes sauvages. *La Terre et la Vie*, **3**, 451–62.

Niethammer, J. (1962). Die Säugetiere von Korfu. *Bonner Zoologische Beiträge*, **13**, 1–49.

Niethammer, J. & Krapp, F. (1978–1986). *Handbuch der Säugetiere Europas*, vols. 1, 2-1, 2-2. Wiesbaden: Akademische-Verlag.

Osborn, D. (1962). Rodents of the subfamily Microtinae from Turkey. *Journal of Mammalogy*, **43**, 515–29.

Osborn, D. (1963). New distributional records of bats from Turkey. *Mammalia*, **27**, 210–17.

Osborn, D. (1964). The hare, porcupine, beaver, squirrels, jerboas and dormice from Turkey. *Mammalia*, **28**, 573–92.

Palacios, F. (1979). Analisis cromosómica, carga de DNA y electroforésis de las liebre españoles. *Doñana Acta Vertebrata*, **6**, 203–15.

Palombo, M. R. (1985). I grandi mammiferi pleistocenici delle isole del Mediterraneo: tempi e vie di migrazione. *Bulletino Società Paleontologica Italiana*, **24**, 201–24.

Panouse, J. B. (1951). Les Chauves-Souris du Maroc. *Travaux Institut Scientifique Chérifien*, **1**, 1–120.

Panouse, J. B. (1957). Les mammifères du Maroc. Primates, carnivores, pinnipèdes, artiodactyles. *Travaux Institut Scientifique Chérifien*, **5**, 1–206.

Poitevin, F., Catalan, J., Fons, R. & Croset, H. (1986). Biologie évolutive des populations ouest-européennes de crocidures. I. Critères d'identification et répartition géographique de *Crocidura russula* et *Crocidura suaveolens*. *Revue Ecologie*, **41**, 299-315.

Pomel, A. (1894). *Monographie des Vertébrés Fossiles de l'Algérie*. Algiers: Publications Service Geologique de l'Algérie.

Quézel, P. (1981). Floristic composition and phytosociological structure of sclerophyllous mattorals around the Mediterranean. In *Ecosystems of the World*, vol. 11, *Mediterranean-type Shrublands*, ed. F. di Castri, D. W. Goodall & R. L. Specht, pp. 107–21. Amsterdam: Elsevier.

Reumer, J. W. F. & Sanders, E. A. C. (1984). Changes in the vertebrate fauna of Menorca in prehistoric and classical times. *Zeitschrift Säugetierkunde*, **49**, 321–5.

Rode, D. (1947). Les mammifères de l'Afrique du nord. I. Primates, insectivores, chiroptères. *La Terre et la Vie*, **94**, 120–42.

Rode, D. (1948). Les mammifères de l'Afrique du nord. II. Rongeurs. *La Terre et la Vie*, **95**, 125–50.

Romer, A. S. (1928). Pleistocene mammals of Algeria. *Logan Museum Bulletin* No. 26.

Saint Girons, M. C. & Petter, F. (1965). Les Rongeurs du Maroc. *Travaux Institut Scientifique Cherifien (Zoologie)*, **31**, 1–55.

Sanges, M. & Alcover, J. A. (1980). Notícia sobre la microfauna vertebrada holocenica de la grotta Su Guanu o Gonagosula (Oliena, Sardenya). *Endins*, **7**, 57–62.

Schilling, D., Singer, D. & Diller, H. (1986). *Guide des Mammifères Sauvages d'Europe*. Neuchâtel: Delachaux & Niestlé.

Schmidt, K. B. (1954). Faunal realms, regions and provinces. *Quarterly Review of Biology*, **29**, 322–31.

Spitzenberger, F. (1968). Zur Verbreitung und Systematik türkischer Soricinae (Insectivora, Mammalia). *Annalen Naturhistorische Museum Wien*, **72**, 273–89.

Spitzenberger, F. (1970). Zur Verbreitung und Systematik türkischer Crocidurinae (Insectivora, Mammalia). *Annalen Naturhistorische Museum Wien*, **74**, 233–52.

Spitzenberger, F. (1978). Die Säugetierfauna Zyperns. I: Insectivora und Rodentia. *Annalen Naturhistorische Museum Wien*, **81**, 401–41.

Spitzenberger, F. (1979). Die Säugetierfauna Zyperns. II. Chiroptera, Lagomorpha, Carnivora und Artiodactyla. *Annalen Naturhistorische Museum Wien*, **82**, 439–65.

Storch, G. (1970). Holozäne Kleinsaugerfunde aus der Ghar-Dalam Höhle, Malta. *Senckenbergiana Biologica*, **51**, 135–45.

Tchernov, E. (1984). Faunal turnover and extinction rates in the Levant. In *Quaternary Extinctions*, ed. P. S. Martin & R. G. Klein, pp. 528–52. Tucson: The University of Arizona Press.

Trouessart, E. L. (1922). *Distribution géographique des Animaux*. Paris: Doin.

Van den Brink, F. H. & Barruel, P. (1967). *Guide des Mammifères Sauvages de l'Europe Occidentale*. Neuchâtel: Delachaux & Niestlé.

Vigne, J. D. (1990). Biogeographical history of the mammals on Corsica (and Sardinia) since the final Pleistocene. In *Biological Aspects of Insularity*, ed. A. Azzaroli, in press. Rome: Academia dei Lincei.

Vigne, J. D. & Alcover, J. A. (1985). Incidence des relations historiques entre l'homme et l'animal dans la composition actuelle du peuplement amphibien, reptilien et mammalien des îles de Méditerranée occidentale. *Actes 110th Congrès National Sociétés Savantes (Sciences)*, **2**, 79–91.

Vögel, P., Maddalena, T. & Catzeflis, F. (1986). A contribution to the taxonomy and ecology of shrews (*Crocidura zimmermani* and *C. suaveolens*) from Crete and Turkey. *Acta Theriologica*, **31**, 537–45.

Voous, K. H. (1960). *Atlas of European Birds*. London: Nelson.

West, R. G. (1977). *Pleistocene Geology and Biology*. London: Longman.

Willan, R. G. M. (1973). Tunisian wildlife. *Oryx*, **12**, 74–6.

Zimmermann, K., Von Wettstein, O., Siewert, H. & Pohle, H. (1942–1949). Die Wildsäuger von Kreta. *Zeitschrift Säugetierkunde*, **17**, 1–72.

18

Introduced mammals in California

W. Z. LIDICKER JR

There are only 12 species of non-domesticated mammals now definitely living in wild populations that have been introduced by humans into the State of California. An additional three species are native to restricted ranges in the state, but have been introduced extensively beyond these ancestral distributions. Of these 15 introduced species, all but three commensal rodents were released intentionally. Six are from Europe, two are from Asia, one is from North Africa, and the rest are North American natives. There are also seven species of domesticated mammals with feral populations in California, and a few more introduced species of uncertain status. A total of 26 species are discussed in this chapter out of a total of 220 mammalian species for California as a whole.

This review does not pretend to be definitive. The situation in California is dynamic, with accidental and intentional introductions occurring regularly, and with the status of some species being unknown because of inadequate or non-existent monitoring programs. The chapter will, however, firstly summarise the history and status of the 15 introduced species of mammals; secondly, it will consider whether there are patterns of habitat use that characterise this assemblage; and thirdly, it will explore any possible insights that these mammals may suggest into the important and basic questions of what factors result in successful invasions of new areas. Are there correlations with diet, fecundity, use of disturbed habitats, absence of closely related species, or body size of potential competitors? This chapter will also take account of whatever can be learnt about how to predict which species are likely to become pests to humans and/or threats to native fauna and flora.

Species accounts

Didelphis virginiana *(Virginia opossum)*

The opossum was introduced undoubtedly on multiple occasions and became well-established early in this century (Grinnell, 1933). One documented introduction occurred about 1910 in the vicinity of San Jose with animals brought from Tennessee (Grinnell, 1915). It now occurs widely in the Central Valley and in coastal areas, extending north into Oregon and Washington. Opossum distribution reaches at least 1200 m in elevation. The species is most abundant in suburban and agricultural habitats.

Oryctolagus cuniculus *(European rabbit)*

This species is established only on Santa Barbara Island (von Bloeker, 1967) and the Farralon Islands off the central Californian coast. Rabbits were introduced to the Farralons as a food source by lighthouse keepers, probably about a century ago. A persistent threat of mainland establishment exists because of domestic rabbit escapes and release by well-meaning pet owners.

Vulpes vulpes *(red fox)*

The red fox is native to California, occurring in the mountains of the northern and eastern parts of the State from 1370 to 3500 m elevation (Grinnell, 1933; Grinnell *et al.*, 1937). The native populations belong to the distinctive subspecies *V. v. necator*. In about 1885 a population of this species became established in the Sacramento Valley in the vicinity of Marysville Buttes (Grinnell, 1933; Seymour, 1960). This population was almost certainly introduced from eastern North America and represents foxes of the subspecies *V. v. fulvus*. These valley foxes have now spread widely (Gray, 1977) in the Central Valley and in recent years have extended to coastal areas of central and northern California. They inhabit agricultural areas as well as oak woodland, chaparral and mixed grassland–brush habitats.

Sciurus carolinensis *(eastern gray squirrel)*

Tree squirrels from eastern North America were introduced into California following the gold rush period in the 1850s and the development of a settled middle class (Byrne, 1979). Few historical details remain of what most certainly involved multiple introductions. Since 1933 when importation of squirrels into California became illegal, most introductions involved movements of individuals within the state. Importations almost certainly began about 100 years ago, but are not documented by specimens until 1938 for the eastern gray squirrel. A careful morphological analysis of current populations by Byrne (1979) resulted in the conclusion that at least three subspecies of gray squirrels were used as source populations.

Currently, eastern gray squirrels occur in the San Francisco Bay area, especially the San Francisco Peninsula, extending as far south as Santa Cruz

County. They are also established in some urban parks of the Central Valley and along the Calavaras River. They are restricted to heavily wooded areas and urban parks. Very little interaction with the native western gray squirrel (*S. griseus*) seems to have occurred, but where the eastern gray squirrel co-exists with the much larger fox squirrel (*S. niger*), the latter has been displaced.

Sciurus niger *(fox squirrel)*

Documentation of early introductions of the fox squirrel is more adequate than for the eastern gray. The fox squirrel is known to have been introduced to the San Fernando Valley of southern California before 1904 and to the Fresno area about 1900. The earliest known specimens, however, are dated 1921 from the San Francisco Peninsula. All Californian populations of this species can be assigned to the widespread subspecies *rufiventer* (Byrne, 1979). Current distribution includes most of coastal California north to Mendocino County, and those parts of the Central Valley where there are orchards and/or riparian habitats. It also occurs in some urban parks where *S. carolinensis* is absent.

Fox squirrels occur in more open sites than do eastern gray squirrels and are common in many suburban localities; they have become pests in orchards and walnut groves. In some places, such as in a few riparian sites and in the coast range on the eastern side of San Francisco Bay, fox squirrels have apparently displaced or prevented the expansion of western gray squirrels (Byrne, 1979).

Ondatra zibethica *(muskrat)*

This mammal provides the second case of a species, native to California and inhabiting a very restricted area, that has been subsequently introduced widely. The original distribution of the muskrat in California (Grinnell *et al.*, 1937) involved a population living along the Colorado River (*O. z. bernardi*) and another series of populations extending from northern Nevada into the north-easternmost fringes of the state (*O. z. mergens*).

Subsequent enlargement of the muskrat's range in California has resulted both from expansion of native populations and from introductions from elsewhere. The former has occurred in Yuma County where Colorado River populations have expanded into the Imperial Valley along with extensive irrigation development (Grinnell, 1914). Numerous introductions were made elsewhere within California starting in the early 1920s, especially to the Central Valley and various coastal counties. The sources of these introductions are for the most part unknown, but undoubtedly most came from various parts of eastern and central North America. In one case planted stock is known to have come from Lassen County within the native range of *O. z. mergens* (Dixon, 1929). Muskrats now occur throughout northern California in aquatic habitats, and in isolated patches in southern and coastal regions. By 1960, an estimated 100000 were trapped annually for their pelts (Seymour, 1960). In many areas muskrats are considered a pest because their burrows damage levee banks and irrigation channels (Grinnell, 1914; Grinnell *et al.*, 1937).

Mus musculus *(house mouse)*

This European importation occurs ubiquitously in California. Mostly it is commensal with humans, but feral populations exist in agricultural areas, annual grasslands, grass-brush mixtures and in saltmarshes. When the house mouse occurs with California voles (*Microtus californicus*), the former becomes a fugitive species (DeLong, 1966; Lidicker, 1966). House mice have probably been in California for more than 200 years.

Rattus rattus *(roof rat) and* R. norvegicus *(Norway rat)*

These two species of rats arrived from Europe certainly prior to 1856 (Grinnell, 1933) and possibly much earlier than that. They both occur in major urban centres and in the Central Valley region. Feral populations of the roof rat occur in riparian situations extending eastward into the Sierra Nevadan foothills. Norway rats occur ferally in rice fields and throughout the Sacramento River delta. In the San Francisco Bay area, they occur in saltmarshes, along dykes and on many islands such as Brooks Island (Lidicker, 1973).

Sus scrofa *(wild pig)*

Domestic pigs first arrived in California with the Spanish in 1769 (Barrett, 1977) and feral populations no doubt originated shortly thereafter. After 1850, domestic pigs were frequently released by ranchers to forage in oak woodlands (Shaw, 1940), and most wild pig populations in California are descended from these free-ranging swine (Barrett, 1978). In 1925, European wild pigs were introduced to Carmel Valley in Monterey County (Barrett, 1977). From there they have spread south through the Santa Lucia Mountains, and stock from this source has been re-introduced elsewhere in California. As a consequence of this history of introductions, wild pig populations in California vary greatly in the extent to which European wild boars are mixed with domestic pigs; the majority, however, is pure domestic stock (R. H. Barrett, personal communication). The least amount of domestic admixture occurs in populations from Monterey and San Benito counties.

Wild pigs are now common over extensive areas of California, particularly along the Sierra Nevadan foothills and in the Coast Ranges. The species became classified as a State game species in 1957. By 1975, the estimated annual kill by hunters exceeded 16700 pigs (Barrett, 1977). Currently, pigs have a major negative impact on crops, rangelands and native biota. California presently faces a difficult management problem because of wild pig populations.

Dama dama *(fallow deer) and* Axis axis *(axis deer)*

These two species are treated together because their histories in California are almost identical. The fallow deer is originally from Europe and the axis deer is a native of India. Both were introduced to the Point Reyes area in the 1940s by a private citizen who obtained surplus animals from the San Francisco Zoo (Wehausen & Elliott, 1982). Populations have persisted and grown, and are now managed by the staff of the Point Reyes National Seashore within whose boundaries the species are largely confined. Their demography and food habits have been studied by Wehausen & Elliott (1982) and by Elliott & Barrett (1985).

Cervus elaphus *(elk or wapiti)*

Elk are native to California with two subspecies formerly occupying large parts of the state. The Roosevelt elk (*C. e. roosevelti*) occurred over the northern coastal zone, including the Trinity Mountains. The tule elk (*C. e. nannodes*) was distributed throughout the Central Valley and the adjacent inner coast ranges (Dasmann, 1965). By 1900, the species was nearly extirpated from California. Some recovery has occurred subsequently and many transplantations have been made. In addition, Rocky Mountain elk (*C. e. nelsoni*) have been introduced into California and now persist at three foci (Dasmann, 1965; R. H. Barrett, personal communication; D. R. McCullough, personal communication): the Hearst Ranch area of San Luis Obispo County and adjacent Monterey County, the Shasta Lake area (Shasta County), and the Tejon Ranch (Kern County). A privately managed herd occurs on Santa Barbara Island.

Hemitragus jemlahicus *(Himalayan tahr)*

This Asian relative of the goat is listed by Williams (1979) as established in California. It occurs only in the vicinity of the Hearst Ranch in San Luis Obispo County from which it was released (Barrett, 1966).

Ammotragus lervia *(Barbary sheep)*

This hardy North African species became feral about 1953 in the vicinity of the Hearst Ranch, San Luis Obispo County. It has been studied by Barrett (1980) and appears to be established successfully there with the potential to displace native species (Barrett, 1967).

Discussion

Fifteen introduced species represent 6.9 per cent of the 216 species of mammals recorded for California (Williams, 1979; Mooney *et al.*, 1986). Since three of these mammals are Californian natives, however, the percentage of introduced species is reduced to 5.5 per cent. Of these 15, six have very restricted local distributions (European rabbit and five ungulates). Of the remaining nine species, five are largely commensal or suburban in distribution (viz. Virginia opossum, eastern gray squirrel, house mouse, Norway rat, roof rat), although all have invaded non-commensal habitats to some extent. The remaining four are widespread and are expanding their distributions (fox squirrel, muskrat, red fox, wild pig) and, except for the red fox, already have become problems in various ways. In the cases of the muskrat and wild pig, economic losses are balanced to some extent by their value as fur-bearers and objects of sport hunting respectively.

To this list of 15 species, Williams (1979) has added two introduced species which are questionably established in California. These additions are the nutria (*Myocaster coypus*), a rodent native to South America, and the sambar deer (*Cervus unicolor*), originally from India and now at large in the Hearst Ranch area. In a report on the southern Californian islands, von Bloeker (1967) listed two other introduced species not yet mentioned, viz. the European hare (*Lepus europaeus*) present on East Anacapa Island, and the snow deer (*Capreolus pygargus*) on Santa Rosa Island. Their current status is unknown. (A third species, the American bison (*Bison bison*) definitely occurs on Santa Catalina Island, but is not included here because it is a fenced and managed population.)

Finally, we must add at least seven species of domestic mammals that have established feral populations in California. These include two equids, the horse (*Equus caballus*) and the burro or donkey (*E. asinus*). They are restricted to a few islands and to desert mountain ranges where they do considerable damage to the vegetation, and possibly compete with native species such as bighorn sheep (*Ovis canadensis*). Of the two equids, the burro has been more successful, with numbers estimated at 2000 to 5500 almost thirty years ago (McKnight, 1961). Feral cattle (*Bos taurus*), sheep (*Ovis aries*) and goats (*Capra hircus*) do not persist in the presence of large predators (Mooney *et al.*, 1986). Consequently, they survive in the feral state mainly on islands (McKnight, 1961). Goats are notorious for their habitat destruction on the Channel Islands where large feral populations occur on San Clemente, Santa Cruz, and Santa Catalina (McKnight, 1961; von Bloecker, 1967; Coblentz, 1980). Sheep also can have major impacts on native vegetation (Hobbs, 1980). Lastly, there are two domestic carnivores which are increasingly important as feral species, namely domestic cats (*Felis catus*) and dogs (*Canis familiaris*). Dog packs occur widely, menacing wildlife

(McKnight, 1961; Mooney *et al.*, 1986) and even humans. Less conspicuous, but more insidious, is the impact of feral cats. Pearson (1985) has documented the major effects this species can have on native rodent populations in the San Francisco Bay region.

Adding the two questionably established species (Williams, 1979), the two of unknown status on the Channel Islands (von Bloecker, 1967) and the seven domestics to the twelve non-natives in the original list gives a total of 23 species. This value is 10.5 per cent of a total of 220 species (Williams, 1979, does not include domestic dogs and cats, European hares, or snow deer in his list). Including the red fox, muskrat and elk in the total would raise the percentage value to 11.8

It remains to consider whether this review of introduced mammals in California can add any insights to the important question of what makes a successful coloniser (Ehrlich, 1986; Mooney & Drake, 1986). Certainly, the most successful and widespread species are omnivorous and have impressive fecundities (opossum, house mouse, Norway and roof rats, red fox, wild pig). Some of these and others (tree squirrels) have profited directly from suburban and agricultural developments. It is clear, however, that disturbed habitats are not always a key factor, as shown by the cases of the red fox, muskrat and the six ungulates. Three of the most successful species (red fox, fox squirrel and wild pig) inhabit a variety of habitats. Communities in regions of mediterranean climate in California are significantly involved with introduced mammals mainly because these areas are so often disturbed. No other particular relationship seems to exist, however, between mediterranean climates and success of introduced mammals. Three species are occasionally found in chaparral, but none is restricted to it. No introduced species (with the possible exception of the Rocky Mountain elk) inhabits heavily forested regions.

It is not apparent that large size is an important factor in the success of any of the cases discussed here (see Ehrlich, 1986). Continent of origin also does not seem to offer predictive value for colonising success. Dispersal ability is, however, clearly important, as all the widespread introduced species are known to be good dispersers. Among the two introduced deer on the Point Reyes peninsula, the fallow deer is the better disperser and has achieved a more extensive range than has the axis deer (Wehausen & Elliott, 1982).

The two species that are native Californians with greatly expanded current distributions are particularly informative as to what leads to success. The best guess for the muskrat is that desert and mountain barriers prevented access to central and coastal parts of California. Once these barriers were broached, rapid colonisation followed. On the other hand, native red fox could easily have dispersed out of the mountains into the Central Valley. But it took a different genetic stock (from eastern North America) to become successful at the lower elevations. The inevitable conclusion from this survey of the introduced mammals in California is that whilst factors such as omnivory, high fecundity, use of disturbed habitats, dispersal, etc., all contribute to the probability of successful colonisation, they are not sufficient for success. Chance, the availability of 'open niches', and non-generalisable aspects of viability (i.e. 'toughness') may make the critical difference in successful colonisations.

Acknowledgements. For advice and critical discussion I thank R. H. Barrett, D. R. McCullough and D. F. Williams.

References

Barrett, R. H. (1966). History and status of introduced ungulates on Rancho Piedra Blanca. M.S. thesis, University of Michigan.

Barrett, R. H. (1967). Some comparisons between the Barbary sheep and the desert bighorn. *Transactions of the Desert Bighorn Council*, **11**, 16–26.

Barrett, R. H. (1977). Wild pigs in California. In *Research and Management of Wild Hog Populations. Proceedings of a Symposium*, ed. G. W. Wood, pp. 111–13. Georgetown, South Carolina: B. W. Baruch Forest Science Institute, Clemson University.

Barrett, R. H. (1978). The feral hog on the Dye Creek Ranch, California. *Hilgardia*, **46**, 283–355.

Barrett, R. H. (1980). History of the Hearst Ranch Barbary sheep herd. In *Symposium on the Ecology and Management of Barbary Sheep*, ed. C. D. Simpson, pp. 46–50. Lubbock: Texas Tech University.

Byrne, S. (1979). The distribution and ecology of the non-native tree squirrels *Sciurus carolinensis* and *Sciurus niger* in northern California. Ph.D. thesis, University of California, Berkeley.

Coblentz, B. E. (1980). Effects of feral goats on the Santa Catalina Island ecosystem. In *The California Islands: Proceedings of a Multi-disciplinary Symposium*, ed. D. M. Power, pp. 167–70. Santa Barbara: Santa Barbara Museum of Natural History.

Dasmann, W. P. (1965). *Big Game of California*. Sacramento: California Department of Fish & Game.

DeLong, K. T. (1966). Population ecology of feral house mice: interference by *Microtus*. *Ecology*, **47**, 481–4.

Dixon, J. (1929). Artificial distribution of fur-bearing mammals. *Journal of Mammalogy*, **10**, 358–9.

Ehrlich, P. R. (1986). Which animal will invade? In *Ecology of Biological Invasions of North America and Hawaii*, ed. H. A. Mooney & J. A. Drake, pp. 79–95. New York: Springer-Verlag.

Elliott, H. W. III & Barrett, R. H. (1985). Dietary overlap among axis, fallow, and black-tailed deer and cattle. *Journal of Range Management*, **38**, 435–9.

Gray, R. L. (1977). Extension of red fox distribution in California. *California Fish & Game*, **63**, 58.

Grinnell, J. (1914). An account of the mammals and birds of the lower Colorado River valley. *University of California Publications in Zoology*, **12**, 51–294.

Grinnell, J. (1915). The Tennessee possum has arrived in California. *California Fish & Game*, 1, 114–16.

Grinnell, J. (1933). Review of the Recent mammal fauna of California. *University of California Publications in Zoology*, **40**, 71–234.

Grinnell, J., Dixon, J. S. & Linsdale, J. M. (1937). *Fur-bearing Mammals of California*. Berkeley: University of California Press.

Hobbs, E. (1980). Effects of grazing on the northern population of *Pinus muricata* on Santa Cruz Island, California. In *The California Islands: Proceedings of a Multi-disciplinary Symposium*, ed. D. M. Power, pp. 159–65. Santa Barbara: Santa Barbara Museum of Natural History.

Lidicker, W. Z. Jr (1966). Ecological observations on a feral house mouse population declining to extinction. *Ecological Monographs*, **36**, 27–50.

Lidicker, W. Z. Jr (1973). Regulation of numbers in an island population of the California vole, a problem in community dynamics. *Ecological Monographs*, **43**, 271–302.

McKnight, T. (1961). A survey of feral livestock in California. *Association of Pacific Coast Geographers Yearbook*, **23**, 28–42.

Mooney, H. A. & Drake, J. A. (ed.) (1986). *Ecology of Biological Invasions of North America and Hawaii*. New York: Springer-Verlag.

Mooney, H. A., Hamburg, S. P. & Drake, J. A. (1986). The invasions of plants and animals into California. In *Ecology of Biological Invasions of North America and Hawaii*, ed. H. A. Mooney & J. A. Drake, pp. 250–72. New York: Springer-Verlag.

Pearson, O. P. (1985). Predation. In *Biology of New World* Microtus, ed. R. H. Tamarin, pp. 535–66. *American Society of Mammalogy Special Publication* No. 8.

Seymour, G. (1960). *Furbearers of California*. Sacramento: California Department of Fish & Game.

Shaw, E. B. (1940). Geography and mast feeding. *Economic Geography*, **16**, 233–49.

Von Bloecker, J. C. Jr (1967). Land mammals of the southern California islands. In *Proceedings of the Symposium on the Biology of the California Islands*, ed. R. N. Philbrick, pp. 245–63. Santa Barbara: Santa Barbara Botanic Garden.

Wehausen, J. D. & Elliott, H. W. III (1982). Range relationships and demography of Fallow and Axis deer on Point Reyes National Seashore. *California Fish & Game*, **68**, 132–45.

Williams, D. F. (1979). Checklist of California mammals. *Annals of the Carnegie Museum*, **48** (23), 425–33.

19

Ecology of a successful invader: the European rabbit in central Chile

F. M. JAKSIC & E. R. FUENTES

The ecology of the European rabbit (*Oryctolagus cuniculus*), introduced to different parts of the world, is relatively well known (see Lockley, 1964, for Great Britain; Myers, 1970, for Australia; Gibb *et al.*, 1978, for New Zealand; Rogers, 1979, 1981, for France). Comparatively little, however, is known about the ecology of rabbits in their native lands (but see Soriguer, 1979, for Spain) or in those parts of South America where it has become established, as, for instance, in the mediterranean-climate region of central Chile.

That European rabbits were not present in the central Chilean matorral prior to 1845 is evident from Gay (1847) who noted the climatic and physiognomic similarities between southern Spain and central Chile. Gay (1847) enthusiastically recommended introduction of the rabbit. European rabbits were introduced eventually and became established in central Chile about the turn of this century. Whether they were purposely released or escaped to the wild from cages is not known. Albert (1902) listed the species of native and introduced animals in Chile in 1900 and rabbits were not known from central Chile at that time. Although European hares (*Lepus capensis*) were listed by Albert, rabbits were not; in fact, rabbits were apparently not abundant in central Chile as late as 1940 (Osgood, 1943). By the early 1960s, however, rabbits were already considered a pest (Greer, 1965). Reportedly, they raided agricultural plots, forestry plantations and grazing lands, thus interfering with important human concerns (Ferriere *et al.*, 1983). From their unknown release and/or establishment locations, probably around Santiago (33° 28' S), rabbits expanded their distribution both to the north and south of the country (Pefaur *et al.*, 1968; Pefaur, 1969). To the north, their expansion has so far reached the Limari valley, near Ovalle (30° 37' S), but not the Elqui valley, near La Serena (29° 53' S) (Fuentes & Campusano, 1985). To the south, their distribution apparently continues to expand, as evergreen hygrophyllous forests are cleared for agriculture. So far, the southernmost limit seems to be around Valdivia or Osorno (*c*. 40° 30' S). Chilean populations of rabbits crossed from the Chillán eastern mountains (*c*. 36° 25' S) into neighbouring Argentina between 1945 and 1950 (Howard & Amaya, 1975) and have since expanded both north and south of Neuquen province (Bonino & Amaya, 1984).

European rabbits were also introduced to Tierra del Fuego Island (*c*. 54° S) at the southernmost tip of South America. An account of the course of their introduction can be found in Jaksic & Yanez (1983). The stock involved in this invasion seems to be different. Whereas central Chilean rabbits were derived from Spanish stocks (Housse, 1953), those in Tierra del Fuego were brought from France by way of the Malvinas (Falkland) Islands (Jaksic & Yanez, 1983). A vast biogeographical experiment is now underway, as the southern (French) stock of rabbits is expanding its distribution northward in Argentina, while the central Chilean stock expands southward (see Bonino & Amaya, 1984).

Interesting as the above situation may be, there is little documentation available on the ecology of European rabbits in Tierra del Fuego and continental surroundings except for Jaksic & Yanez (1983), Amaya & Bonino (1980) and Bonino & Amaya (1984). In comparison, much is known of the ecology of central Chilean populations. This chapter will synthesise all the information available and will attempt to answer the question of why did rabbits become a pest in central Chile, whereas in their native lands of southern Spain they have never attained pest status. What may be so different between Spain and Chile that caused the originally innocuous rabbits to become such invasive creatures in their new home? An answer to these questions clearly demands a comparative approach to the ecology of rabbits in their native and colonised lands.

The rabbit as a native: the southern Spanish case

Rabbits in Spain have never been considered a pest, but rather a reliable commodity by small game hunters, fur traders and peasants. Spaniards actually worry because rabbits have not reached their historic levels of abundance prior to myxomatosis (Soriguer, 1979, 1980*b*). Indeed, this viral disease was never introduced deliberately to Spain to control rabbits, but came with infected rabbits from southern France (Valverde, 1967; Soriguer, 1979).

Soriguer (1979) has produced the most comprehensive study so far on the ecology of rabbits in south-western Spain. Although he did not consider central Chile, he did compare his findings with those of other researchers in different parts of the world where rabbits had been introduced. Soriguer (1979) found that the fecundity of Spanish rabbits was the lowest reported; each female produces only between 11 and 16 newborns yearly. In addition, the mortality caused by both myxomatosis and predation was the highest reported for the species. Soriguer wondered how rabbits could survive under such conditions and still be able to maintain population densities of between 0.8 and 9.7 individuals per hectare (data from two years at a single site). According to Soriguer (1979), the answer lay in the Spanish rabbit's early sexual maturity (between 3.5 and 4 months old), reached at a very small body size (mean = 900 g). These values were the lowest ever reported for any rabbit population in the world (Soriguer, 1979, 1980*a*, 1981).

Soriguer (1979) proposed that the short life expectancy of rabbits in Spain has two main causes. One is that myxomatosis became an endemic disease that operates in a density-dependent way. When rabbit populations increase, and start depressing their food resources, an increasingly important segment of the population becomes poor in condition, and weakened rabbits become more susceptible to the disease. Further, rabbits being social organisms, the frequency of body contacts increases in dense populations, and this accelerates the rate of transmission of myxomatosis (Soriguer, 1980*b*). The second cause is that Spanish rabbits are the staple prey of a horde of predators, including diurnal and nocturnal birds of prey, mammalian carnivores, and snakes, among others (Soriguer, 1979; Delibes & Hiraldo, 1981; Jaksic & Delibes, 1987). No fewer than 30 species of Falconiformes, Strigiformes, Carnivora and Serpentes have been documented to prey on rabbits (Jaksic & Soriguer, 1981). Rabbits in Spain somehow reduce their predation risks through behavioural mechanisms, including habitat and microhabitat selection. On the broader scale, Spanish rabbits have been shown to select dense scrub (for shelter) with adjacent grassland for grazing (Rogers & Myers, 1979, 1980; Soriguer & Rogers, 1981). Rabbits rarely venture into the open and restrict their foraging to the vicinity of sheltering structures such as burrows, rock outcrops and shrubs. Apparently, Spanish rabbits have to compromise forage needs with predation risks. Open pastures are plentiful but risky places, whereas in the scrub there is not much to eat but at least it offers a safe place.

In summary, rabbits in Spain are kept in check directly by myxomatosis and predation, and indirectly by the limited food available in safe places.

The rabbit as an invader: the central Chilean case

In contrast to the demographic characteristics reported by Soriguer (1979) for rabbits in Spain, rabbits in Chile have two distinguishing characteristics (Zunino & Vivar, 1985): females produce on the average 19 newborns per year; and also, sexual maturity is reached between 4 and 6 months old, with a mean weight of 1195 g. In their study area near Valparaiso (33° 11' S), Zunino & Vivar (1983) detected densities between 2.3 and 25.6 rabbits per hectare over the year. In the Juan Fernandez Archipelago (*c.* 30° 37' S) Saiz *et al.* (1982) documented mean densities of between 15.3 and 51.6 rabbits per hectare in different parts of two islands. It appears, then, that rabbits in Chile produce more newborns per female, mature later, at a heavier weight, and reach higher densities than in Spain. Further, Chilean rabbits attain larger sizes; among 520 rabbits examined by Soriguer (1979) at his Spanish study site, only 11 per cent exceeded 1200 g, whereas among 735 rabbits examined by Zunino & Vivar (1985) at their Chilean study site, 43 per cent exceeded that weight. Apparently, then, the life expectancy of rabbits in Chile is greater than in Spain.

Three main causes may be invoked for the increased life expectancy of rabbits in central Chile. First of all myxomatosis is nonexistent. Density-dependent population regulation seems to be effected by emigration from densely populated areas. Although rabbits are considered slow dispersers, empirical information from Argentina (Howard & Amaya, 1975) shows that they may spread at a mean rate of 8 linear kilometres per year, and at maximum rates of 16 km per year. Secondly, predation upon rabbits in central Chile is very low, a point that will be elaborated further below. Thirdly, low predation has resulted in a habitat shift: instead of choosing dense scrub, and the vicinity of sheltering areas, rabbits in Chile are found in open scrub, foraging away from potential refuges. This point will also be enlarged on below.

Low levels of predation upon central Chilean rabbits have been documented repeatedly (Jaksic *et al.*, 1979*a, b*; Jaksic & Yanez, 1980; Jaksic & Soriguer, 1981; Simonetti & Fuentes, 1982; Jaksic & Ostfeld, 1983; Jaksic, 1986). The information available indicates that native predators consume mainly native small mammals. In particular, the caviomorph rodent *Octodon degus* has been shown to be the staple prey of most central Chilean predators (Jaksic *et al.*, 1981; Jaksic, 1986; Jaksic & Delibes, 1987), whereas rabbits are either not preyed upon or preyed on only among their juvenile ranks (Simonetti & Fuentes, 1982). Although central Chilean predators have been claimed to be generalists/opportunists (Jaksic *et al.*, 1981; Jaksic, 1986; Jaksic & Delibes, 1987), preying approximately on the rank order that prey abundances are available, the fact that the abundant rabbits are scarcely preyed upon poses an interesting question.

Jaksic *et al.* (1979*a*) proposed that a low level of predation upon rabbits was the consequence of a lack of adaptation of local predators to hunt for a recently introduced prey. Indeed, the escape response of rabbits (zig-zag sprints, backward dashes, and leaps, all combined) may prove bewildering to native predators 'accustomed' to the simpler escape behaviour (straight dash to shelter) of native rodents (Jaksic, 1986). It could be argued, however, that central Chilean native mammal prey are smaller on the average than introduced rabbits; adult *Octodon degus* weigh about 230 g, five to six times less than an adult rabbit (1300 g). Perhaps rabbits escape 'by size' (Jaksic, 1986). Indeed, that juvenile (small) rabbits are more preyed upon than adults (Simonetti & Fuentes, 1982) seems to support this contention.

Two lines of evidence indicate that escape by size, however, is unlikely. Firstly, Jaksic & Soriguer (1981) and Jaksic & Ostfeld (1983) addressed this point specifically by statistically comparing the body sizes of predators in both Chile and Spain in relation to rabbit consumption. These authors found that despite the fact that related and similarly-sized predators were found in these two mediterranean-type regions, rabbit consumption was three to four times higher in Spain. Secondly, Fuentes & Simonetti (1982) found that the rabbit-sized vizcacha (*Lagidium viscacia*) restricts its activities to the vicinity of sheltering rock-outcrops, and that despite its relatively large size (mean adult weight = 1600 g) it falls prey to local predators. Fuentes & Simonetti (1982) consequently claimed that the vizcacha's restricted habitat use is a response for reducing predation risks.

It appears, then, that central Chilean rabbits fall within the handling capabilities of native predators but the latter nonetheless disregard this introduced prey in favour of native prey. Indeed, it may not be necessary to invoke insufficient time for co-adaptive adjustments between native predators and introduced rabbits (Fuentes & Jaksic, 1980). Simple optimal-foraging arguments may depict central Chilean predators as making an optimal choice between the large (energetically profitable) but difficult to catch (energetically expensive, coupled with low capture success) rabbits, and the smaller but easier to catch native rodents (Jaksic, 1986).

Whatever the ultimate cause of the relatively low predation upon rabbits in Chile, it has resulted in a clear behavioural shift in habitat selection, both at the macro- and micro-habitat levels. Whereas in Spain rabbits concentrate their foraging activities in the vicinity of potential shelters (shrubs, burrows), much as do central Chilean *Octodon degus* and Californian rabbits (Bartholomew, 1970; Jaksic *et al.*, 1979*b*; Jaksic & Ostfeld, 1983), introduced rabbits in Chile behave otherwise. They have been shown to use the open areas between shrubs more than the protected areas under shrubs (Jaksic *et al.*, 1979*b*). Further, Simonetti & Fuentes (1982) have shown that there is an ontogenetic process leading to this behaviour. Juvenile rabbits do concentrate their activities close to sheltering places, but as they grow and become adults, they extend their activity onto open places. This process is in line with the relatively higher predation reported on juveniles, and the relative immunity of adults, thereby suggesting a causal link between predation pressure and habitat choice.

On a broader scale, Jaksic & Soriguer (1981) showed that rabbits in Spain choose dense scrub, with a peak of activity and/or abundance at about 75 per cent shrub cover. Rabbits in Chile, however, are more active/abundant at about 25 per cent shrub cover. This difference in habitat selection may be explained as a compromise between predation risk and food acquisition. Certainly, less shrub cover means more pasture for rabbits, but at least in Spain, less cover also means less protection against predators. In Chile, however, less cover means more forage for the rabbits, without an increase in predation risk comparable to that in Spain. This may explain their use of open scrub, away from shelters, their larger body size, increased life expectancy and, ultimately, their status as an invasive animal in Chile.

Humans are not without blame in transforming a supposedly innocuous animal into a pest. Inefficient as central Chilean predators may be, they help to remove some fraction of the rabbit population. But humans have persecuted predators either because they reputedly eat poultry (most of the larger-sized hawks, see Jaksic & Jimenez, 1986), because their pelt has commercial value (the cats and foxes, see Iriarte & Jaksic, 1986) or out of sheer ignorance (owls are considered birds of ill-omen, see Jaksic & Jimenez, 1986). As a result, predators in central Chile are considerably less abundant than their relatives in the similar mediterranean region of California (Miller, 1980). If Chilean predators were more abundant, perhaps rabbits would not venture so far into the open areas. Humans have also created more favourable habitat for rabbits. Through firewood extraction and woodland clearing for agriculture and livestock raising, humans have opened the originally dense matorral, thereby allowing entrance and establishment of rabbits. Indeed, Simonetti (1983) has shown that goats interfere with native rodents but not with rabbits, actually favouring the latter. Another factor is that roads are favoured corridors for dispersing rabbits (Howard & Amaya, 1975).

In summary, low predation, habitat shift and increased forage seem to constitute the three main causes of the status of rabbits as pests in central Chile.

Ecological consequences of the rabbit invasion

The economic damage produced by introduced rabbits is well known. Both in Chile and Argentina they interfere with livestock raising (cattle and sheep) by consuming pasture (Amaya & Bonino, 1980; Ferriere *et al.*, 1983; Bonino & Amaya, 1984). Their raids on horticultural plots are legendary (e.g. see Housse, 1953; Greer, 1965) and their killing of pine seedlings strongly affects forestry activities (Pine *et al.*, 1979; Ferriere *et al.*, 1983).

The ecological consequences of the introduction of rabbits are less well known (Fuentes, 1981; Fuentes & Simonetti, 1982) but some striking effects have been documented. Jaksic & Fuentes (1980) showed experimentally that rabbits are an important agent accounting for the restriction of native perennial herbs to areas underneath the canopy of sheltering shrubs. According to these authors, the finding of Keeley & Johnson (1977) that mediterranean Chile and California differ in the spatial distribution of native herbs owes more to the grazing activities of rabbits than to microclimatic differences. The important implication here is that before the introduction of rabbits, the herb layers in Chile and California may have been more similar or 'convergent' than nowadays.

Rabbits are also avid browsers when herbs disappear. Fuentes *et al.* (1983) documented experimentally that rabbits are not only more destructive of shrub seedlings than native rodents but that they also are more lethal. Given the rabbits' extensive foraging ranges, they are able to locate and readily browse to death seedlings of several native shrub species. Fuentes *et al.* (1983) speculated that rabbits may be arresting the process of secondary succession in the Chilean matorral and, as a long-term consequence, may be further opening the already sparse matorral.

Nothing is known about interactions between rabbits and the native fauna. As stated above, rabbits have not been an important addition to the prey base of native predators. On the other hand, they do not seem to compete for forage with folivorous rodents, given their complementary spatial distributions (between-shrubs compared to under-shrubs). Although this complementarity may be thought of as competition-induced habitat partitioning, the fact is that native small mammals are always in the scrub, and under shrubs, regardless of the presence of rabbits, and vice versa. Rabbits may interfere with rodents that select dense scrub, however, provided that the long-term effect of rabbits may be the opening up of those areas. But at the same time they may favour other folivorous rodents such as *Octodon degus* and *Abrocoma bennetti*, which benefit from the existence of more 'edges' between matorral and pasture formations. Rabbits are known to colonise and enlarge burrow systems of native animals (Housse, 1953; Howard & Amaya, 1975), probably evicting them. But at the same time they favour other native animals that require large burrows as enlarged by rabbits, e.g. the strigid owl *Athene cunicularia* (Housse, 1953).

In summary, the ecological consequences of the rabbit invasion have so far been explored with reference to the vegetation and they are striking. Research is badly needed with reference to interactions between rabbits and native faunal elements.

Concluding remarks

Gay's (1847) recommendation of introducing Spanish rabbits to central Chile was not good advice; rabbits have proved to be disastrous both for the economy and the ecology of the country. In Gay's defence it should be stated that in those days there was no way to predict how the rabbits and their reputed predators were going to behave. In a more positive vein, the introduction of rabbits in central

Chile could be interpreted as a 'natural experiment' that has allowed a better understanding of how native predators interact with both native and introduced small mammals, and what are some of the outcomes of this interaction. The experiment is still underway and scientific advantage should be taken of this unique opportunity: will the predators eventually 'learn' to hunt efficiently for rabbits, and how will the rabbits react? The current level of knowledge does not allow even the crudest extrapolation. Only one thing is certain and should be taken as well-meaning advice: don't introduce rabbits if you don't have them!

Acknowledgements. The authors' research reported in this chapter has been funded at several stages by grants from the Man and the Biosphere Program of UNESCO, the National Science Foundation of the United States of America, the Direccion de Investigacion de la P. Universidad Catolica of Chile and the Fondo Nacional de Ciencia y Tecnologia of Chile. The most recent support has come from grants DIUC 093/87 and 094/87 and FONDECYT 1161 and 748.

References

Albert, F. (1902). *La Caza en El Pais*. Seccion de Ensayos Zoolojicos i Botanicos, Ministerio de Industria, Santiago, pp. 1–13.

Amaya, J. N. & Bonino, N. A. (1980). El conejo silvestre europeo (*Oryctolagus cuniculus*) en Tierra del Fuego. *Instituto Nacional de Tecnologia Agropecuaria, Buenos Aires, IDIA*, **387–8**, 14–33.

Bartholomew, B. (1970). Bare zone between California shrub and grassland communities: the role of animals. *Science*, **170**, 1210–12.

Bonino, N. A. & Amaya, J. N. (1984). Distribucion geografica, perjuicios y control del conejo silvestre europeo *Oryctolagus cuniculus* (L.) en la Republica Argentina. *Instituto Nacional de Tecnologia Agropecuaria, Buenos Aires, IDIA*, **429–32**, 25–50.

Delibes, M. & Hiraldo, F. (1981). The rabbit as prey in the Iberian Mediterranean ecosystem. In *Proceedings of the First World Lagomorph Conference, Guelph, Ontario*, ed. K. Myers & C. D. MacInnes, pp. 614–22. Guelph: University of Guelph.

Ferriere, G., Cerda, J. & Roach, R. (1983). El conejo silvestre en Chile. *Corporacion Nacional Forestal (Chile), Boletin Tecnico*, **8**, 1–35.

Fuentes, E.(1981). El matorral en perspectiva biologica. *Monografias Biologicas (Universidad Catolica de Chile)*, **1**, 27–43.

Fuentes, E. R. & Campusano, C. (1985). Pest outbreaks and rainfall in the semi-arid region of Chile. *Journal of Arid Environments*, **8**, 67–72.

Fuentes, E. R. & Jaksic, F. (1980). Consideraciones teoricas para el control biologico del conejo europeo en Chile central. *Medio Ambiente (Chile)*, **4**, 45–9.

Fuentes, E. R. & Simonetti, J. A. (1982). Plant patterning in the Chilean matorral: are the roles of native and exotic mammals different? In *Proceedings of the Symposium on Dynamics and Management of Mediterranean-type Ecosystems*, ed. C. E. Conrad & W. C. Oechel, pp. 227–33. USDA Forest Service, Pacific

Southwest Forest and Range Experiment Station, Berkeley, California, General Technical Report PSW-58.

Fuentes, E. R., Jaksic, F. M. & Simonetti, J. A. (1983). European rabbits versus native rodents in central Chile: effects on shrub seedlings. *Oecologia*, **58**, 411–14.

Gay, C. (1847). Historia fisica y politica de Chile. *Zoologia*, **1**, 19–182.

Gibb, J. A., Ward, C. P. & Ward, G. D. (1978). Natural control of a population of rabbits, *Oryctolagus cuniculus* (L.) for 10 years in the Kourarau enclosure. *New Zealand DSIR Bulletin*, **223**, 1–89.

Greer, J. K. (1965). Mammals of Malleco Province, Chile. *Michigan State University, Publications of the Museum, Biological Series*, **3**, 49–152.

Housse, R. (1953). *Animales Salvajes de Chile en su Clasificacion Moderna*. Santiago: Ediciones de la Universidad de Chile.

Howard, W. E. & Amaya, J. N. (1975). European rabbit invades western Argentina. *Journal of Wildlife Management*, **39**, 757–61.

Iriarte, J. A. & Jaksic, F. M. (1986). The fur trade in Chile: an overview of seventy-five years of export data (1910–1984). *Biological Conservation*, **38**, 243–53.

Jaksic, F. M. (1986). Predation upon small mammals in shrublands and grasslands of southern South America: ecological correlates and presumable consequences. *Revista Chilena de Historia Natural*, **59**, 209–21.

Jaksic, F. M. & Delibes, M. (1987). A comparative analysis of food-niche relationships and trophic guild structure in two assemblages of vertebrate predators differing in species richness: causes, correlations and consequences. *Oecologia*, **71**, 461–72.

Jaksic, F. M. & Fuentes, E. R. (1980). Why are native herbs in the Chilean matorral more abundant beneath bushes: microclimate or grazing? *Journal of Ecology*, **68**, 665–9.

Jaksic, F. M. & Jimenez, J. E. (1986). The conservation status of raptors in Chile. *Birds of Prey Bulletin*, **3**, 95–104.

Jaksic, F. M. & Ostfeld, R. S. (1983). Numerical and behavioral estimates of predation upon rabbits in mediterranean-type shrublands: a paradoxical case. *Revista Chilena de Historia Natural*, **56**, 39–49.

Jaksic, F. M. & Soriguer, R. C. (1981). Predation upon the European rabbit (*Oryctolagus cuniculus*) in mediterranean habitats of Chile and Spain: a comparative analysis. *Journal of Animal Ecology*, **50**, 269–81.

Jaksic, F. & Yanez, J. (1980). Quien controla las poblaciones de conejos introducidos? *Medio Ambiente (Chile)*, **4**, 41–4.

Jaksic, F. M. & Yanez, J. L. (1983). Rabbit and fox introductions in Tierra del Fuego: history and assessment of the attempts at biological control of the rabbit infestation. *Biological Conservation*, **26**, 367–74.

Jaksic, F. M., Fuentes, E. R. & Yanez, J. L. (1979*a*). Two types of adaptation of vertebrate predators to their prey. *Archivos de Biologia y Medicina Experimentales (Chile)*, **12**, 143–52.

Jaksic, F. M., Fuentes, E. R. & Yanez, J. L. (1979*b*). Spatial distribution of the Old World rabbit (*Oryctolagus cuniculus*) in central Chile. *Journal of Mammalogy*, **60**, 207–9.

Jaksic, F. M., Greene, H. W. & Yanez, J. L. (1981). The guild structure of a community of predatory vertebrates in central Chile. *Oecologia*, **49**, 21–8.

Keeley, S. C. & Johnson, A. W. (1977). A comparison of the pattern of herb and shrub growth in comparable sites in Chile and California. *American Midland Naturalist*, **97**, 120–32.

Lockley, R. M. (1964). *The Private Life of the Rabbit*. New York: Avon Books.

Miller, S. (1980). Human influences on the distribution and abundance of wild Chilean mammals: prehistoric–present. Ph.D. thesis, University of Washington, Seattle.

Myers, K. (1970). The rabbit in Australia. In *Proceedings of the Advanced Study Institute on Dynamics of Numbers in Populations*, ed. P. J. den Boer & G. R. Gradwell, pp. 478–506. Wageningen: Centre for Agricultural Publishing & Documentation.

Osgood, W. H. (1943). The mammals of Chile. *Field Museum of Natural History, Zoological Series*, **30**, 1–268.

Pefaur, J. (1969). Consideraciones sobre el problema de la conservacion de los mamiferos silvestres chilenos. *Boletin de la Universidad de Chile*, **93–4**, 4–10.

Pefaur, J., Hermosilla, W., di Castri, F., Gonzalez, R. & Salinas, F. (1968). Estudio preliminar de mamiferos silvestres chilenos: su distribucion, valor economico e importancia zoonotica. *Revista de la Sociedad de Medicina Veterinaria (Chile)*, **18**, 3–15.

Pine, R. H., Miller, S. D. & Schamberger, M. L. (1979). Contributions to the mammalogy of Chile. *Mammalia*, **43**, 339–76.

Rogers, P. M. (1979). Ecology of the European wild rabbit, *Oryctolagus cuniculus* (L.), in the Camargue, southern France. Ph.D. thesis, University of Guelph, Ontario.

Rogers, P. M. (1981). Ecology of the European wild rabbit, *Oryctolagus cuniculus* (L.) in Mediterranean habitats. II. Distribution in the landscape of the Camargue, S. France. *Journal of Applied Ecology*, **18**, 355–71.

Rogers, P. M. & Myers, K. (1979). Ecology of the European wild rabbit, *Oryctolagus cuniculus* (L.) in Mediterranean habitats. I. Distribution in the landscape of the Coto Doñana, Spain. *Journal of Applied Ecology*, **16**, 691–703.

Rogers, P. M. & Myers, K. (1980). Animal distributions, landscape classification and wildlife management, Coto Doñana, Spain. *Journal of Applied Ecology*, **17**, 545–65.

Saiz, F., De la Hoz, E., Toro, H., Zuniga, L., Vasquez, E., Cossio, F., Leon, J. & Wendt, M. C. (1982). *Proposicion de un Metodo de Control Integrado del Conejo en el Archipielago de Juan Fernandez*. Valparaiso: Edicion de la Universidad Catolica de Valparaiso.

Simonetti, J. A. (1983). Effects of goats upon native rodents and European rabbits in the Chilean matorral. *Revista Chilena de Historia Natural*, **56**, 27–30.

Simonetti, J. A. & Fuentes, E. R. (1982). Microhabitat use by European rabbits (*Oryctolagus cuniculus*) in central Chile: are adult and juvenile patterns the same? *Oecologia*, **54**, 55–7.

Soriguer, R. C. (1979). Biologia y dinamica de una poblacion de conejos (*Oryctolagus cuniculus* L.) en Andalucia occidental. Doctoral Thesis, Universidad de Sevilla.

Soriguer, R. C. (1980*a*). El conejo *Oryctolagus cuniculus* (L.), en Andalucia Occidental: parametros corporales y curva de crecimento. *Doñana Acta Vertebrata*, **7**, 83–90.

Soriguer, R. C. (1980*b*). Mixomatosis en una poblacion de conejos en Andalucia Occidental. Evolucion temporal, epidemia invernal y resistencia genetica. In *Primera Reunion Iberoamericana de Zoologia de Vertebrados, La Rabida*, ed. J. Castroviejo, pp. 241–50. La Rabida: Consejo Superior de Investigaciones Cientificas.

Soriguer, R. C. (1981). Estructura de sexos y edades en una poblacion de conejos (*Oryctolagus cuniculus* L.) de Andalucia Occidental. *Doñana Acta Vertebrata*, **8**, 225–36.

Soriguer, R. C. & Rogers, P. M. (1981). The European wild rabbit in mediterranean Spain. In *Proceedings of the First World Lagomorph Conference, Guelph, Ontario*, ed. K. Myers & C. D. MacInnes, pp. 600–13. Guelph: University of Guelph.

Valverde, J. A. (1967). *Estructura de una Comunidad de Vertebrados Terrestres*. Madrid: Consejo Superior de Investigaciones Cientificas.

Zunino, S. & Vivar, C. (1983). '*Oryctolagus cuniculus*': Uso y control del recurso. Final report (mimeograph) to Corporacion Nacional Forestal, V. Region, Valparaiso, pp. 1–55.

Zunino, S. & Vivar, C. (1985). Ciclo reproductor de los conejos en Chile central. I. Madurez y relacion sexual. *Anales de Museo de Historia Natural de Valparaiso (Chile)*, **16**, 101–10.

20

Mammals introduced to the mediterranean region of South Africa

R. C. BIGALKE & D. PEPLER

In contrast to the marked impact exerted by introduced plants (Wells, this volume), introduced mammals are unimportant in the South African mediterranean zone. This relative unimportance is partly because of historical accident but also because the characteristics of fynbos, the dominant vegetation type in the region, have influenced the establishment of introduced mammals.

Commensal rodents

Commensal rodents were the first comers. The house or black rat, *Rattus rattus*, is known from Iron Age sites in Zambia (AD 1500–1600), the northern Transvaal (about AD 1000) and Natal (8th century) (Avery, 1985). Whether they travelled southward to reach the mediterranean-climate region or whether there was a fresh introduction at the Cape from European ships – which seems likely – is not known. The species is well established as a commensal in all but the drier parts of South Africa, as well as in many other parts of Africa (de Graaf, 1981). Feral populations were reported in forests at Pirie and in Pondoland, eastern Cape Province, by Shortridge (1934). Smithers (1983) stated that where, in restricted local areas, they have established themselves in the field, black rats have done so where there is thick underbrush in forested areas or other types of substantial cover. There are no such records from the mediterranean zone of South Africa. There is no evidence of black rats affecting vegetation or the populations of indigenous birds or mammals, as they have done, for example, in New Zealand, the Galapagos and Hawaii (Lever, 1985). The relative importance of physiological and ecological requirements, competition and predation (many wild predators take the species; see Kingdon, 1974; de Graaf, 1981) in restricting black rats to commensalism is not known.

The brown rat, *Rattus norvegicus*, is confined to coastal ports, larger coastal towns and their immediate vicinity in southern and eastern Africa (Kingdon,

1974; de Graaf, 1981). It was probably introduced by European ships and is known to have been in the Cape Colony by 1832 (Avery, 1985). This rat is reported to inhabit the shoreline south of Durban (Natal) and near Hout Bay (Cape Town) but there is no evidence of interactions with indigenous faunal communities. The distribution of brown rats may be limited by climatic factors. The species is reported to acclimatise less readily than *R. rattus* to hot countries because of its presumed origin in the cold climate of western China (de Graaf, 1981).

The origin and date of arrival of the third commensal rodent, *Mus musculus*, the house mouse, in southern Africa is unknown. It is widespread in the mediterranean zone of South Africa and elsewhere but the only feral populations known occur in Botswana (Smithers, 1971). If the house mouse's failure to become established in the wild is because of competition, as Rosevaar (1969) suggested for West Africa, the Botswana exception would repay further study.

Grey squirrel

The only other introduced rodent present in South Africa is the grey squirrel *Sciurus carolinensis*. It is of particular interest in the context of this volume since it was introduced deliberately to the mediterranean region of South Africa and its subsequent history is fairly well documented. A few pairs were obtained, almost certainly from Britain, by C. J. Rhodes, who released them on his estate 'Groote Schuur', on the outskirts of Cape Town, probably during the 1890s (Lever, 1985). Their subsequent rapid spread was documented by Bigalke (1937) and Davis (1950). Millar (1980) provided the most recent information. Squirrels now occupy patches of suitable habitat within an area of some 7000 km^2. The boundaries of their range extend to about 70 km from Cape Town (Figure 20.1). Extensions outside this area to the environs of the towns of Swellendam (190 km from Cape Town) and Ceres (100 km) are the result of deliberate and misguided introductions in 1957 and 1968 respectively.

Squirrels have not invaded indigenous ecosystems but are confined to urban, agricultural or afforested environments. Lever (1985) summarised Millar's conclusions on their habitat requirements as follows: 'The grey squirrel in South Africa is restricted to areas of mixed (exotic) timber plantations with higher than average rainfall, and to orchards and vineyards; it is generally absent from pastoral land, exotic *Eucalyptus* plantations, bush and veld, and only occurs in urban surroundings with an adequate supply of tree cover and watercourses. The existence of at least one of the three staple seed-bearing food species – *Pinus pinaster*, *P. pinea* and *Quercus robur* – seems essential to its survival' (Lever, 1985, p. 326). Apart from damaging fruit and vegetable crops, squirrels are blamed for preying on eggs and nestlings of indigenous birds, especially in urban gardens. The extent of their influence on bird populations is unknown.

The grey squirrel's failure to establish itself in fynbos after almost a century at the Cape requires explanation. Two possibilities suggest themselves. Fynbos ecosystems either do not provide an arboreal or scansorial squirrel niche or else they do, but *Sciurus carolinensis* is not adapted to fill it. In the second case the absence of indigenous squirrels must be accounted for. To do so, the zoogeography of African Sciuridae must be examined. Arboreal and scansorial squirrels occupy forest, woodland and savanna. They are essentially tropical in origin and their species richness shows a marked latitudinal gradient declining polewards. The southernmost occurrence is that of *Paraxerus palliatus* at 29° S in forests of northern Natal (Bigalke, 1972). Unlike other tropical groups with relict distributions in forest outliers, Sciurids do not extend further down the coast to the eastern or southern Cape Province, as do for example, *Dendrohyrax arboreus*, the tree hyrax, and *Cephalophus monticola*, the blue duiker. The reason may lie in the evolutionary history of the group. Kingdon (1974) pointed out that fossil squirrels are not known until the late Miocene in Africa. As a result of this 'late

Figure 20.1. Present range of grey squirrel (*Sciurus carolinensis*) in the south-western Cape of South Africa.

start' they show less differentiation than Asian squirrels. Their penetration into all potentially suitable habitats may also have been limited as a result. In any event, Sciurids do not appear to have come into contact with fynbos.

Even if they had, the characteristics of the vegetation make it appear unlikely that they would have become established. The upper stratum of mature fynbos is commonly 2–3 m tall and dominated by broad-leaved sclerophyllous shrubs (Kruger, 1979). Fruit- and nut-bearing plants are not common, many species have small seeds and foliage is rich in secondary compounds. Few of even the biggest shrubs provide holes in their trunks which could serve as dens. The fact that *Sciurus carolinensis* appears well adapted to the south-western Cape climate but has not encroached into fynbos suggests that there is no niche available for a squirrel.

Lagomorphs

The South African mainland does not have a problem with rabbits (*Oryctolagus cuniculus*) because of a remarkable set of circumstances. Within a month of the arrival of the Dutch party at the Cape of Good Hope in 1652, the Commander, Jan van Riebeeck, wrote to the Dutch East India Company suggesting that rabbits be introduced as a source of meat (Skead, 1980). In 1654 nine animals were released on Robben Island, a few kilometres offshore in Table Bay. They died out but were supplemented at various times from 1656 to 1658. By 1659 there was a well-established population. They were not released on the mainland because of an express prohibition by the Council of the Dutch East India Company, which feared that the animals would damage gardens and crops.

From Robben Island (507 ha) rabbits were translocated to Dassen Island (222 ha) in the 1660s. The subsequent history of these and other island populations is discussed by Cooper & Brooke (1982) and Lever (1985). Schapen Island (41 ha) has a population stemming from a reintroduction in the present century, long after the previously established population had become extinct. Rabbits are still present on Jutten (46 ha) and Vondeling (21 ha) Islands but have died out on Marcus (11 ha) and Meeu (7 ha). The sparse vegetation and summer drought appear to be the main factors limiting populations, together with predation by feral cats (*Felis catus*) in the case of Dassen Island. In summer, rabbits feed on algae in the intertidal area of several islands, perhaps mainly for their moisture content. Lever's (1985) conclusion that rabbits would more than likely have been successful invaders of the mainland if they had been introduced, is probably correct.

Ungulates

Three ungulate species have been introduced to the South African mediterranean zone. Domestic pigs (*Sus scrofa*) were released in a number of State forests in the

1920s to control the pine tree emperor moth *Imbrasia cytherea*, larvae of which pupate in the soil. Subsequently, a few Austrian and Bavarian wild boars were brought in to improve the stock (Siegfried, 1962; Lever, 1985). Some populations have become extinct but a few have stabilised (Botha, 1985). They utilise pine plantations and raid surrounding farmland where they are shot whenever possible (D. Pepler, personal observation). In some instances they are subject to leopard predation.

Fallow deer, *Dama dama*, were already kept in the park of Newlands House, Cape Town in 1869 (Thompson, 1968, in Lever, 1985). Subsequently, the herd was transferred to the farm 'Lourensford' at Somerset-West. C. J. Rhodes imported some animals in 1897 for his estate 'Groote Schuur' (Siegfried, 1962; Chapman & Chapman, 1980). This enclosed estate, now government property, still supports a large, managed herd, mainly on planted pastures under plantations of introduced oaks and pines. On 'Lourensford' and the surrounding farms, fallow deer are still present but the population appears to be small. They inhabit a mixture of orchards, pastures and thickets of Australian acacias and introduced pines with a small admixture of indigenous fynbos. Since the deer damage fruit trees by browsing and de-barking, some sporadic shooting takes place and leopard predation is another possible limiting factor. The farms are bordered on one side by uninhabited State forest land with typical mountain fynbos but after almost a century, fallow deer do not appear to have occupied this readily available habitat (R. C. Bigalke, personal observation). Another old established population in the mediterranean zone is that on the van der Byl family estates in the district of Bredasdorp. Hills support fynbos but much of the land is taken up by sheep pastures and wheat fields and these may provide most of the habitat utilised. There are no indications of dispersal onto surrounding farms, although the possibility that dispersing animals are shot as soon as they appear, cannot be discounted (R. C. Bigalke, personal observation).

In contrast to pigs and fallow deer, the Himalayan tahr, *Hemitragus jemlahicus*, is a successful, vigorously invasive introduction. A pair was sent to the zoo at 'Groote Schuur' on the lower slopes of Table Mountain in Cape Town from the National Zoological Gardens in 1935 (Bigalke, 1977). Escapees dispersed onto the mountain and by 1972 the population was estimated to number 330. A shooting campaign to eliminate them was started in 1973, when 31 were shot. A major effort resulted in kills of 228 in 1976 and 191 in 1977 but since then annual bags have been small (Brooke, Lloyd & de Villiers, 1986). These authors believe that the rate of increase has been high enough to compensate for shooting mortality and that there may be as many animals now as there were when shooting started in 1973. Destruction of plant cover and erosion on steep slopes are serious problems caused by the presence of tahr. Macdonald & Jarman (1984) also identified an effect of tahr on nutrient cycling. Dispersal to other mountains in the chain running down the Cape Peninsula has, somewhat surprisingly, not occurred. In other directions dispersal is limited by urban development. If any animals were to succeed in reaching the major ranges of the Cape Fold Belt 50 km away, there is little doubt that the species would also become established there.

Differences between the success of the three introduced ungulates in the mediterranean zone calls for discussion. Pigs, so widespread as feral animals in many parts of the world, have not become established in fynbos. That this may be due mainly to inadequate food resources is suggested by the fact that bushpigs (*Potamochoerus porcus*) in the southern Cape occupy forests but make little use of surrounding fynbos (A. Seydack, personal communication). Poor thermo-regulatory capabilities resulting from their physical characteristics (Fraser, 1980) may also play a role however. Very hot summer weather and cold wet winters are typical of the areas where pigs were released.

Fallow deer are the world's most widely naturalised mammal apart from rats, mice and feral domestic animals (Lever, 1985). Their failure to colonise fynbos indicates that the vegetation rather than lack of adaptability is likely to be a limiting factor. Kruger & Bigalke (1984) point out that fynbos provides 'poor' forage for herbivores. Plants are mainly sclerophyllous, secondary compounds which apparently reduce palatability are common and plant parts generally have low nutrient concentrations. Palatable grasses are also scarce. Pastoralists use fire to maintain fynbos in early successional stages where these disadvantages are minimised. In the absence of such management, domestic stock can derive little benefit from fynbos. Fallow deer feed on grass and the foliage and fruits of shrubs and trees. Their relatively large size requires a considerable daily intake although, since they are not amongst the really large herbivores, quality must be relatively high. Only early post-fire successional stages in fynbos could perhaps provide adequate food and these are not available unless a deliberate mosaic pattern of burning is followed. The environs of western Cape farms where the animals are long established have probably not been managed in this way.

The success of tahr may stem partly from its smaller size and lower absolute food requirements. The disturbed nature of much of the vegetation on Table Mountain probably provides better forage than fynbos in good condition. Food selection by tahr in the difficult environment which they have so successfully colonised would repay further study.

Summary

The South African mediterranean region has few introduced mammals. Two rats and the house mouse have remained commensal, probably at least partly because no unoccupied niches are available. The grey squirrel is successful in human settlements in towns and on farms and in forested areas where it has access to introduced seed- and fruit-bearing trees. It has not established itself in fynbos probably because of the absence of a squirrel niche.

Rabbits have been successful on most offshore islands where they were introduced. The sparse and much modified vegetation of the islands cannot be classified as fynbos. The domestic pig is present in some forest plantations. Its lack of success as a feral species is thought to be related to food shortage and perhaps to climatic factors. Fallow deer have not colonised fynbos because of its poor forage qualities. The success of tahr on one mountain may be linked to the atypical vegetation of the area but has not been explained satisfactorily. The mediterranean-climate area of South Africa has not been exposed to many introduced mammals. It has resisted most of those which have arrived.

References

Avery, D. M. (1985). The dispersal of brown rats *Rattus norvegicus* and new specimens from 19th century Cape Town. *Mammalia*, **49**, 573–6.

Bigalke, R. (1937). The naturalization of animals, with special reference to South Africa. *South African Journal of Science*, **33**, 45–63.

Bigalke, R. C. (1972). The contemporary mammal fauna of Africa. In *Evolution, Mammals and Southern Continents*, ed. A. Keast, F. C. Erk & B. Glass, pp. 141–94. Albany: State University of New York Press.

Bigalke, R. (1977). The Himalayan tahr on Table Mountain. *Zoologica Africana*, **12**, 504.

Botha, S. A. (1985). The feral pig: an unsuccessful alien in Africa. Poster presentation, National Synthesis Symposium on the Ecology of Biological Invasions, Stellenbosch. Pretoria: CSIR.

Brooke, R. K., Lloyd, P. H. & de Villiers, A. L. (1986). Review of invasive alien terrestrial vertebrate species in South Africa. In *The Ecology and Management of Biological Invasions in Southern Africa*, ed. I. A. W. Macdonald, F. J. Kruger & A. A. Ferrar, pp. 63–74. Cape Town: Oxford University Press.

Chapman, N. G. & Chapman, D. I. (1980). The distribution of fallow deer: a worldwide review. *Mammal Review*, **10**, 61–138.

Cooper, J. & Brooke, R. K. (1982). Past and present distribution of the feral European rabbit on southern African offshore islands. *South African Journal of Wildlife Research*, **12**, 71–5.

Davis, D. H. S. (1950). Notes on the status of the American grey squirrel (*Sciurus carolinensis* Gmelin) in the south-western Cape (South Africa). *Proceedings of the Zoological Society of London*, **120**, 265–8.

de Graaf, G. (1981). *The Rodents of Southern Africa*. Pretoria: Butterworth.

Fraser, A. F. (1980). *Farm Animal Behaviour*. London: Ballière Tindall.

Kingdon, J. (1974). *East African Mammals*, vol. 2, part B (*Hares and Rodents*). London: Academic Press.

Kruger, F. J. (1979). South African heathlands. In *Ecosystems of the World*, vol. 9A,

Heathlands and Related Shrublands, ed. R. L. Specht, pp. 19–80. Amsterdam: Elsevier.

Kruger, F. J. & Bigalke, R. C. (1984). Fire in fynbos. In *Ecological Effects of Fire in South African Ecosystems*, ed. P. de V. Booysen & N. M. Tainton, pp. 67–114. Berlin: Springer-Verlag.

Lever, C. (1985). *Naturalized Mammals of the World*. London: Longman.

Macdonald, I. A. W. & Jarman, M. L. (ed.) (1984). Invasive alien organisms in the terrestrial ecosystems of the fynbos biome, South Africa. South African National Scientific Programmes Report No. 85. Pretoria: CSIR.

Millar, J. C. G. (1980). Aspects of the ecology of the American grey squirrel *Sciurus carolinensis* Gmelin in South Africa. M.Sc. thesis, Department of Nature Conservation, University of Stellenbosch.

Rosevaar, D. R. (1969). *The Rodents of West Africa*. London: British Museum (Natural History).

Shortridge, G. C. (1934). *The Mammals of South-West Africa*. London: Heinemann.

Siegfried, W. R. (1962). A report on introduced vertebrates in the Cape Province. Cape Provincial Administration, Department of Nature Conservation, Annual Report No. 19, pp. 80–7.

Skead, C. J. (1980). *Historical Mammal Incidence in the Cape Province*, vol. 1, *Eastern and Northern Cape*. Cape Town: Department of Nature & Environmental Conservation.

Smithers, R. H. N. (1971). The mammals of Botswana. *Memoirs of the National Museum of Rhodesia*, **4**, 1–340.

Smithers, R. H. N. (1983). *The Mammals of the Southern African Subregion*. Pretoria: University of Pretoria.

21

Mammals introduced to southern Australia

T. D. REDHEAD, G. R. SINGLETON, K. MYERS & B. J. COMAN

Many mammals have been introduced to southern Australia in the 200 years of European settlement of the region. During one of the most active periods for animal introduction, between 1860 and 1880, more than 60 species of vertebrates were released into the Australian environment. More were to follow later with the advent of a broader immigration policy and the development of aquaculture and aviculture. Some of these introductions failed, whilst others prospered. Some introduced vertebrates are now problems in northern Australia. The majority of the 47 species of mammals brought into Australia (Myers, 1986) were introduced, however, into southern, mediterranean-climate regions (Table 21.1).

Of the larger pest species, only the goat (*Capra hircus*), pig (*Sus scrofa*) and water buffalo (*Bubalus bubalis*) have been studied in detail, mainly in relation to regional pest situations (see e.g. Henzell & McCloud, 1984, on goats; Taylor & Friend, 1984, on water buffalo). Several smaller species of pest mammal have been the subjects of intensive research because of their impact on the southern Australian sheep and grain industries; they include the mouse (*Mus domesticus*) and rabbit (*Oryctolagus cuniculus*) and one of their principal predators, the European red fox (*Vulpes vulpes*). In this chapter we shall attempt to elucidate the importance of the latter introduced mammals in mediterranean regions in Australia by examining their invasive characteristics.

Herbivores

Deer

Approximately 16 species or subspecies of deer or deer-like mammals were introduced into Australia; the fate of these has been discussed more specifically by Myers (1986). Although facing few problems in Australia in relation to climatic adaptation or predation, the deer as a group appear to have failed because

Table 21.1. *Mammals introduced to southern Australia*

Common name	Scientific name	Reason/route of introduction	Result	Origin
Eastern grey squirrel	*Sciurus carolinensis*	Aesthetic	Died out	North America
Three-striped palm squirrel	*Funambulus pennanti*	Zoo release	Localised in Sydney and Perth	India
Black rat	*Rattus rattus*	Commensal	Widespread	S.E. Asia
Brown rat	*Rattus norvegicus*	Commensal	Widespread	Asian steppes
House mouse	*Mus domesticus*	Commensal	Widespread	Central Asia
Ferret	*Mustela putorius furo*	Hunt rabbits	Did not establish	Mediterranean
Gold-spotted mongoose	*Herpestes javanicus auropunctatus*	Control rabbits and rats	Died out	India, Java, Sumatra
Feral dog	*Canis familiaris*	Commensal	Widespread	
Red fox	*Vulpes vulpes*	Hunting	Widespread	Europe
Domestic cat	*Felis catus*	Commensal	Widespread	Europe
European rabbit	*Oryctolagus cuniculus*	Hunting	Widespread	Western Mediterranean
European hare	*Lepus capensis*	Hunting	Widespread	Europe
Feral horse	*Equus caballus*	Draught, transport	Widespread	Europe
Feral donkey	*Equus asinus*	Draught	Mostly NW, monsoonal	North Africa
Feral pig	*Sus scrofa*	Meat	Widespread	Eurasia
Arabian camel	*Camelus dromedarius*	Draught, transport	Arid central west	North Africa, Arabia
Indian spotted mouse deer	*Tragulus meminna*	Hunting	Died out	India
Musk deer	*Moschus moschiferus*	Hunting	Died out	Central and east Asia
Fallow deer	*Dama dama*	Hunting	Local herds	Mediterranean and central Europe
Hog deer	*Axis porcinus*	Hunting	Large herds in Gippsland	India–Asia
Axis deer (chital)	*Axis axis*	Hunting	Small local herds	India
Indian and Sumatran sambar	*Cervus unicolor*	Hunting	Local populations, expanding	India, Sumatra
Barasingha	*Cervus duvauceli*	Hunting	Died out	India
Formosan and Japanese sika	*Cervus nippon*	Hunting	Died out	Japan
Red deer	*Cervus elaphus*	Hunting	Local herds	Europe
Wapiti	*Cervus elaphus canadensis*	Hunting	Died out	North America

Table 21.1. (*cont.*)

Common name	Scientific name	Reason/route of introduction	Result	Origin
Chinese water deer	*Hydropotes inermis*	Hunting	Died out	China
Roe deer	*Capreolus capreolus*	Hunting	Died out	Europe
Feral sheep	*Ovis ammon musimon*	Meat, wool	Widespread	Mediterranean
Feral goat	*Capra hircus*	Meat, milk	Widespread	Mediterranean
Indian black buck	*Antilope cervicapra*	Hunting	Local herds	India
Feral cattle	*Bos taurus*	Meat	Widespread	Europe

Source: Based on Myers, 1986, after Marshall, 1966; Rolls, 1969; Long, 1972; Frith, 1975; Bentley, 1978.

southern Australian vegetation was not adequate as browse, and obligate herding species (those with a fixed seasonal rut) competed directly with domestic stock. The two most successful species, the sambar (*Cervus unicolor*) and hog deer (*Cervus porcinus*), are non-herding grazers, with no fixed rut, living in habitats not used for grazing and which afford shelter from predators.

Other large herbivores

The many other large mammals introduced into Australia for draught, transport, meat, milk and wool, show different patterns of colonisation success rate from those of the deer (Myers, 1986). Almost all the large introduced herbivores survive in habitats similar to those of their original environment. For instance, the wild pig, an unspecialised feeder on roots, herbs, grasses, fruit and animals, from swampy forests of Europe and Asia, likewise inhabits the same environments in Australia and eats the same food types. The wild goat is almost extinct in Asia Minor and the Greek Islands, where it once inhabited forest and brush on mountains and slopes; it seeks out similar environments in Australia. Even the domestic sheep displays strong evolutionary relationships with its Mediterranean origin in Corsica and Sardinia. Sheep farming predominates in the essentially mediterranean-climate region of southern Australia.

Rodents and lagomorphs

Two squirrel species, both food specialists, failed to cope with the Australian environment. The eastern grey squirrel (*Sciurus carolinensis*), from the North American softwood forests, died out (Seebeck, 1984), and the Indian palm squirrel (*Funambulus pennanti*) remains isolated at one location in Perth where it depends essentially on human hand-outs (Myers, 1986). In contrast, the house mouse (*Mus domesticus*), the black rat (*Rattus rattus*) and the Norway rat (*Rattus norvegicus*) have become widely established. *Mus domesticus* originated in central Asia and has colonised virtually the whole continent. Of the two rats, *R. rattus* is the more successful and widespread, with *R. norvegicus* confined to environments with permanent water.

The lagomorphs (*Lepus capensis* and *Oryctolagus cuniculus*) were introduced for sport and food in the mid-1800s and both species erupted in spectacular fashion. The hare, an inhabitant of cool European woodlands and steppes, moved northwards at approximately 60 km/year from its release points near Melbourne (Jarman & Johnson, 1977) to reach the eastern Queensland border about 1900. It failed to penetrate the tropics and has not colonised dry regions of Australia. Today the hare is restricted to areas of south-eastern Australia. The rabbit, from the western Mediterranean, quickly swept over the southern half of Australia (Myers, 1986) and rapidly threatened the sheep and wheat industries and caused significant environmental degradation.

It is necessary to look at modern studies of *Mus* and *Oryctolagus* in order to understand these events more clearly.

Mus domesticus

House mice were probably introduced to Australia during early European settlement in the late 1700s. However, the oldest specimen in an Australian museum was not lodged until 1841 (Mahoney & Richardson, 1988). The absence of an earlier record possibly reflects the commensal nature of house mice and their successful accompaniment of Europeans during colonisation of the globe – the early European mammalogists in Australia may have been too familiar with mice to bother recording their presence.

Nothing is known of the rates or routes of colonisation of Australia by mice. It is presumed that colonisation by mice extended inland from the ports, as they 'hitchhiked' with humans during the movement of livestock and grain around the country. The dependence on ships for carrying immigrants, livestock and domestic goods to Australia has presented countless opportunities for unintentional imports of mice. In more recent years, quicker transit times (see Wace, 1985), increased trade and greater diversity of countries trading with Australia may have increased the opportunities for mice to enter the country. On the other hand, use of containers and better hygiene practices may lessen those opportunities.

World wide, the house mouse complex comprises seven major species (Table 21.2). In Australia, *Mus domesticus* is the only species of house mice that has been recorded (G. R. Singleton, unpublished data). It is somewhat surprising that other species are not present. For example, *M. castaneus* and *M. musculus* are

Table 21.2. *Some species of house mice throughout the world*

Species	Region
Mus domesticus	Widely distributed, including Western Europe, North & South America, and Australia
Mus musculus	Eastern Europe, Central Asia
Mus bactrianus	Middle East to India
Mus spretus	South Europe, North Africa
Mus abbotti	South-east Europe, Middle East
Mus castaneus	South and South-east Asia
Mus hortulanus	Eastern Europe
Mus molossinus	East Asia

both highly commensal in southern Asia (Marshall, 1981), from where large numbers of camels (*Camelus dromedarius*), and associated food and baggage, were imported to Australia between 1860 and 1920. Similarly, opportunities for *M. castaneus*, and perhaps *M. musculus wagneri* and *M. molossinus*, to enter Australia could have been numerous during the later half of the nineteenth century, when large numbers of Chinese arrived in Australia to work the newly discovered gold fields.

Although *Mus domesticus* live in fields in southern France, Corsica, Sardinia and Greece, they occur primarily in houses in Europe (Sage, 1981) and also in southern Africa (see Bigalke & Pepler, this volume). In Australia, *M. domesticus* lives both a commensal and feral existence, and has successfully colonised deserts, offshore islands, and agricultural land. In the case of the latter, mice have become an important agricultural pest and we shall now focus on the ecology of the invasion of mice in agricultural habitats.

Significant domestic and industrial losses, caused by house mice living commensally in major cities are not apparent. It is in the grain-growing regions of eastern and southern Australia, where populations irrupt aperiodically to form plagues, that house mice are of major economic concern (see Figure 21.1). The history of mouse plagues in New South Wales and Victoria has been well documented (Saunders & Giles, 1977; Singleton & Redhead, 1989). They have occurred on average every 7 to 9 years (though the intervals between plagues are highly variable) since the first plague around the turn of the century. The economic costs of mouse plagues have not been well documented, but estimates range from A$50 to A$100 million for each of the recent outbreaks in the 1980s. During plagues, mice cause substantial losses in crops (especially soon after sowing and when grain has formed), in domestic houses, to farm machinery, to livestock producers, to town businesses (especially to produce merchants and grocers), to grain stored on-farm and to grain stored in bulk storages at railheads. Mice in plague numbers present a health threat to humans and their domestic stock, and cause high levels of stress to people who have to share their houses with hordes of mice for periods of up to six months. Also, the poisons used to combat plagues present an environmental problem through pollution of waterways and the possible poisoning of some non-target species.

It is not clear why the cereal-belt of south-eastern Australia is faced with the sporadic occurrence of mouse plagues. We know of only one other region – the Xinjiang Autonomous Region of China – which experiences similar plagues of house mice (Yan Zhitan & Zhong Mingming, 1984). It is interesting that the species of house mouse which forms plagues there is *M. musculus* (Marshall, 1981; T. Redhead, unpublished observation).

Globally, there is a great diversity of rodent species using diverse life forms and strategies to exploit successfully a wide range of habitats (see Happold, 1983, for diversity of rodents in tropical savannas throughout the world). Two broad classes of rodents can be recognised: those with high fecundity and short generation time, and those with low fecundity and long generation time. The latter group, with a low rate of population increase and high survival (*K*-strategists), are associated with habitats where there is little variation in resources between years. The former group (*r*-strategists) can undergo rapid population increase when conditions become favourable. *Mus domesticus* in rural Australia is an *r*-strategist species (Berry, 1977) which, living in habitats characterised by large and irregular variation in environmental factors such as rainfall, forms plagues.

In the cereal-belt in south-eastern Australia, with its mediterranean-type climate, rainfall is highly variable. Moreover, a major consequence of modification of the environment for summer and/or winter cropping is the short-term availability of massive amounts of grain. These grains, together with grass seeds, seeds of annual forbs, and seeds of improved pasture species (such as *Medicago* spp.), provide the main food sources for small mammals. Since the clearing of vast tracts of land for cropping, none of the native rodent species has increased in abundance. Indeed, very few persist in the grain-belt, even at low levels of abundance. Few small mammals native to Australia are granivores (Watts & Braithwaite, 1978; Morton, 1979; Cockburn, 1981) and there has been insufficient time for the few granivorous native species of small mammal to adapt and evolve to the drastically changed landscape. Land clearing for extensive cropping has occurred only in the past 80 to 120 years. Outside the grain-belt, in the arid interior of Australia, the native rodents *Notomys* sp. and *Rattus villosissimus* form plagues after periods of favourable climatic conditions (Newsome & Corbett, 1975).

The house mouse is a successful *r*-strategist with a cosmopolitan distribution. Why then do plagues of mice rarely occur in other parts of the world except China? The reason is not clear but one or more of the following hypotheses warrant consideration:

1. Elsewhere there are highly fecund native rodents, and these competitively exclude or compete with house mice (see e.g. Bigalke & Pepler, this volume)
2. Elsewhere house mice have more predators and pathogens which prevent the formation of mouse plagues. In the Victorian Mallee, the principal predators of house mice are the introduced fox (*Vulpes vulpes*), the introduced cat (*Felis catus*), brown snakes (*Pseudonaja textilis*) and native raptors such as the black shouldered kite (*Elanus axillaris*), brown falcon (*Falco berigora*), swamp harrier (*Circus assimilis*), Australian kestrel (*Falco cenchroides*) and the barn owl (*Tyto alba*)

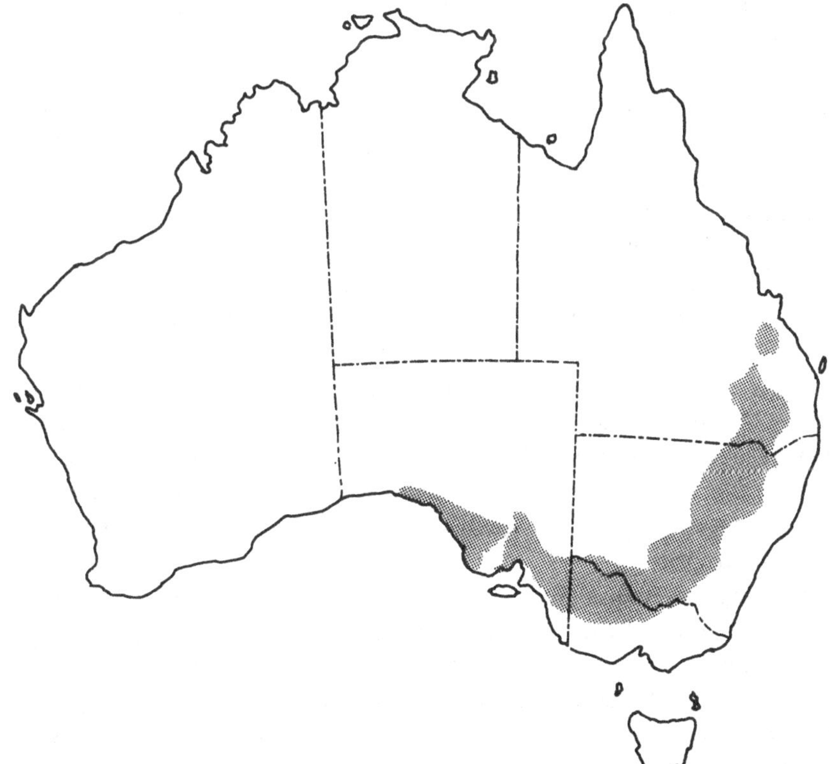

Figure 21.1. Region (shaded) of south-eastern Australia that is prone to mouse plagues.

3. Elsewhere climates are less variable and suitable conditions necessary for the generation of a plague never arise
4. Elsewhere winters are so harsh that few individuals survive through to spring. According to Redhead (1982) a necessary precursor of mouse plagues is winter breeding by mice in fields. In environments where the winter is always severe, this could never occur
5. Elsewhere farming practices and land use patterns may prevent large numbers of mice from surviving winter, or from breeding in winter.

Although house mice are well adapted to respond to favourable conditions a mouse plague does not occur overnight. In the dry Mallee wheatlands and the irrigated cereal crops of southern Australia, above-average autumn rainfall has been identified as a key trigger of plagues which occurred 12 to 20 months later (Redhead, 1982; Singleton, 1989). To explain the formation of a mouse plague Redhead (1982) developed a tri-phasic model. Aspects of that model, known also as the 'food-quality, spacing-behaviour model', were examined by Bomford (1987) and Bomford & Redhead (1987). The model postulates that mouse plagues are triggered by external environmental events, specifically above-average rainfall in autumn, and the consequences of that event impinge first on mouse populations in patches of refuge habitat. The effect of the above-average amounts of rain is to extend the period for which high-quality food is available to mice in the refuge patches. Consequently, some mice continue to breed in winter, and Phase 1 of a plague is then under way. Mice in these populations with winter breeding are in good condition, with high growth rates of individuals, and high weight-to-length ratios (but not obesity). Also during Phase 1, many young mice migrate from the refuge patches into other areas which, because of the good autumn rains, are now able to support mouse populations.

The second phase commences in the summer, 6 or 7 months after the plague trigger. The average number of young in each litter is about 50 per cent higher than at the commencement of breeding in the early summer of other years, and there are few, if any, non-breeding adult females. This high productivity leads to a mouse plague (Phase 2 peak) in autumn and winter, some 12 to 15 months after the plague trigger.

In some years, such as 1980, a further plague peak (Phase 3 peak) may occur, roughly two years after the plague trigger. Following heavy autumnal rains in 1983, a Phase 2 peak was experienced in autumn and winter of 1984. There was no general Phase 3 peak in 1985, except on some irrigated farms, and it appears that survival rate after the Phase 2 peak determines whether a Phase 3 peak will eventuate.

The tri-phasic model incorporating food-quality and spacing-behaviour appears to be generally applicable. It is also recognised, however, that distribution of mice during plagues is heterogeneous and the timing of plague peaks depends on local conditions, including the type and stage of crop present.

Oryctolagus cuniculus

The invasion of south-eastern Australia by the rabbit can be attributed to a number of possible factors:

1. It is pre-adapted physiologically to the climate of much of the southern half of the continent
2. It carries a depauperate parasite fauna when compared with rabbits in Spain, where it originated
3. It was presented with a suitable and abundant food source because of alteration of the landscape by humans and domesticated stock which changed a predominantly perennial vegetation into one containing a large proportion of annuals
4. It was almost completely free of predators and serious disease during its primary irruption because of grazing mammals and human persecution of native predators, i.e. introduced predators (the fox and the cat) and disease (myxomatosis) became part of the system too late to be able to influence events
5. It colonised extensive existing burrow systems of native grazing mammals in some arid environments
6. It is an excellent ecological generalist.

As well the rabbit has penetrated some environments, without human assistance, on a large scale (Myers, 1986).

We find it most useful to think of the rabbit in Australia in terms of Caughley's (1977) strategic model of plant–herbivore–predator relationships (see Myers, 1986, Figure 6). European settlers set up the first trophic level by altering the environment, then added the herbivores and finally the predators to form second and third trophic levels.

Introduced predators

Five species of predatory mammals have been introduced into Australia. Two of these, the tropical mongoose (*Herpestes javanicus auropunctatus*) and the ferret (*Mustela putorius furo*) failed to become established.

The other three, the dingo, fox and cat, have been very successful. The fox and cat, supreme generalists, followed the rabbit northwards and westwards about 20 years later.

*The fox (*Vulpes vulpes*)*

Although there are newspaper accounts of the introduction of foxes into Australia as early as 1855, it is likely that the first successful releases took place in southern Victoria in 1871 (Rolls, 1969). Curiously, one of these releases took place within a short distance of Geelong, in southern Victoria, where Thomas Austin had, a few years earlier, released the English rabbit. This must rank as one of the few examples where a predator and its natural prey were introduced contemporaneously.

The subsequent spread of foxes is not well documented. The fox's present distribution, which covers all of mainland Australia except the tropical north (Figure 21.2), was achieved in the space of 100 years. Data from early bounty payment records suggest that the fox spread most rapidly across the inland saltbush and Mallee country and more slowly in the forested ranges near the coast (Jarman, 1986). In Western Australia, however, the early spread seems to have been along the southern coastline of the state with a succession of sightings from Eucla in 1912 to Geraldton in 1925 (Long, 1972) (see Figure 21.2).

If we assume that all Australian foxes originated from early introductions to southern Victoria, then data from early sightings elsewhere on the continent suggest annual dispersal distances of up to 160 km/year. Although some studies have recorded movements of this magnitude from countries other than Australia, they are the exception rather than the rule. Thus, the first colonising animals were either partially shifted about by humans or, alternatively, behaved in a manner different from their longer-established counterparts in Europe and elsewhere. Recent data on cub dispersal in southern Victoria (Coman, unpublished data) indicates an average annual dispersal distance of 11 km with exceptional movements up to 30 km.

It seems clear that the early spread and establishment of fox populations was closely linked to the spread of rabbits. Several studies of food habits in Australia have highlighted the importance of rabbit in fox diet (e.g. McIntosh, 1963; Coman, 1973) and the data on early spread of foxes show that foxes spread more quickly where rabbits were present. Indeed, Long (1972) has noted that, in Western Australia, the fox appeared to follow approximately the same path of invasion as the rabbit, though several years later.

The interrelationship between foxes and rabbits is dramatically illustrated in Figure 21.3. In Victoria, a State-wide bounty payment scheme for fox scalps was initiated in 1949. Numbers returned for payment quickly rose to a high level but then fell dramatically in 1952–3 when myxomatosis reduced the rabbit population to a very low level. Significantly though, scalp numbers did rise again within a few years and this is good evidence of the ability of foxes to switch to alternative food sources.

Although population densities of foxes in a range of Australian habitats are not available, it appears that the animal reaches its highest densities in southern agricultural areas where the fragmentary habitats and secure food supplies mimic, to some extent, the situation in Europe. In central Victoria, densities as high as 4 foxes/km^2 have been recorded in summer (Coman, unpublished data). In the arid zone, densities are probably lower and more variable over time. It is reasonable to expect that, in line with huge fluctuations in the availability of prey species in the arid zone, the smaller predators show a similar if less spectacular, boom and crash cycle.

The impact of fox predation on indigenous fauna in Australia has been the subject of much conjecture but very little scientific evidence. Dietary studies for foxes taken from heavy bush country (e.g. Coman, 1973; Brunner *et al.*, 1975) indicate that small native mammals comprise the bulk of the intake. In many instances, however, predation is probably having little effect on long-term

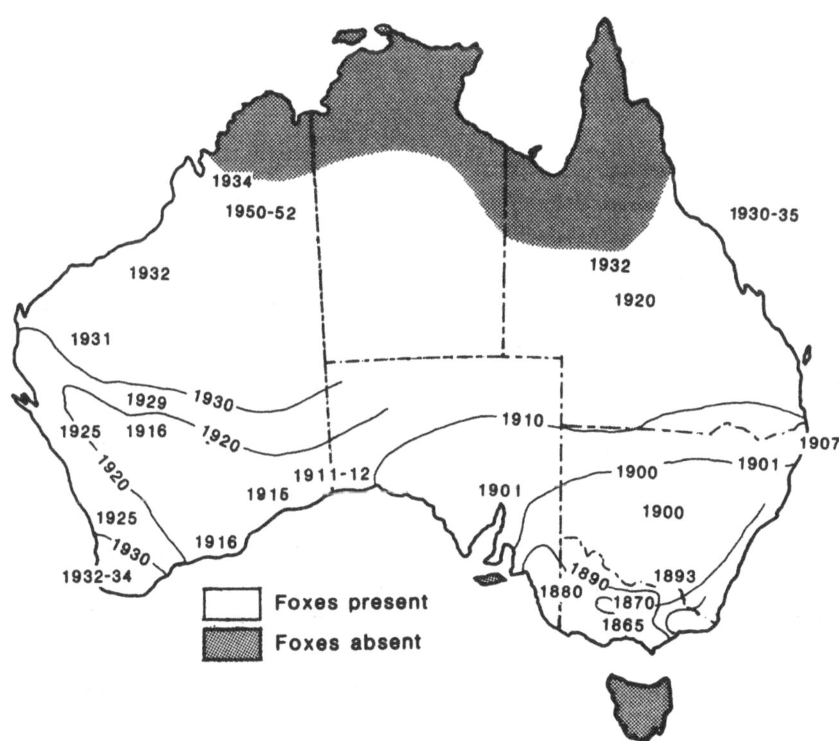

Figure 21.2. Approximate distribution of the red fox in Australia (after Coman, 1985) and probable dates of the first arrival of the red fox in districts (adapted from Jarman, 1986).

population size. In mainland Australia, it is difficult to partition the impact (on indigenous fauna) between predators (such as cats, dogs and foxes), land clearing for agriculture, grazing by sheep and cattle and changes in vegetative cover because of alteration in the timing and intensity of wildfires. In recent years, there has been one study in Western Australia which implicates foxes as a major factor in population decline for remnant populations of rock wallabies (*Petrogale lateralis*) (J. Kinnear, personal communication). Despite high annual recruitment through breeding, wallaby numbers have either decreased or remained static. In the period 1982–6, intensive fox control was undertaken in two of these areas and in both cases, there were substantial increases in the wallaby population. Unprotected wallaby populations nearby either declined or increased marginally.

In recent times there has been a flourishing export trade in fox pelts and large numbers of foxes are harvested in the southern Australian states each year, but pelt prices fluctuate markedly. An investigation of age structure for foxes killed by hunters in Victoria (Coman, 1988) indicates that over half of the animals sampled were less than one year old and very few survived beyond four years. This suggests a high annual turnover – a situation common to the red fox in Europe (Haltenorth & Roth, 1968).

Conclusions

Colonisation, however defined, is an integral part of the biology of all species. The processes and qualities which permit an organism to invade a new environment are the same as those which give it the ability to expand its range in its home environment. As Barrett & Richardson (1986) stated, there is no single optimal genetic solution to the challenges facing the colonist, and there is no optimal ecological solution either.

The brief examination we have presented in this overview of the introduction of vertebrates into southern Australia leads to the following five general conclusions:

1. Amongst mammals, as has been found in invading species of plants and birds in Australia (Newsome & Noble, 1986), the probability of colonisation is highly influenced by similarity between the climate of origin and climate of introduction (see Swincer, 1986). Unlike many invertebrates, with their short generation times, most mammals must already possess a genetic background capable of supporting the physiological tolerances and ecological plasticity demanded of them.

2. Introduced predators now depend heavily on introduced herbivores as a source of food. The studies described reveal the existence of a system completely different from that which prevailed when Europeans colonised Australia. Most

of the smaller species of native mammals which formed the staple items of diet of endemic predators are no longer present in sufficient density to support their populations and they have been forced to turn to rabbit and mouse for their continued existence. Thus the introduced rabbit and mouse not only form the basic food source of the introduced fox and cat but essentially support a whole system of predators in mediterranean Australia.

3. A large number of introduced mammals in Australia owe their success and their continued existence to widespread human alterations to environment, either as invaders of human-made environments, or as colonisers of new environments from centres of high density created by human settlements. Again, there are strong parallels with the history of invasions by plants and birds in Australia (Newsome & Noble, 1986). For Australia, where most of the full ecological equivalents have been removed, or had their status altered, it is thus rather questionable to try and discuss problems relating to the potential of communities for invasion. It is likely to be far more profitable to compare the histories of establishment of the same species in different environments (where the data exist) or on different continents. The European rabbit maintains large populations on offshore islands in North America but, despite more than 140 deliberate introductions, has never become established on the mainland which is inhabited by a full complement of closely related lagomorph species and their predators and diseases.

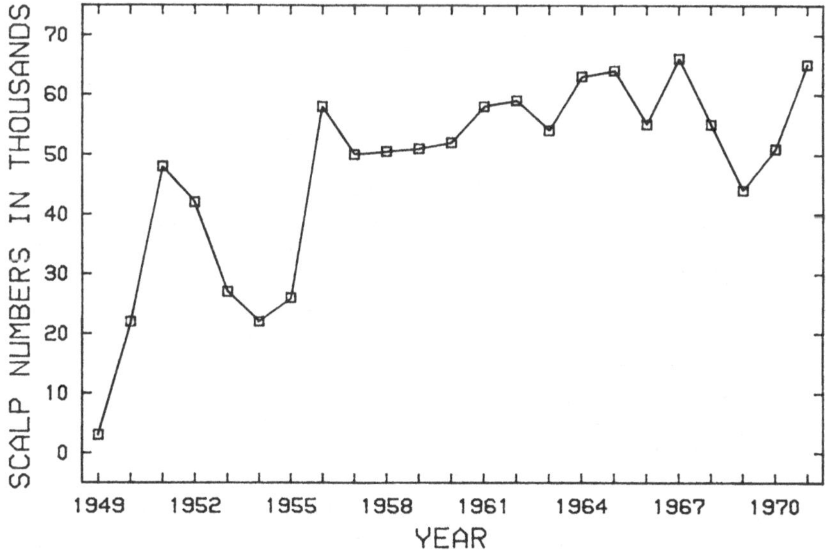

Figure 21.3. Bounty payments for fox scalps in Victoria from 1949 to 1971.

4. Few ecological specialists have become established among the many mammals introduced into Australia, although many have failed. Those that succeeded did so because of strong human support. Mammals like roe deer and mouse deer are so difficult to keep in captivity that the numbers released could not have been large. Species like the fallow deer, on the other hand, which thrive in captivity, were released in large numbers. Vertebrate species which have become established in Australia were almost without exception pre-adapted to the Australian environment and the resources it offered.

5. The Australian experience is a prime example of the supreme ability of humans themselves as a mammalian coloniser, extremely competent in predation, competition and environmental conditioning, including the use of fire (Myers, 1986). As human populations continue to rise, all species which compete with humans or serve as a resource, and all environments which are of human use, will change in status. But as far as the mammals are concerned, decisions on which species humans want to have live with them may, in the end, belong principally to the cultural rather than the biological sphere.

Acknowledgements. We are grateful to Drs D. Spratt and O. Pouliquen for their comments on the manuscript. We thank Greg Young for preparing the figures.

References

Barrett, S. C. H. & Richardson, B. J. (1986). Genetic attributes of invading species. In *The Ecology of Biological Invasions: An Australian Perspective*, ed. R. H. Groves & J. J. Burdon, pp. 21–33. Canberra: Australian Academy of Science.

Berry, R. J. (1977). The population genetics of the house mouse. *Science Progress (Oxford)*, **64**, 341–70.

Bomford, M. (1987). Food and reproduction of wild house mice. II. A field experiment to examine the effect of food availability and food quality on breeding in spring. *Australian Wildlife Research*, **14**, 197–206.

Bomford, M. & Redhead, T. (1987). A field experiment to examine the effect of food quality and population density on reproduction of wild house mice. *Oikos*, **48**, 304–11.

Brunner, H., Lloyd, J. W. & Coman, B. J. (1975). Fox scat analysis in a forest park in south-eastern Australia. *Australian Wildlife Research*, **2**, 147–54.

Caughley, G. (1977). *Analysis of Vertebrate Populations*. Chichester: Wiley.

Cockburn, A. (1981). Diet and habitat preference of the silky desert mouse, *Pseudomys apodemoides* (Rodentia). *Australian Wildlife Research*, **8**, 475–97.

Coman, B. J. (1973). The diet of red foxes, *Vulpes vulpes* L., in Victoria. *Australian Journal of Zoology*, **21**, 391–401.

Coman, B. J. (1985). Australian predators of livestock. In *Parasites, Pests and Predators*, ed. S. M. Gaafar, W. E. Howard & R. E. Marsh, pp. 411–25. Amsterdam: Elsevier.

Coman, B. J. (1988). The age structure of a sample of red foxes (*Vulpes vulpes* L.) taken by hunters in Victoria. *Australian Wildlife Research*, **15**, 223–9.

Haltenorth, T. & Roth, H. H. (1968). Short review of the biology and ecology of the red fox. *Säugetierkundliche Mitteilungen*, **16**, 339–52.

Happold, D. C. D. (1983). Rodents and lagomorphs. In *Tropical Savannas*, ed. F. Bourlière, pp. 363–99. Amsterdam: Elsevier.

Henzell, R. P. & McCloud, P. I. (1984). Estimation of the density of feral goats in part of arid South Australia by means of the Petersen estimate. *Australian Wildlife Research*, **11**, 93–102.

Jarman, P. (1986). The red fox – an exotic, large predator. In *The Ecology of Exotic Animals and Plants*, ed. R. L. Kitching, pp. 43–61. Brisbane: Wiley.

Jarman, J. A. & Johnson, K. A. (1977). Exotic mammals, indigenous mammals and land-use. *Proceedings of the Ecological Society of Australia*, **10**, 146–66.

Long, J. L. (1972). Introduced birds and mammals in Western Australia. Agriculture Protection Board of W.A. Technical Series No. 1 (DAIS 09/72-500).

Mahoney, J. A. & Richardson, B. J. (1988). Muridae. In *Zoological Catalogue of Australia*, vol. 5, *Mammalia*, ed. D. W. Walton, pp. 154–92. Canberra: Australian Government Publishing Service.

Marshall, J. T. (1981). Taxonomy. In *The Mouse in Biomedical Research*, vol. 1, ed. H. L. Foster, J. D. Small & J. G. Fox, pp. 17–26. New York: Academic Press.

McIntosh, D. L. (1963). Food of the fox in the Canberra district. *CSIRO Wildlife Research*, **8**, 1–20.

Morton, S. R. (1979). Diversity of desert-dwelling mammals: a comparison of Australia and North America. *Journal of Mammalogy*, **60**, 253–64.

Myers, K. (1986). Introduced vertebrates in Australia, with emphasis on the mammals. In *The Ecology of Biological Invasions: An Australian Perspective*, ed. R. H. Groves & J. J. Burdon, pp. 120–36. Canberra: Australian Academy of Science.

Newsome, A. E. & Corbett, L. K. (1975). Outbreaks of rodents in semi-arid and arid Australia: causes, preventions, and evolutionary considerations. In *Rodents in Desert Environments*, ed. I. Prakash & P. K. Ghosh, pp. 117–53. The Hague: Junk.

Newsome, A. E. & Noble, I. R. (1986). Ecological and physiological characters of invading species. In *Ecology of Biological Invasions: An Australian Perspective*, ed. R. H. Groves and J. J. Burdon, pp. 1–20. Canberra: Australian Academy of Science.

Redhcad, T. D. (1982). Reproduction, growth and population dynamics of house mice in irrigated and non-irrigated cereal farms in New South Wales. Ph.D. thesis, Australian National University.

Rolls, E. C. (1969). *They All Ran Wild*. Sydney: Angus & Robertson.

Sage, R. D. (1981). Wild mice. In *The Mouse in Biomedical Research*, vol. 1, ed. H. L. Foster, J. D. Small & J. G. Fox, pp. 39–90. New York: Academic Press.

Saunders, G. R. & Giles, J. R. (1977). A relationship between plagues of the house mouse, *Mus musculus* (Rodentia: Muridae) and prolonged periods of dry weather in south-eastern Australia. *Australian Wildlife Research*, **4**, 241–7.

Seebeck, J. H. (1984). The eastern grey squirrel, *Sciurus carolinensis*, in Victoria. *Victorian Naturalist*, **101**, 60–5.

Singleton, G. R. (1989). Population dynamics of an outbreak of house mice (*Mus domesticus*) in the Mallee wheatlands of Australia – hypothesis of plague formation. *Journal of Zoology (London)*, **219**, 495–515.

Singleton, G. R. & Redhead, T. (1989). House mouse plagues. In *Mediterranean Landscapes in Australia: Mallee Ecosystems and Their Management*, ed. J. C. Noble & R. A. Bradstock, pp. 418–33. Sydney: Surrey Beatty & Sons.

Swincer, D. E. (1986). Physical characteristics of sites in relation to invasions. In *The Ecology of Biological Invasions: An Australian Perspective*, ed. R. H. Groves & J. J. Burdon, pp. 67–76. Canberra: Australian Academy of Science.

Taylor, J. A. & Friend, G. R. (1984). Ground surface features attributable to feral buffalo *Bubalus bubalis*. I. Their distribution relative to vegetation structure and plant life form. *Australian Wildlife Research*, **11**, 303–9.

Wace, N. (1985). Australia – the isolated continent. In *Pests and Parasites as Migrants; an Australian Perspective*, ed. A. Gibbs & R. Meischke, pp. 3–22. Canberra: Australian Academy of Science.

Watts, C. H. S. & Braithwaite, R. W. (1978). The diet of *Rattus lutreolus* and five other rodents in southern Victoria. *Australian Wildlife Research*, **5**, 299–303.

Yan Zhitan & Zhong Mingming (1984). The prediction to fluctuations in house mouse (*Mus musculus*) population studies of mechanisms. *Acta Theriologica Sinica*, **4**, 139–46.

Part IIIc Birds

The majority of bird species that have become invasive have done so because of their commensalism with humans and their associations with human activities. This is as true for birds native to the Mediterranean Basin as for those of other regions of mediterranean climate, as the first chapter in this section shows. The avifauna of the Mediterranean Basin is itself diverse. When introduced to other regions, bird species have not always been invasive; in fact, many introductions have failed. Only about 5 per cent of the avifaunas of mediterranean California and Chile are potentially invasive and an even lower percentage is potentially invasive of the truly mediterranean shrublands termed chaparral and matorral. This conclusion seems also to apply to birds in South African fynbos and southern Australian heaths, except for human-modified sites where birds such as starlings and sparrows are successful. The success of such species is very much because of the creation of environments such as croplands which provide food sources of 'European-type' plants and of areas with a greatly reduced diversity of native species.

In most regions of mediterranean climate a few birds have become spectacularly invasive. The numbers of species which have done so are, however, a lower percentage than that able to be predicted from a knowledge of the ecological characteristics of other groups of invasive organisms such as plants and, to a lesser extent, mammals.

22

Invasions and range modifications of birds in the Mediterranean Basin

J. BLONDEL

One of the most striking features of the Mediterranean Basin is the extremely complicated geography and geotopography of the islands and lands encircling this 'sea-among-the-lands' (Mediterranean). The landmasses which encircle the sea are part of three continents: Europe, south-western Asia and Africa. This latter feature is supposed to be a diversification factor in the biogeographical origin of the biotas. Because of the topographical diversity of the region, with high mountains (up to 4000 m) coming near to the sea, we can find only a few kilometres apart, but at different elevations along the slopes of the mountains, such Siberian faunal elements as Tengmalm's owl *Aegolius funereus* side by side with thermophilous Mediterranean or semi-arid species such as the Sardinian warbler *Sylvia melanocephala* or the black-eared wheatear *Oenanthe hispanica*. This admixture of species and faunal elements represents a striking telescoping of faunas which, as will be shown later, reflects the past history of the region. As a result, the Mediterranean Basin, which comprises about 2 970 000 km^2, is a land of exceptionally diverse climates and habitats, ranging from the arid steppes of northern Africa and the Middle East to the moist forests of fir and beech in the mountains of the continents and some larger islands.

The richness in bird species in the region is extremely high: 347 species of birds regularly breed there, as compared with no more than 419 species in the whole of Europe which extends over about three times the area (Voous, 1960; Blondel, 1985). An examination of the biogeographical affinities of 209 land birds out of these 347 breeding species shows that they belong to three main categories: firstly, a boreal set of forest species which are widespread in both deciduous and mixed forests of the Palaearctic (this category contains 74 species and is dominant in the Mediterranean region, including the Mediterranean parts of North Africa); secondly, birds of Palaearctic grasslands and of southern and south-eastern steppes (the latter speciated at the southern margins of the

Mediterranean within the large belt of dry and open habitats which encircles the region from the shores of the Atlantic Ocean to the steppes of central Asia; 92 species belong to this category); and thirdly, a small endemic set which speciated within the limits of the Mediterranean Basin (this category comprises no more than 43 species, i.e. 13 per cent of the total avifauna).

Although many tropical groups, such as the Cracidae, Psittacidae, Musophagidae, Coliidae, Trogonidae, Bucerotidae and some others, were widespread in Eurasia during the Miocene as well as the vultures (Cathartiidae) of the New World during the Pliocene (Brodkorb, 1971), all these groups became extinct before the beginning of the Pleistocene. The relationships of the modern avifaunas between the Mediterranean and continents other than Eurasia are thus rather scanty. On the other hand, there are close affinities between Mediterranean and Asian avifaunas. This is especially true of the faunas of the steppes of south-western Asia. On many occasions in the past and up to the present, the steppe landscapes acted as a screen of 'steppe Palaearctic' between the 'forest Palaearctic' and tropical Asia. More than 100 species of birds come from this part of Asia and are a strong component of the bird faunas of the eastern part of the Mediterranean Basin.

To sum up, the Mediterranean bird fauna belongs to three main faunal units: (i) boreal sylvatic, (ii) southern and eastern steppic and semi-arid; (iii) Mediterranean, with the latter being very small for historical reasons (see later). These units have evolved in different geographical areas and their occurrence in the region is a long story closely linked to the climatic vicissitudes of the Pleistocene followed just afterwards by human impacts. Human actions on Mediterranean landscapes have been drastic for at least 10000 years in the eastern Mediterranean and for more than 6000 years in the western. Anthropogenic pressures have, to varying degrees, had an impact on the structure and function of most Mediterranean habitats. For this reason, any interpretation of the present distributional patterns of the biotas must take into account human actions on Mediterranean landscapes.

As far as the problem of invasions in the Mediterranean Basin is concerned, one could hypothesise, mindful of the caveat that disturbed areas are especially prone to invasions (Elton, 1958; Baker & Stebbins, 1965), that such human modifications to landscapes have produced many opportunities for alien birds to invade the new habitats so created. Facts do not support this hypothesis. If the term 'invasion' is restricted either to species which move spontaneously to new areas or else to introduced species that have been brought intentionally or unintentionally to areas outside their natural ranges of distribution, there have been extremely few cases of invasions by alien birds in the Mediterranean Basin.

At first, this seems surprising because dispersal abilities and adaptive properties of birds should have produced many opportunities for invasion and range expansion. Moreover, several attempts to introduce foreign birds as game or as ornamental species have failed. As is the case with many other organisms, especially plants, there are many fewer bird invaders in the Palaearctic Mediterranean region than in other mediterranean-type ecosystems in the world, i.e. South Africa (Macdonald *et al.*, 1986), California (Mooney & Drake, 1986), Chile, and especially Australia (Groves & Burdon, 1986).

The study of biological invasions is complicated because this invasion process can be investigated at several scales of time and space and not only from distributional or biogeographical points of view. From an ecological, i.e. functional viewpoint, native species can be invaders within their distributional range, especially as a result of population explosions which may change the structure and functioning of communities and eventually lead populations to invade new habitats within their natural range. Thus local and regional modifications to distributional patterns, as well as population explosions of species as a response to human induced changes in ecological conditions, are all relevant to a study of invasive processes. Although these indigenous invasions are fundamentally different in their origins from invasions by introduced species, these processes may adversely affect elements of natural ecosystems important for conservation and ecosystem management.

In this chapter I wish to address four main points. Firstly, the past history and development of bird faunas in the Mediterranean region will be examined briefly in order to state what should be the patterns of distribution of the species in the present bioclimate, i.e. before the impact of humans started to modify Mediterranean ecosystems. Secondly, I shall discuss the few cases of invasions, either spontaneous or those deliberately produced by humans. Thirdly, modifications to the distributional ranges of birds as a result of human-induced changes will be mentioned. Finally, I shall mention briefly some consequences at the community level of the demographic outbreaks of some species which benefit from human activities. Only breeding species will be considered. The process of seasonal 'invasions' by migrating birds, especially the birds which winter in the region but which breed further north, will not be considered; neither will some irregular and unpredictable movements such as irruptions or eruptions of birds, even though some of them may be spectacular and occur on a grand scale. Incidentally, it must be pointed out that migratory species are not necessarily more liable to invade new breeding areas than sedentary species (O'Connor, 1986). Similarly, many drift migrants, which are displaced from their normal passage routes by winds, never succeed in breeding in the new areas they reach by chance. For instance, drift migrations regularly bring to Europe large numbers of migrant individuals from North America (Sharrock & Sharrock, 1976) but these birds never succeed in establishing themselves as breeding populations.

Distributional patterns of the bird faunas in the Mediterranean Basin before the Neolithic

As I have discussed this problem at length elsewhere (Blondel, 1982, 1985, 1986, 1988) only a brief summary will be given here. The fauna of the cold, temperate and warmer Mediterranean regions of Europe belongs to Holarctic faunal types (Voous, 1963) and has taken part or has been subjected to the geographic and climatic history of the Eurasian continent, at least during the Pleistocene. This history, especially the climatic vicissitudes of the Pleistocene (Steinbacher, 1948; Moreau, 1954; Blondel, 1982, 1985, 1986, 1988) greatly influenced the ecological characteristics and the geographical distribution of the fauna. The main events of this history are briefly summarised as:

1. During the last 2 million years there have been a series of alternating expansions and contractions of glacial and arctic conditions with consequent shifts in most ecological zones. The processes of expansion of vegetation belts during the periods of climatic improvement were rather rapid since estimates from pollen records of post-glacial rates of spread give 400 to 2000 m per year for tree species in North America and Europe (Davis, 1981; Huntley & Birks, 1983). This alternation is much more complicated than formerly believed (Birks, 1986). In particular, it is now recognised from palaeobotanical and palaeontological records that the diversity of conditions of temperature and moisture within the Mediterranean during the climatic phase (either glacial or interglacial), as well as the geotopographical diversity of the region, allowed for the coexistence at a regional scale of all the vegetation belts and their associated faunas of Europe.

2. During the interglacial periods such as the one at the present time, Mediterranean forests at low and middle altitudes were mostly dominated by broad-leaved trees (*Quercus pubescens*), not evergreen trees (such as *Q. ilex*) as occurs today. The dramatic extension of evergreen vegetation is mainly a secondary feature arising because of human impact (Pons, 1981, 1984).

3. Mediterranean-type shrublands never disappeared from the region during the Pleistocene, even during the most severe climatic conditions (Pons, 1981). They persisted as shrubland habitats distributed patchily which allowed for the survival and differentiation of the few species characteristic of this habitat type.

4. Consequently, during the most severe climatic phases, all the forest types of Europe and their associated faunas had to find a refugium in the Mediterranean region which was larger than it is at present because of a lowering of the sea level of between 100 and 150 m.

5. The present extension of mediterranean-type shrublands is a modern feature arising from human deforestation which continued from the early Neolithic (9000 BP). Before this impact, forests were much more widespread than at present except in local patches where climatic, edaphic and topographical conditions allowed for only a shrubby vegetation.

These events explain why there have been so few speciation processes in the region, despite the fact that the Mediterranean region extends over three continents and despite its high geographical and topographical diversity which is propitious for biological isolates. On many occasions during the Pleistocene there has been an accumulation in the Mediterranean of all the faunal types of Europe. The faunal elements of central and northern Europe expanded from their Mediterranean refugia to the north at each climatic improvement *without leaving the region* and came back again at each climatic deterioration. In these circumstances there has never been any kind of geographical isolation between the forests of the Mediterranean and those of central Europe, which should have been a prerequisite for allopatric speciation. From a biogeographical point of view, the bird fauna of Mediterranean forests cannot be different from, and definitely belongs to, the same general faunal stock of the western Palaearctic forest blocks as a whole, since these forest blocks were never fragmented during the Pleistocene. The bird faunas of Mediterranean forests are not different, just impoverished, because the Mediterranean region is at the south-western margin of Eurasia. This history explains why out of the 347 species of birds which presently breed in the Mediterranean Basin (Blondel, 1986, 1988), no more than 47 belong to a Mediterranean endemic fauna.

Birds as invaders in the Mediterranean Basin

I shall consider as invaders only the species which succeeded in establishing themselves as regular breeders, i.e. which built viable, self-sustaining populations in the wild without any human intervention. Thus, species which have been intentionally introduced for food, such as the chicken *Gallus gallus*, or for ornament, such as some species of pheasants (*Phasianus* spp.) and the peacock (*Pavo cristatus*), will not be discussed. In contrast to many other organisms, such as plants, mammals, insects, parasites and diseases, human dispersal has not been an important factor for bird invasions in the Mediterranean Basin.

There is hardly any species of bird from other continents which regularly occurs as an established breeder in Mediterranean habitats, either lowland or montane. It is a general feature that extremely few species succeeded in invading Europe. For instance, in Britain, despite attempts to introduce up to 100 species, no more than four completely alien birds have become naturalised over wide areas (Lack, 1976). Two of them are native to the nearby mainland: the red-legged partridge *Alectoris rufa* and the little owl *Athene noctua*. One species came from Asia, the pheasant *Phasianus colchicus* and one from North America, the Canada goose *Branta canadensis*. In France, there have been many attempts to introduce species as game birds. The quail *Colinus virginianus* and the pheasant *Syrmaticus reevesii* have been introduced many times; some local but very small populations breed in the wild (Cruon & Nicolau-Guillaumet, 1985). Other unsuccessful attempts at introduction of quail as game birds are *Francolinus francolinus* from the Middle East as well as *Callipepla californica* and *Colinus virginianus* from North America to Corsica (Thibault, 1983). The only truly naturalised species is the pheasant *Phasianus colchicus*, the natural distributional range of which extends from Caucasia to the China Sea and Formosa. This bird was introduced by the Romans some two millennia ago.

The only example of a deliberate and successful introduction of species to a new area within the Mediterranean Basin is that of the partridges of the genus *Alectoris* to Mediterranean islands. Four species of partridge occur in this region and their distribution is allopatric, with *A. rufa* in the western part of Mediterranean Europe (Spain to Italy), with *A. graeca* in the central part (Italy to Greece and Bulgaria), with *A. chukar* in the eastern part (Turkey and the Middle East), and with *A. barbara* in North Africa. There is a narrow belt of hybridisation between *A. rufa* and *A. graeca* (Bernard-Laurent, 1984) and between *A. graeca* and *A. chukar* (Dragoev, 1974) where these species come in contact. It has been established from palaeontological records that there were no *Alectoris* on Mediterranean islands before humans invaded them some 10000 to 8000 years ago (Mourer-Chauviré, 1975; Vilette, 1983). There is direct evidence from historical records that at least two or three and presumably all four species have been introduced by humans several times to each of the larger islands of the Mediterranean as is the case for Corsica (Thibault, 1983). There is at the present time only one species of *Alectoris* left on each of these islands: *A. rufa* on Corsica, *A. barbara* on Sardinia, *A. graeca* on Sicily and *A. chukar* on Crete and Cyprus. The distribution of these partridges on the Mediterranean islands has become allopatric just as it is on the mainland and each species inhabits the island which is nearest the part of the mainland to which it is native. Each species seems to occupy on the island a larger spectrum of habitats than it does on the mainland (Thibault, 1983), which is an insular feature (Blondel, 1986). The story of the partridges on Mediterranean islands is a nice example of competitive exclusion superimposed on very particular patterns of colonisation. A similar story has occurred for all the non-flying mammals of the Mediterranean islands which have been introduced by humans since they invaded the islands (see for instance, Alcover, 1979, for the Balearics, and Vigne, 1983, for Corsica).

Finally, two rather anecdotal cases may be cited here. The first is that of very

Table 22.1. *Examples of European bird species which have expanded their distributional range since the beginning of the century*

Species
Anthus pratensis
Aythia ferina
A. fuligula
Bubulcus ibis
Carpodacus erythrinus
Cettia cetti
Cisticola juncidis
Dryocopus martius
Emberiza aureola
Hippolais polyglotta
Phoenicurus ochruros
Phylloscopus bonelli
P. trochiloides
Picoides syriacus
Podiceps nigricollis
Regulus ignicapillus
Serinus canaria
Streptopelia decaocto
Turdus pilaris

Source: Harrison, 1982, and others

small free-living populations of the ring-necked parakeet *Psittacula krameri* which breed occasionally in European cities, e.g. in Nice and Brussels (P. Devillers, personal communication). The second example is that of the Asiatic red avadavar *Amandava amandava* which developed a rather large free-breeding population (more than 1000 birds) in the Spanish province of Extramadura (de Lope *et al.*, 1984). This cage bird now occurs in many countries on several continents but it does not seem to become a pest anywhere.

Range modifications of species are a common feature in biogeography and it is not easy to decide whether or not some distributional shifts must be interpreted as invasive processes. There are hundreds of cases of species which either expand or reduce their distributional range. For most of them the causes of these changes are unknown and cannot be obviously related to any habitat modification. More than 20 species are known in Europe to exhibit such changes (Table 22.1). For example, the ranges of several warbler species appear to be extending northwards in western Europe. Although it has been suggested that these changes may be a response to a general post-glacial amelioration of the climate (Kalela, 1952), the rapidity of range extension for many species belies this view (Cody, 1985). Habitats change faster in response to human impact than to climatic shifts and genetic changes in expanding populations may facilitate a rapid range expansion. Cetti's warbler *Cettia cetti* took 50 years to reach Britain from the Mediterranean (Bibby, 1982). The firecrest *Regulus ignicapillus* has colonised Britain and spread over the country in the last 20 years. Other species with largely southern European ranges, such as Bonelli's warbler *Phylloscopus bonelli* and the melodious warbler *Hippolais polyglotta*, are also moving northwards in continental Europe (Fouarge, 1969; Paquet, 1978). Moreover, there are many cases of birds which are becoming more and more abundant in northern Europe, such as, for instance, the willow warbler *Phylloscopus trochilus* in Finland (Järvinen & Väsinänen, 1978) and blackcaps, which show an increasing tendency to overwinter in continental Europe. Overall, the biogeographic picture of these warblers is dynamic: species composition in a given region is continuously changing over the long term (Cody, 1985).

One example will illustrate the process of invasion by a bird species. The spread of the collared dove *Streptopelia decaocto* is the most impressive. With those of the serin *Serinus canaria* (Mayr, 1926), the muskrat *Ondatra zibethicus* (Skellam, 1951) and the grey squirrel *Sciurus carolinensis* (Reynolds, 1985), the collared dove is one of the best known examples of an animal invasion (Nowak, 1971). The collared dove is a representative of the Indian–African faunal element (Voous, 1960). Until the late 1920s the range of this species extended from southern and central Asia to the Balkans in Bulgaria, northern Greece and eastern Yugoslavia. At the end of the 1920s the bird showed a quite unexpected and explosive process of invasion in western Europe. It reached Holland in the late forties, northern France in 1952 and the British Isles in 1955. The colonisation of Mediterranean Europe has been much slower. Spain was colonised in the 1970s and it was not until 1976 that the first breeding pairs were recorded in Corsica (Thibault, 1983). On the whole, the rate of spread of the collared dove was very high, since it occupied an average of 74300 km^2 each year (Nowak, 1971). The process of spread was natural and active but the causal mechanism remains largely unknown, although some hypotheses have been put forward by Mayr (1951) and Nowak (1965, 1971). The habitat of this dove is mainly anthropogenic, and because it mostly lives in human settlements the species' spread has been a step-wise colonisation from city to city.

Range modifications of birds as a result of human-induced changes

The Mediterranean Basin has been so dramatically modified by humans for several millennia that there have been great changes in levels of abundance and distributional patterns for most species of birds. I shall summarise only some general trends. At the scale of the Mediterranean Basin as a whole, the most important changes are a dramatic reduction of the habitats for forest-dwelling species. As explained in a previous section, the history of the Mediterranean region is also that of a progressive shrinkage of forest habitats and of a symmetrical extension of matorral (shrubland)-type habitats and farmlands. Extremely few species have succeeded in invading matorrals, and since matorrals are only secondary habitats characterised by few speciation processes, their bird communities are very poor. Deforestation and the degradation of most lowland and semi-montane habitats in the Mediterranean Basin, arising as a result of human pressures, have led to a spatial extension of both steppic and Mediterranean species which were formerly much less numerous and much more patchily distributed. Consequently, there has been a significant withdrawal of boreal and temperate forest species as well as a loss of some larger species (especially raptors) because of the breaking up of favourable habitats into small isolated woodlots. Recent trends in landscape management, however, illustrate how distributional patterns are influenced by land use.

Rural France is characterised by a general desertion of the country which started at the end of the nineteenth century and was accelerated after each world war. Because of the increase in human population and a very severe exploitation of the vegetation during the Second World War, human pressure on natural habitats was very severe between the years 1940 and 1950 and this probably rivalled that which prevailed over several millennia since the Neolithic or Bronze age (A. Pons, personal communication). But this pressure, which lasted only a decade or so, stopped almost at once just after the war because of the use of fossil fuel instead of firewood for machines and domestic purposes. As a result, there is at the present time, a generalised return of vegetative cover with more and more old coppices becoming taller and taller. Simultaneously, plant ecologists have noticed that this biological reversion is accompanied by a strong subduing of the Mediterranean character of the vegetation. When they analysed the avifaunal changes occurring in the Camargue and its surroundings over the last century, Blondel & Isenmann (1981) found that out of 30 changes, 6 were extinctions and 24 were new acquisitions. Many of these new species are middle-European forest birds which progressively have come back to their former habitats as humans stopped cutting wood. Furthermore, in many cases, the dramatic destruction of forests by wood cutting, fire and overgrazing, produced such catastrophic erosion of soils that important reforestation programs started in Mediterranean France as early as the end of the last century are still in progress. As a result, many Mediterranean mountains which were almost bare of woody vegetation a century ago are now covered with forests. Although artificial, many of these forest stands are now inhabited by species of middle-European habitats which have driven out the species typical of more open landscapes and which had secondarily invaded these degraded habitats. Thus these forest birds were indirectly reintroduced by humans to their former habitats (Blondel, 1976).

The rate of vegetation improvement is fast enough to be noticed by the biologist. This is true not only for plants but also for birds: during the last 20 years, no less than six species (viz. *Buteo buteo*, *Pernis apivorus*, *Dryocopus martius*, *Turdus philomelos*, *Sitta europaea* and *Parus palustris*) which were unknown as breeding birds in Provence have enlarged their distributional range towards the south.

On the other hand, many kinds of highly artificial and heterogeneous anthropogenic habitats (e.g. farmlands and orchards) have been invaded by native species which were rare or only locally distributed before human impact. Urban settlements have existed for at least 3500 years in the Mediterranean region, especially in the eastern Mediterranean where there were flourishing pre-Greek civilisations in the Middle East, Turkey, Cyprus, Crete, etc. The building materials of these ancient cities as well as their spatial organisation and the way of life of their inhabitants were probably very favourable for a progressive adaptation of many species to live in the immediate vicinity of cities, more or less as commensals. To be a successful human commensal requires a species to be inquisitive, familiar, gregarious, sociable, cunning, omnivorous but never harmful to human interests. Some large species were very useful as municipal scavengers and were protected for this reason. This must be true for such species as ibises, storks, kites, vultures, falcons (*Falco tinnunculus*, *F. naumanni*), some owls (*Tyto alba*, *Athene noctua*, *Otus scops*), swifts, swallows, *Phoenicurus ochruros*, *Emberiza striolata*, sparrows, several finches, *Sturnus unicolor*, *S. vulgaris*, crows and some others. Many of these species are 'synanthropic' (Harding & Sutton, 1985), and some of them, such as the starling and the house sparrow, are among the most aggressive invaders in the world. For most of the species which succeeded in adapting to cities in the Mediterranean Basin this commensalism has been a starting point for huge demographic and spatial expansions through direct transportation by humans, intentional or not, as well as by spontaneous step-wise progression from city to city. Unfortunately very little is known of the role of ancient Mediterranean urbanisation on the life histories and the range modifications of these species. But close relationships, commensal with humans, such as that which has been described by Meininger *et al.* (1980) for the crow *Corvus splendens* for which no population is known to live away from human settlements, must have had important repercussions on the distribution and adaptive characters of the species. Erz (1964) suggested that many of the species commonly found in large European cities belong to a pool of more or less human-adapted populations which exchange propagules between themselves in the archipelago of cities and which are almost completely isolated from populations of the natural habitats around their perimeters. Such an hypothesis, which has been questioned by Tomialojc (1985), still has to be tested by demographic and genetic studies.

Invasions as demographic outbreaks

From an ecological viewpoint large increases in population sizes within communities also belong to the subject of invasions. But it is usually extremely difficult to evaluate at the community level the mechanisms and consequences of such demographic outbreaks. This problem would require a level of treatment which is outside the scope of this chapter. Some aspects of this problem only will be summarised briefly here. As before, many cases of invasions at the community level are because of changes in habitats as a response to human activities. Indeed, there are extremely few instances of birds which manage to invade natural habitats as opposed to human settlements and farmlands. It is well known that natural primaeval habitats, especially primaeval forests, resist invasion by alien species including migrant birds (Moreau, 1972; Herrera, 1978). Most demographic outbreaks have occurred as a result of an improved food supply, especially in winter. Two prerequisites for successful demographic outbreaks are firstly, a release in limiting factors such as the food supply in highly seasonal environments, and secondly, the availability of species which are, so-to-say, pre-adapted to exploit new resources produced by humans, such as crops, refuse, garbage, etc. For these reasons most of the species which produce spectacular outbreaks and eventually become pests are either granivorous species, such as sparrows, doves, the black-faced dioch *Quelea quelea* (Ward, 1965), or omnivorous species, such as crows and gulls. As demonstrated by Lebreton (1981) from models of population dynamics, population changes are more sensitive to adult survival than to an improvement in fecundity, especially in long-lived birds, such as gulls *Larus* sp. For this reason, the superabundant food supply that these species can exploit all the year around, such as garbage and refuse from fishery operations, allows them to manifoldly increase their populations (see for instance Isenmann, 1976–1977; Lebreton & Isenmann, 1976; Cramp & Simmons, 1983). This arises because their adult survival has been dramatically improved by eliminating seasonality in their resources, thanks to the permanent superabundance of food for these birds.

The problem of demographic outbreaks is important in relation to wildlife conservation because invasive species cause considerable problems for management of nature reserves. Since most nature reserves, even the largest ones, are too small in size to be completely free from the influence of their surrounding habitats, demographic-invading species can become a threat to local communities. For this reason programs of containment of pest birds had to be developed in many localities in Europe including the Mediterranean region. For instance, gulls which survive to a large extent all the year around on refuse and garbage have to find quiet sites to breed. Quite often they find such safe sites in reserves and since they are too numerous to feed upon the local natural resources, they become harmful to communities by either harvesting resources which should be exploited by other species, or by importing into the reserve organic material coming from elsewhere, thereby participating in a process of eutrophication of the reserve, or thirdly, by becoming predators on eggs and chicks of other birds. One example of such problems is that of the National Reserve of the Camargue. In order to protect breeding colonies of flamingos, avocets and terns against predation by the Mediterranean herring gull *Larus cachinnans*, programs of containment of the population of this latter species, which increased from a few pairs in the 1930s to about 1000 pairs in the 1960s, became a regular management practice.

Conclusion

As far as birds are concerned, the problem of biological invasions in the Mediterranean Basin must be considered at three scales of space and of biological processes. Firstly, invasion *per se*, that is the appearance of a new alien species, is an extremely rare event and, in any case, of no biological significance for Mediterranean biotas. Secondly, changes in the distribution of native species within the Mediterranean region have occurred many times and for most species as a result of human impacts on landscapes. Schematically, these changes find expression in a decrease in the distribution of forest birds and an increase in that of open and shrubby habitats as well as a large development of the so-called synanthropic species. Thirdly, demographic outbreaks of some species have managed to take advantage of human conditions. Many of them which were rare under natural conditions became pests as a result of habitat changes. The most conspicuous examples are several species of gull (*Larus* spp.), the starling *Sturnus vulgaris* (which vastly increased during the last hundred years; Berthold, 1968), two species of sparrows and, to a lesser extent, the woodpigeon *Columba palumbus* (Tomialojc, 1980), as well as some other species. Unfortunately very little is known so far about the community responses to this kind of invasion. As pointed out by Williamson & Brown (1986), models are required to indicate which invasive species would disrupt the structure of communities.

From an extensive review of the literature concerning invasions, Simberloff (1981) found little support for classical community models derived from niche theory, such as the model of equilibrium island biogeography and the model of limiting similarity of coexisting species. Both models imply that invasive species should interact with native species in the community they invade and eventually displace them. The fact that Simberloff found no effect on the indigenous species for 80 per cent of recorded invaders does not mean that invading species do not threaten the communities. Indeed, invasions may cause problems to natural communities and accordingly, are relevant to the achievement of the aims of the world conservation strategy, namely the maintenance of essential ecological processes and life support systems, the preservation of genetic diversity and the sustainable utilisation of species and ecosystems (IUCN, 1979).

References

Alcover, J. A. (1979). *Els Mamifers de les Baleares*. Palma de Mallorca: Editorial Moll.

Baker, H. G. & Stebbins, G. L. (ed.) (1965). *The Genetics of Colonizing Species*. New York: Academic Press.

Bernard-Laurent, A. (1984). Hybridation naturelle entre Perdrix bartavelle (*Alectoris graeca saxatilis*) et Perdrix rouge (*Alectoris rufa*) dans les alpes maritimes. *Gibier Faune Sauvage*, **3**, 79–96.

Berthold, P. (1968). Die massenvermehrung des stars *Sturnus vulgaris* in fortpflanzungs-physiologischer sicht. *Journal für Ornithologie*, **109**, 11–16.

Bibby, C. J. (1982). Polygyny and breeding ecology of the Cetti's warbler *Cettia cetti*, *Ibis*, **124**, 288–301.

Birks, H. J. B. (1986). Late Quaternary biotic changes in terrestrial and lacustrine environments, with particular reference to north-west Europe. In *Handbook of Holocene Palaeoecology and Palaeontology*, ed. B. E. Berglund, pp. 3–65. New York: Wiley.

Blondel, J. (1976). L'influence des reboisements sur les communautés d'oiseaux, l'exemple du Mont-Ventoux. *Annales des Sciences Forestières*, **33**, 221–45.

Blondel, J. (1982). Caractérisation et mise en place des avifaunes dans le bassin Méditerranéen. *Ecologia Mediterranea*, **8**, 253–72.

Blondel, J. (1985). Historical and ecological evidence on the development of Mediterranean avifaunas. *Acta XVIII Congressus Internationalis Ornithologici, Moscow*, **1**, 373–86.

Blondel, J. (1986). *Biogéographie Evolutive*. Paris: Masson.

Blondel, J. (1988). Biogéographie évolutive á différentes échelles: l'histoire des avifaunes méditerranéennes. *Acta XIX Congressus Internationalis Ornithologici, Ottawa*, **1**, 155–88.

Blondel, J. & Isenmann, P. (1981). *Guide des Oiseaux de Camargue*. Neuchâtel: Delachaux & Niestlé.

Brodkorb, P. (1971). Origin and evolution of birds. In *Avian Biology*, vol. 1, ed. D. S. Fariner & J. R. King, pp. 19–55. New York: Academic Press.

Cody, M. L. (1985). *Habitat Selection in Birds*. New York: Academic Press.

Cramp, S. & Simmons, K. E. L. (1983). *Handbook of the Birds of Europe, the Middle East and North Africa*, vol. 3, *Waders and Gulls*. Oxford: Oxford University Press.

Cruon, R. & Nicolau-Guillaumet, P. (1985). Notes d'ornithologie française. *Alauda*, **53**, 34–63.

Davis, M. B. (1981). Quaternary history and the stability of forest communities. In *Forest*

Succession: Concepts and Applications, ed. D. C. West, H. H. Shugart & D. B. Potkin, pp. 132–53. New York: Springer-Verlag.

de Lope, F., Guerrero, J. & de la Cruz, C. (1984). Une nouvelle espèce à classer parmi les oiseaux de la Péninsule Ibérique: *Estrilda* (*Amandava*) *amandava* (Ploceidae, Passeriformes). *Alauda*, **52**, 312.

Dragoev, P. (1974). On the population of the rock partridge *Alectoris graeca* (Meisner) in Bulgaria and methods of census. *Acta Ornithologica*, **14**, 251–5.

Elton, C. S. (1958). *The Ecology of Invasions by Animals and Plants*. London: Methuen.

Erz, W. (1964). Populationsökologische untersuchungen an der avifauna zweier nordwestdeutscher grosstädte. *Zeitschrift für Wissenschaftliche Zoologie*, **170**, 1–111.

Fouarge, J. (1969). Le pouillot de Bonelli (*Phylloscopus bonelli*) étend-il son aire de nidification vers le nord? *Aves*, **6**, 34–139.

Groves, R. H. & Burdon, J. J. (ed.) (1986). *The Ecology of Biological Invasions: An Australian Perspective*. Cambridge: Cambridge University Press.

Harding, P. T. & Sutton, S. L. (1985). *Woodlice in Britain and Ireland: Distribution and Habitat*. Monks Wood: NERC Biological Records Centre.

Harrison, C. (1982). *An Atlas of the Birds of the Western Palaearctic*. Princeton, N.J.: Princeton University Press.

Herrera, C. M. (1978). On the breeding distribution pattern of European migrant birds: MacArthur's theme re-examined. *Auk*, **95**, 496–509.

Huntley, B. & Birks, H. J. B. (1983). *An Atlas of Past and Present Pollen Maps for Europe: 0–13 000 Years Ago*. Cambridge: Cambridge University Press.

Isenmann, P. (1976–1977). L'essor spatial et démographique de la mouette rieuse en Europe. *L'Oiseau*, **46**, 337–66, **47**, 25–40.

IUCN (1979). *World Conservation Strategy*. Morges: IUCN.

Järvinen, O. & Väsinänen, R. A. (1978). Recent changes in forest bird populations in northern Finland. *Anales Zoologici Fennici*, **15**, 279–89.

Kalela, O. (1952). Changes in the geographic distribution of Finnish birds and mammals in relation to recent changes in climate. *Fennia*, **75**, 41–59.

Lack, D. (1976). *Island Biology Illustrated by the Land Birds of Jamaica*. Oxford: Blackwell Scientific Publications.

Lebreton, J. D. (1981). Contribution à la dynamique des populations d'oiseaux. Modèles mathématiques en temps discret. Thèse Etat, Université Claude Bernard, Lyon.

Lebreton, J. D. & Isenmann, P. (1976). Dynamique de la population camarguaise de mouette rieuse (*Larus ridibundus*): un modèle mathématique. *Revue Ecologie (Terre et Vie)*, **30**, 529–49.

Macdonald, I. A. W., Kruger, F. J. & Ferrer, A. A. (ed.) (1986). *The Ecology and Management of Biological Invasions in Southern Africa*. Cape Town: Oxford University Press.

Mayr, E. (1926). Die Ausbreitung des Girlitz. *Journal für Ornithologie*, **74**, 572–671.

Mayr, E. (1951). Speciation in birds. *Proceedings X International Ornithological Congress, Uppsala*, pp. 484–93.

Meininger, P. L., Mullé, W. C. & Bruun, B. (1980). The spread of the house crow *Corvus splendens* with special reference to the occurrence in Egypt. *Le Gerfaut*, **70**, 245–50.

Mooney, H. A. & Drake, J. A. (ed.) (1986). *Ecology of Biological Invasions of North America and Hawaii*. New York: Springer-Verlag.

Moreau, R. E. (1954). The main vicissitudes of the European avifauna since the Pliocene. *Ibis*, **96**, 411–31.

Moreau, R. E. (1972). *The Palaearctic–African Bird Migration Systems*. New York: Academic Press.

Mourer-Chauvire, C. (1975). Les oiseaux du Pléistocène moyen et supérieur de France. Thèse Etat, Université Claude Bernard, Lyon.

Nowak, E. (1965). Die Türkentaube. Neue Brehm Bücherei No. 353, Wittenberg Lutherstadt.

Nowak, E. (1971). The range expansion of animals and its causes. *Zeszyty Naukowe*, **3**, 1–255.

O'Connor, R. J. (1986). Biological characteristics of invaders among bird species in Britain. *Philosophical Transactions of the Royal Society of London*, **B314**, 583–98.

Paquet, A. (1978). Un couple d'hypolais polyglotte (*Hippolais polyglotte*) cantonné dans l'Entre-Sambre-et-Meuse. *Aves*, **15**, 81–3.

Pons, A. (1981). The history of the Mediterranean shrublands. In *Ecosystems of the World*, vol. 11, *Mediterranean-type Shrublands*, ed. F. di Castri, D. W. Goodall & R. L. Specht, pp. 131–8. Amsterdam: Elsevier.

Pons, A. (1984). Les changements de la végétation de la région méditerranéenne durant le Pliocène et le Quaternaire en relation avec l'histoire du climat et de l'action de l'homme. *Webbia*, **38**, 427–39.

Reynolds, J. C. (1985). Details of the geographic replacement of the red squirrel (*Sciurus vulgaris*) by the grey squirrel (*Sciurus carolinensis*) in eastern England. *Journal of Animal Ecology*, **54**, 149–62.

Sharrock, J. T. R. & Sharrock, E. M. (1976). *Rare Birds in Britain and Ireland*. Berkhamstead: Poyser.

Simberloff, D. (1981). Community effects of introduced species. In *Biotic Crises in Ecological and Evolutionary Time*, ed. M. H. Nitecki, pp. 53–81. New York: Academic Press.

Skellam, J. G. (1951). Random dispersal in theoretical populations. *Biometrika*, **38**, 196–218.

Steinbacher, G. (1948). Der Einfluss der Eiszeit auf die europaische Volgelwelt. *Biologisches Zentralblatt*, **67**, 444–56.

Thibault, J. C. (1983). *Les Oiseaux de la Corse*. Ajaccio: Parc Naturel Régional de Corse.

Tomialojc, L. (1980). The impact of predation on urban and rural woodpigeon (*Columba palumbus*) populations. *Polish Ecological Studies*, **5**, 141–220.

Tomialojc, L. (1985). Urbanization as a test of adaptive potentials in birds. *Acta XVIII Congressus Internationalis Ornithologicii, Moscow*, pp. 608–14.

Vigne, J. D. (1983). Les mammifères terrestres non volants du post-glaciaire de la Corse. Thèse, Université de Paris VI, Paris.

Vilette, J. (1983). Avifaunes de la fin du Pléistocène supérieur et de l'Holocène dans le sud de la France et en Catalogne. Systématique, paléoenvironnement, paléoethnologie. Thèse, Université Claude Bernard, Lyon.

Voous, K. H. (1960). *Atlas of European Birds*. Edinburgh: Nelson.

Voous, K. H. (1963). The concept of faunal elements or faunal types. *Proceedings XIII International Ornithological Congress*, pp. 1104–8.

Ward, P. (1965). Feeding ecology of the black-faced dioch *Quelea quelea* in Nigeria. *Ibis*, **107**, 173–214.

Williamson, M. H. & Brown, K. C. (1986). The analysis and modelling of British invasions. *Philosophical Transactions of the Royal Society of London*, **B314**, 505–22.

23

Invasions in the mediterranean avifaunas of California and Chile

F. VUILLEUMIER

Avian biogeographers often ignore the introduced faunal element in their studies of local or regional avifaunas. They usually analyse the native species of a given fauna and pay little or no attention to the non-native element. This tradition is more noticeable in analyses of continental avifaunas than in insular ones. Such a difference in treatment is because of the fact that many insular avifaunas (Mayr, 1965; Wodzicki, 1965) have a very large component of introduced species (both in terms of species numbers and relative abundance of individuals in some species), whereas continental avifaunas have fewer (Mayr, 1965; Navas, 1987).

This chapter will describe and analyse the introduced species occurring in the continental avifaunas living in the mediterranean bioclimatic zones of California and Chile. To the author's knowledge this analysis is the first attempt to census introduced species in these avifaunas and to make biogeographic comparisons between them. Mooney *et al.* (1986), paraphrasing Small (1974), discussed very briefly the introduced bird species established in California, but did not focus on birds of the mediterranean-climate zone of that state. I know of no paper which reviews the introduced species of birds found in Chile. Navas (1987) discussed the nine introduced species of birds occurring in Argentina.

Introduced species found in a given avifauna can be classified in three categories:

1. Species that were deliberately introduced by humans ('introduced' in the strict sense)
2. Species that were released accidentally, e.g. caged birds kept in zoos or as pets that escaped from confinement (called 'escapees')
3. Species that invaded the fauna being studied without deliberate human help during a process of natural range expansion (called 'expanding').

Examples of these three categories in the mediterranean avifaunas of California or Chile are respectively (i) *Alectoris chukar*, introduced to California;

(ii) *Zosterops japonicus*, an escapee in California; and (iii) *Bubulcus ibis*, a species which is expanding in both California and Chile. In practice, it is often difficult to distinguish unequivocally between introduced species in the strict sense and escapees, because some species that have escaped from captivity also have been released deliberately by humans at times. Also, some introduced species have had spectacular range expansions after their original introduction.

In order to place introduced species (*sensu lato*) in their proper avifaunal context, it has been necessary in the preparation of this chapter to review critically the breeding avifaunas of California and Chile. Thus as a first step, a complete list of all breeding species, including native and non-native ones, was compiled for both areas. Subsequently, as a second step, a list of introduced species was made. These two lists are the basis for comparisons and discussions of the significance of introduced species in the avifaunas of these two mediterranean zones. Appendix 23.1 lists all species breeding in the mediterranean zone of California, and Appendix 23.2 is the list for the mediterranean zone of Chile. These lists may be the first published lists of mediterranean-zone species of birds for California and Chile.

Materials and methods

The list of bird species breeding in California and the mediterranean zone of that State (Appendix 23.1) was compiled from Grinnell & Miller (1944), Miller (1951), Small (1974), the 1983 checklist of the American Ornithological Union (AOU), supplemented by personal field observations made during trips to California in 1971, 1975, 1980 and 1984 (the latter are largely unpublished but some observations were used in Blondel *et al.*, 1984*a*, 1984*b*). The list of introduced species found in California was compiled from the same sources and from Hardy (1973).

The list of bird species breeding in Chile and the mediterranean zone of that country was prepared from Hellmayr (1932), Philippi (1964), Johnson (1965, 1967), Meyer de Schauensee (1966, 1982), Araya & Millie (1986), supplemented by personal field observations made during trips to Chile in 1965, 1967, 1985, 1987 and 1988 (which, as for California, are largely unpublished, but see Blondel *et al.*, 1984*a*, 1984*b*). The list of introduced species found in Chile was compiled from the same sources and from Sick (1968).

Avifaunas of California and Chile

Total avifaunas

A total of 284 species of land and water birds (excluding pelagic birds) breed in the State of California (Appendix 23.1). Nineteen of these 284 breeding species (7 per cent) are introductions ('introduced', 'escapees' or 'expanding'). Thus the

Table 23.1. *Numbers of breeding land and water bird species in the avifaunas of California and Chile*

	California	Chile
Total avifauna		
Total no. species	284 (100%)	263 (100%)
Total no. introduced	19 (7%)	7 (3%)
Total no. native	265 (93%)	256 (97%)
Mediterranean avifauna		
Total no. species	199 (100%)	128 (100%)
Total no. introduced	18 (9%)	7 (5%)
Total no. native	181 (91%)	121 (95%)

total number of native breeding species in California is 265 (93 per cent of total). Mooney *et al.* (1986) stated that 525 species occur in California, a figure based on Small (1974) that includes all species of breeding birds, as above, and also pelagic species, migrants, occasional visitors and accidentals. Small (1974, p. 33) wrote that '302 species breed or have bred in California of which 277 are regular nesters'. Note that five families in the Californian avifauna are not native and are composed of introduced species only: Psittacidae (five species of parrots and parakeets), Sturnidae (starling), Zosteropidae (Japanese white-eye), Passeridae (house sparrow) and Estrildidae (African fire-finch).

A total of 263 species of land and water birds, excluding pelagic species, breed in Chile (Appendix 23.2). Seven of these 263 species (3 per cent) are introductions ('introduced', 'escapees' or 'expanding'). Thus the total number of native breeding birds in Chile is 256 species (97 per cent of the total). Note that two families in Chile are not native and composed of introductions, namely Phasianidae (ring-necked pheasant, California quail) and Passeridae (house sparrow).

With 284 and 263 species of breeding birds (265 and 256 native species), respectively, the avifaunas of California and Chile are similar in species richness (Table 23.1). This is to be expected as each region is similarly diverse, both ecologically and topographically, and has the same range of habitats, including high mountain grasslands, deserts, mediterranean shrublands, cultivated valleys and temperate rainforests.

Mediterranean avifaunas

The mediterranean bioclimatic zone of California supports 199 species of breeding land and water birds (pelagic species excluded). Eighteen of these 199 species (9 per cent) are introductions. Thus the total number of native breeding species in the mediterranean zone of California is 181 (91 per cent of the total mediterranean breeding avifauna).

The mediterranean zone of Chile supports 128 species of breeding land and water birds (pelagic species excluded). Seven of these 128 species (5 per cent) are introductions. Thus the total number of native breeding species in the mediterranean habitats of Chile is 121 (95 per cent of the total mediterranean breeding avifauna).

There are 71 more species (total fauna) or 60 more species (native fauna) in the mediterranean bioclimatic zone of California than in the equivalent zone of Chile (Table 23.1). This large difference (33–36 per cent) is expected on the basis of the general conclusion reached by Cody (1970) and by Vuilleumier (1985) that Chilean regional avifaunas show insular characteristics, including low species diversity. This fact seems to be because of, in part, the greater geographical isolation of the Chilean fauna, as it is cut off from faunal exchange with other faunas by the Andes.

Bird species introduced to California

The list below certainly includes the most important species, but it is probably incomplete. Some other species are or have been released from time to time in California, but their populations are exceedingly small (e.g. *Pycnonotus jocosus* is listed by Small (1974) but not by AOU (1983)). Many species have been introduced and released in southern California (Hardy, 1973), which, as for southern Florida (Owre, 1973), has a climate that is very conducive to at least temporary survival of a number of tropical or subtropical species.

1. *Bubulcus ibis* (cattle egret) (mediterranean zone; expanding). This species, not mentioned by Grinnell & Miller (1944), now breeds in north-western and central California (Small, 1974; AOU, 1983), including the mediterranean zone. The cattle egret has undergone a spectacular, world-wide expansion in the twentieth century (Handtke & Mauersberger, 1977) and has apparently reached the Americas from Africa, making its first landfall in South America. It arrived in California unaided directly by humans, although it clearly benefits from human surroundings (cattle and cattle pastures). The cattle egret is a water bird and does not occur in mediterranean shrublands. I conclude that its impact on the Californian native avifauna is insignificant.

2. *Perdix perdix* (gray partridge) (not in mediterranean zone; introduced). Numerous introductions of this Eurasian species have been made (Grinnell & Miller, 1944), but it may not survive in California (Small, 1974; AOU, 1983). Its impact on mediterranean avifaunas is thus nil, since it does not occur in the mediterranean zone at present.

3. *Alectoris chukar* (chukar) (mediterranean zone; introduced). This Eurasian species has been introduced since 1932 (Grinnell & Miller, 1944) and is now established throughout the State (Small, 1974) locally 'south to extreme northern Baja California' (AOU, 1983, p. 134). It lives in 'arid foothills with rocky areas, along the desert and valley edges' (Small, 1974, p. 64). Drinking water is artificially supplied (Small, 1974). The ecological role of the chukar in mediterranean habitats has not been assessed. This species does not seem very abundant and would not seem to have an important ecological function in the mediterranean avifauna.

4. *Phasianus colchicus* (ring-necked pheasant) (mediterranean zone; introduced). Attempts to introduce this south-central Palaearctic species have been made since 1885 (Grinnell & Miller, 1944) and thousands have been liberated over the succeeding years. It is well established locally (Small, 1974; AOU, 1983). The ring-necked pheasant occurs in 'brushy fields, agricultural land, stubble fields' (Small, 1974, p. 64). It is probably largely restricted to cultivated areas in the mediterranean zone of California. Its ecological role in the local avifauna is not clear, but may not be very important.

5. *Pavo cristatus* (common peafowl) (mediterranean zone; introduced or escapee, or both). This tropical Asian species is not mentioned by Grinnell & Miller (1944). It is now locally established in a semi-domesticated state (Small, 1974; AOU, 1983). Hardy (1973) discussed a small feral population in southern California, where birds appear to be 'thoroughly wild and completely independent of man for food' (p. 507). Because of the extremely small and localised population in the mediterranean zone, this species clearly plays an insignificant ecological role in this avifauna.

6. *Meleagris gallopavo* (wild turkey) (mediterranean zone; introduced). This species, native to North America, was once widespread on this continent. In California attempts at implantation go back to 1877 (Grinnell & Miller, 1944). The turkey is established locally at present (Small, 1974; AOU, 1983) and is found in 'oak woodlands, riparian woodland, pine forests' (Small, 1974, p. 65). Local populations in the woodlands of the mediterranean zone are small and are not likely to have much impact on the avifauna.

7. *Columba livia* (rock dove) (mediterranean zone; introduced). This Palaearctic and oriental species is now abundant in a semi-domesticated state and occurs mainly in urban, suburban and farming areas (Grinnell & Miller, 1944; Hardy, 1973; Small, 1974). Because of its largely urban and suburban habitat in the mediterranean zone of California, this species is unlikely to have a significant impact on the avifauna, in spite of its large population size.

8. *Streptopelia risoria* (ringed turtle-dove) (mediterranean zone; introduced). The origin of this species is unclear. Introductions were made from domestic

stock, which 'may have been derived from either *S. roseogrisea* of Africa or *S. decaocto* of Eurasia, these two forms considered conspecific by some authors. For the present, it seems best to retain the usage of *S. risoria*' (AOU, 1983). In California it seems to be restricted to city parks in the Los Angeles area (Grinnell & Miller, 1944; Hardy, 1973; Small, 1974; AOU, 1983). Because of the small size and geographical restriction of the population to an urban or semi-urban area, this species probably has no impact on the mediterranean avifauna.

9. *Streptopelia chinensis* (spotted dove) (mediterranean zone; introduced or escapee or both). This eastern Palaearctic and oriental species was 'introduced, or escaped, probably prior to 1917, when already "common" in northern Hollywood, a suburb of Los Angeles' (Grinnell & Miller, 1944, pp. 567–8). This species now occurs 'as far north as Santa Barbara, as far south as San Diego, and inland to the Salton Sea' (Hardy, 1973, p. 507; see also Small, 1974; AOU, 1983). Hardy (1973, p. 507) adds 'Eastward its occurrence seems restricted by the deserts and the species' requirements of large trees, especially eucalyptus. In its optimal suburban habitat it seems to outnumber the Mourning Dove (*Zenaidura macroura*). There is no evidence that it competes with the latter species'. In spite of its relatively widespread distribution and its possible co-occurrence with *Zenaidura macroura*, a native dove, it seems unlikely that this species, found mostly in 'cities, towns, and suburbs" (Small, 1974, p. 87) has much ecological impact on the avifauna. It is of interest to note that introduced trees (viz. *Eucalyptus* spp.) are what this introduced species prefers.

10. *Psittacula krameri* (rose-ringed parakeet) (mediterranean zone; introduced). Grinnell & Miller (1944) did not mention this species, which originates from Africa and southern Asia. Its distribution is restricted to southern California, although its status as an established species is not clear. For instance, Hardy (1973) stated that it was extirpated, whereas AOU (1983, p. 267) stated that 'small introduced groups have also persisted in . . . southern California'. If still extant, this population is too small and localised to have any impact on the avifauna of the mediterranean zone.

11. *Nandayus nenday* (black-hooded parakeet) (mediterranean zone; escapee). This South American species was not mentioned by Grinnell & Miller (1944). Hardy (1973) and AOU (1983) suggest that a small population may exist in southern California. This species cannot have any impact on the mediterranean zone avifauna.

12. *Brotogeris versicolorus* (canary-winged parakeet) (mediterranean zone; introduced or escapee, or both). A South American species, not mentioned by Grinnell & Miller (1944), this bird is very restricted geographically in southern California (Hardy, 1973; Small, 1974; AOU, 1983). Because of the very small population this species has no effect on the mediterranean avifauna.

13. *Amazona viridigenalis* (red-crowned parrot) (mediterranean zone; introduced or escapee, or both). A Mexican species, not mentioned by Grinnell & Miller (1944), this species has now a very small and geographically restricted population in southern California (Hardy, 1973; AOU, 1983) that cannot have any impact on the ecology of the mediterranean avifauna.

14. *Amazona oratrix* (= *A. ochrocephala*) (yellow-headed parrot) (mediterranean zone; introduced or escapee, or both). Another Mexican species of parrot, not mentioned by Grinnell & Miller (1944), but found now in southern California (Hardy, 1973; Small, 1974; AOU, 1983). As it is restricted to urban and suburban areas this species can only have a negligible ecological impact on the mediterranean avifauna.

15. *Sturnus vulgaris* (starling) (mediterranean zone; introduced and expanding). The Palaearctic starling was introduced to New York City in 1890 and spread westward since that time, reaching California much later, since Grinnell & Miller (1944, p. 572) wrote that 'so far as known up to 1944, reached California but once'. According to Small (1974, p. 118) starlings 'arrived in California about 1942 as winter visitors from Oregon'. After a tremendous population and range expansion, it is now widespread and common in 'open country, especially agricultural lands, ranches, farms, cultivated land, cities, towns, suburbs, parks', but not in rainforests or deserts (Small, 1974, p. 118), and 'south to northern Baja California' (AOU, 1983, pp. 585–6, but see also the early paper by Wing, 1943, on the spread of the species prior to its having reached California). The impact of the starling on the mediterranean avifauna of California may be ecologically significant. Small (1974, p. 118) stated that the species 'threatens [the] survival of much more timid hole-nesters as bluebirds, titmice, nuthatches, swallows, wrens, woodpeckers, and even American Kestrels'.

16. *Zosterops japonicus* (Japanese white-eye) (mediterranean zone; escapee). A species native to China, Japan and the Philippines, not mentioned by Grinnell & Miller (1944). According to AOU (1983, pp. 588–9) 'A pair of some race of the *Z. palpebrosus* complex escaped from the San Diego Zoo in the early 1970s, and a small but apparently increasing population is now present in the San Diego area, although attempts are being made to control its establishment and spread'. Because of its localised and urban or suburban population, this species is unlikely to have any impact on the mediterranean avifauna.

17. *Cardinalis cardinalis* (eastern cardinal) (mediterranean zone; introduced). This species, native to North America (including south-eastern California), has been repeatedly introduced elsewhere in that State, including the mediterranean-zone areas (Grinnell & Miller, 1944; Hardy, 1973; AOU, 1983), where it is uncommon and restricted to 'riparian woodland thickets' (Small, 1974, p. 130). The geographical restriction of this population is such

that the species is unlikely to have any effect on the mediterranean zone fauna.

18. *Passer domesticus* (house sparrow) (mediterranean zone; introduced and expanding). This western Palaearctic and south-east Asian species was introduced to New York in 1850, and again several times thereafter through to 1867 (AOU, 1983). Wing (1943) described the early spread of this species westward after its original introduction in the eastern USA. Grinnell & Miller (1944) documented the spread of the species in California and stated that its habitat was 'most notably the vicinity of buildings and their associated trees'; by 1915 the species 'had spread to virtually all sections of the State' (p. 573). According to Small (1974), it reached California from adjacent states. It is now widespread and very common, especially in urban and suburban areas, but also in 'gardens, parks, ranches, stables' (Small, 1974, p. 126). Its impact on the mediterranean avifauna does not seem to be very profound because it is not common in mediterranean shrublands.

19. *Lagonosticta rubricata* (African fire-finch) (mediterranean zone; escapee). This tropical African species was not mentioned by Grinnell & Miller (1944). According to the AOU (1983, p. 788), 'Successful breeding of escaped pairs of this widespread African species was reported at Pacific Grove, Monterey County, California, in 1965 and 1966 [. . .] but no population became established'. If the population did not succeed, then of course this species cannot have any impact on the mediterranean avifauna.

Bird species introduced to Chile

Because there are few ornithologists in Chile, the list of introduced species is probably less complete than that for California, but it is likely that the main species are included.

1. *Bubulcus ibis* (cattle egret) (mediterranean zone; expanding). Not mentioned by Hellmayr (1932) or Philippi (1964). Johnson (1972) summarised the spread of the cattle egret in South America and cited only one record for Chile – in 1968 at Antofagasta. Araya & Millie (1986, p. 98) give the range in Chile 'from Arica to Navarino Island; vagrant to the Chilean part of Antarctica'. The general range of the species in South America was summarised by Meyer de Schauensee (in 1966, not stated to occur in Chile; whereas in 1982, stated to occur only in northern Chile (probably Johnson's record)). The cattle egret reached South America apparently unaided by humans, and has since expanded its range on this continent considerably (see Handtke & Mauersberger, 1977). Chile appears to have been the last South American country to be invaded. The ecological role of the species in the Chilean mediterranean zone avifauna is

unclear, but, as in California, it is not likely to be important, since it is a water bird.

2. *Cairina moschata* (muscovy duck) (mediterranean zone; escapee or introduced). Hellmayr (1932) wrote that this species of tropical American duck did not 'occur in Chile in a wild state' (p. 315) but listed under the name '*Anas iopareia*', 'a hybrid between the Muscovy Duck (*Cairina moschata*) and some domesticated race [of duck]'. Neither Philippi (1964), Johnson (1965, 1972) nor Meyer de Schauensee (1966, 1982) mentioned this species, but Araya & Millie (1986) stated that there were two cases of 'asilvestramiento' in Chile, one in Talca and another in Curico. Since this duck is frequently kept in captivity the Chilean records may represent feral individuals or populations that have originated from escapees, although deliberate introduction(s) cannot be ruled out. The insignificant size of these populations (if, indeed, there are any) clearly shows that the ecological impact of this species is nil on mediterranean avifaunas.

3. *Phasianus colchicus* (ring-necked pheasant) (mediterranean zone; introduced). This Asian species has been introduced repeatedly to Chile, as early as 1886 or 1887 in Coquimbo (from England) (Hellmayr, 1932). 'In 1897, they had largely increased in numbers without penetrating, however, more than fifteen miles inland' (Hellmayr, 1932, p. 424). This stock 'later went into decline and finally died off altogether' (Johnson, 1965, p. 279). Other introductions were made to Cautin from Germany in 1914, and the birds 'soon began to multiply' (Johnson, 1965, p. 279). Another release was made 'on the island of Pichi Colcuma in the middle of Lake Ranco. These also did well and today number perhaps 1000 birds' (Johnson, 1965, p. 279). Johnson goes on to say, however, that the pheasant did not establish itself outside these areas. Philippi (1964) cited Cautin and Valdivia, as did Araya & Millie (1986), who added that the pheasant was found also near Illapel, in the province of Choapa. The ring-necked pheasant can only be considered a marginal member of the Chilean mediterranean avifauna.

4. *Callipepla californica* (California quail) (mediterranean zone; introduced). The California quail, a native of western North America, has been introduced to Chile and to the Juan Fernandez Islands (Hellmayr, 1932; Johnson, 1965). It now occurs in Chile from Atacama (Araya & Millie, 1986) and Coquimbo (Philippi, 1964) to Puerto Montt (Sick, 1968), as well as on the Argentine side of the Andes from Cordoba and San Juan south to Neuquen and Rio Negro (Sick, 1968; Navas, 1971; Olrog, 1979; Navas, 1987; personal observation, Rio Negro, 1965). This species appears to be exported annually from Chile to Argentina and Brazil (Inskipp, 1975).

In the thirties *Callipepla californica* was already common in Chile and according to Hellmayr (1932, p. 424) had 'become perfectly acclimatized throughout the central provinces of Chile, where it is now found in large numbers'. Hellmayr (1932) also stated that 'Chilean specimens agree with the brown-backed, dark-flanked race of the humid coast region of California, which Grinnell [...] has shown to be entitled to the name of *L. californica brunnescens*' (p. 424). Johnson (1965) indicated that the northern limit of this species was the desert, and the southern one the high-rainfall area, and that in spite of intense hunting pressure 'there is no evidence of a decline in numbers' (p. 278). More recently, however, Erazo & Valenzuela (1985) thought that the species was decreasing in numbers because of hunting. Nevertheless, census results in Cody (1970, 1974) and Erazo & Valenzuela (1985) suggest that *Callipepla californica* has a relatively high density in mediterranean habitats of central Chile. It seems therefore clear that this mediterranean zone bird from California has done very well in the mediterranean zone of central Chile, and that it is occupying an area with a vegetation and climate that are very similar to those of its native country.

Callipepla californica is an ecologically important member of the mediterranean avifauna of Chile. It is interesting to point out that the only native species with which it could compete directly is the tinamou *Nothoprocta perdicaria*, with which it coexists. A comparative study of these two species would be especially rewarding, since both are sought about equally by hunters, who must exert a tremendous pressure, influencing their population dynamics and perhaps their habitat selection. Note that in Argentina, *Callipepla californica* does not seem to have any competitors and that, according to Navas (1987, p. 29), it 'has found an empty ecological niche' and 'has become a new food item for native predators'.

5. *Columba livia* (rock pigeon) (mediterranean zone; introduced, expanding). This Palaearctic species was not mentioned by Hellmayr (1932). Philippi (1964, p. 96) stated that it was found 'al estado completamente silvestre en la isla de Masatierra (Juan Fernandez)', but gave no details about its distribution on the mainland. Again, Johnson (1967) only cited *C. livia* as from Masatierra Island. Araya & Millie (1986, p. 232) state that the species is common 'in parks and gardens of our cities', but do not give the range in Chile. Wild-type birds, according to them, are to be found not only on Masatierra Island but also at Vega del Chanaral on the mainland. According to Sick (1968, p. 302; but writing, however, about all of South America) the species is found 'in all towns'. In fact, in spite of its wide distribution and great abundance, not much is known about the species' distribution in South America in general (for Argentina, see Navas, 1987), or in Chile in particular, and especially in the mediterranean zone of that country. It is clearly common there, and generally restricted to towns, and so does not seem to have much of an impact on the mediterranean avifaunas.

6. *Molothrus bonariensis* (shiny cowbird) (mediterranean zone; introduced or expanding). According to Hellmayr (1932) this species appears to have been rare in the late 1800s, but to have become more common in the first decades of the 1900s; in the late twenties and early thirties its range was from Coquimbo to Malleco. Hellmayr (1932, pp. 88–9) goes on to say 'Both Carlos Reed and Rafael Barros believe it to be very unlikely that the Cowbird, avoiding, as it does, the higher mountain ranges, crossed the Andes unaided from Mendoza, where it is known to be abundant, and advance the theory that its present Chilean population may have originated from liberated cage-birds which are frequently imported from Argentina'. Hellmayr, unfortunately, did not give his own opinion about the validity of this theory.

The shiny cowbird is native to tropical South America, eastern Panama, and the Lesser Antilles (Meyer de Schauensee, 1966, 1982). Meyer de Schauensee (1966) gave the Chilean range as 'Atacama (Copiapo) south to Llanquihue (probably introduced from Argentina)', presumably following Hellmayr (1932). The species' range appears to be greater, however, and extends more or less continuously from Atacama to Chiloe and at Chile Chico, Aysen (*fide* Philippi, 1964); from Atacama to Aysen (*fide* Araya & Millie, 1986). Neither Philippi (1964) nor Araya & Millie (1966) mention Hellmayr's or Meyer de Schauensee's introduction theory. Johnson (1967) believes that the introduction theory is more likely than the pure range expansion (invasion) theory. In any case, this species is now very common in the mediterranean habitats of central Chile and plays an important ecological role there because it is a brood parasite of other bird species. Johnson (1967) found a total of 72 parasitised nests, as follows: *Diuca diuca* (44 nests out of 72, or about 61 per cent), *Agelaius thilius* (13/72 or about 18 per cent), *Zonotrichia capensis* (10/72 or about 14 per cent). The following species were found with one parasitised nest each (a little over 1 per cent): *Passer domesticus*, *Sturnella loyca*, *Metriopelia melanoptera*, *Pyrope pyrope* and *Hymenops perspicillata*. All these species breed in the mediterranean bioclimatic zone, although not all are characteristic of the mediterranean matorral and other shrublands, and one of them, *Passer domesticus*, is an introduction. *Molothrus bonariensis* seems to be relatively abundant; Erazo & Valenzuela (1985) ranked it about sixteenth out of 25 species they censused. Whether or not *Molothrus* was introduced originally, it is clear that its range in Chile has expanded, so that it is now a common and ecologically important component of the mediterranean zone avifauna.

7. *Passer domesticus* (house sparrow) (mediterranean zone; introduced and expanding). This western Palaearctic species was apparently first introduced to South America at Buenos Aires in 1872 and in Chile in 1904 (Hellmayr, 1932; Sick, 1959; Johnson, 1967; Navas, 1987). In Chile it was introduced on at least

two occasions. Johnson (1967, pp. 341–2) stated 'Since [its original introductions in Chile and] following its usual tactics of close association with man, it has consistently extended its range both north and south and may be said to have "colonized" the entire country from Tierra del Fuego to Arica. In doing so it has ousted the indigenous Rufous-collared Sparrow [*Zonotrichia capensis*] and Diuca Finch [*Diuca diuca*; both species are mediterranean birds in Chile] from many of their former haunts around the towns and forced them to withdraw to the countryside'. Although the house sparrow lives near humans, Johnson (1967) also mentioned a nest of *Passer domesticus* in a nest of *Bubo virginianus* (presumably in a more rural environment, but unfortunately he did not say where this observation was made). Philippi (1964) and Araya & Millie (1986) indicate the range of the house sparrow to be from Arica to Tierra del Fuego, as well as Easter Island and the Juan Fernandez Archipelago. Johnston & Selander (1973) discussed geographical variation in *Passer domesticus* in South America; their Chilean sample was 100 birds from Santiago, in the mediterranean zone. The house sparrow is clearly an important element of the avifauna of towns and villages of the mediterranean zone; whether it has indeed displaced two native finches, as Johnson suggested, is unclear to me, since the habitats of these two birds to a large extent do not overlap with those of *Passer domesticus*. Thus I would conclude that the house sparrow is not a very significant ecological component of Chile's mediterranean shrublands.

Discussion

The introduced avifauna of California consists of 19 species, 18 of which (95 per cent) breed or occur in the mediterranean zone; and the introduced avifauna of Chile consists of 7 species, all of which live in the mediterranean zone (Table 23.2). Although at first sight the introduced faunal component of mediterranean Chile appears to be much smaller than the Californian one, upon closer inspection the difference is less pronounced. Indeed, of the 18 introduced Californian species, no fewer than 11 can be considered marginal to the mediterranean zone (rare, restricted in distribution, or of dubious implantation or establishment), and only seven are well or relatively well implanted and/or relatively widespread in the mediterranean zone (viz. *Bubulcus ibis*, *Alectoris chukar*, *Phasianus colchicus*, *Columba livia*, *Streptopelia chinensis*, *Sturnus vulgaris* and *Passer domesticus*). In Chile, five species out of seven are similarly well established and/or widespread in the mediterranean zone (viz. *Bubulcus ibis*, *Callipepla californica*, *Columba livia*, *Molothrus bonariensis* and *Passer domesticus*).

Thus, only seven of 199 breeding mediterranean land and water bird species in California (5 per cent) and only 5 of 128 breeding mediterranean land and water bird species in Chile (4 per cent) are introductions that are worth

Table 23.2. *Geographic distribution, habitats and ecological impact of bird species introduced to the mediterranean bioclimatic zones of California and Chile*

Species	Distribution	Habitats	Ecological impact
California			
Bubulcus ibis	w	Marshes	Small
Alectoris chukar	w	Shrublands	Small
Phasianus colchicus	w	Cultivation	Small
Pavo cristatus	vr	Parkland	Nil
Meleagris gallopavo	l	Woodlands	Negligible
Columba livia	w	Urban/suburban	Small
Streptopelia risoria	vr	Urban parks	Negligible
Streptopelia chinensis	w	Urban/suburban	Small
Psittacula krameri	vr	Urban/suburban	Nil
Nandayus nenday	vr	Urban/suburban	Nil
Brotogeris versicolorus	vr	Urban/suburban	Nil
Amazona viridigenalis	vr	Urban/suburban	Nil
Amazona oratrix	vr	Urban/suburban	Nil
Sturnus vulgaris	w	Varied	Significant
Zosterops japonicus	vr	Urban/suburban	Nil
Cardinalis cardinalis	vr	Parkland	Negligible
Passer domesticus	w	Urban/suburban	? Small
Lagonosticta rubricata	vr	Suburban	Nil
Chile			
Bubulcus ibis	w	Marshes	Small
Cairina moschata	vr	?	Nil
Phasianus colchicus	vr	Cultivation	Nil
Callipepla californica	w	Shrublands	Significant
Columba livia	w	Urban/suburban	Small
Molothrus bonariensis	w	Varied	Significant
Passer domesticus	w	Urban/suburban	Small

Distribution: w, widespread; l, localised; vr, very restricted

considering in terms of their potential impact as invaders of the native faunas. These percentages are very low. An examination of species lists (Appendixes 23.1 & 23.2) thus shows unequivocally that the mediterranean avifaunas of California and Chile are composed fundamentally of native taxa.

It is noteworthy that, by pooling the Californian and Chilean lists, three of the nine species composing the total list of biologically potentially significant introductions occur in both regions, namely *Bubulcus ibis*, *Columba livia* and *Passer domesticus*. These three species occur in much of the world, however, and not only in the areas being discussed here. Their success in the mediterranean regions of California and Chile therefore has nothing to do with the mediterranean nature of the regions considered, since these three species have adapted well to a wider range of bioclimates besides the mediterranean one. In a sense, then, these three invasive species are ecological and biogeographic ubiquists and tell us little or nothing about the potential of the Californian and Chilean mediterranean zones as habitats for invasive species.

If we remove these three species from further consideration, four species remain for California (viz. *Alectoris chukar*, *Phasianus colchicus*, *Streptopelia chinensis* and *Sturnus vulgaris*). Two of these species (*Alectoris chukar* and *Phasianus colchicus*) live in mediterranean habitats but have low population densities and probably have only a minor ecological impact on the avifauna. A third, *Streptopelia chinensis*, is urban or suburban in its habitat preference and is unlikely to have a significant impact on the avifauna. Only the fourth, *Sturnus vulgaris*, is numerically abundant and widespread both geographically and ecologically, and clearly has an impact, as it competes with and displaces native, hole-breeding species (Small, 1974). In Chile, after a similar process of elimination, we are left with two species (*Callipepla californica* and *Molothrus bonariensis*). These two species are widespread and/or abundant in the mediterranean bioclimatic zone of Chile, including typically mediterranean habitats (shrublands), and clearly belong now in the mediterranean zone avifauna of that country.

The main conclusions to be drawn from this analysis of introduced bird species in California and Chile are (i) only one species of bird (*Sturnus vulgaris*) has invaded and is established in or around the truly mediterranean habitats (shrublands, chaparral) of California; and (ii) only two really successful invasions are found in mediterranean Chile (*Callipepla californica* and *Molothrus bonariensis*).

An ecologist censusing birds in the mediterranean habitats of California is thus likely to encounter only native birds and no introduced ones. In Chile, the great majority of birds censused in mediterranean habitats will belong to native species, but *Callipepla californica* (a species introduced from another continent, now apparently as completely acclimatised to the mediterranean habitats as a native species) and *Molothrus bonariensis* (a South American species that was either introduced or that expanded its range naturally, now apparently acclimatised to, and relatively common in, mediterranean habitats) will appear regularly in the counts.

Two final points remain to be made. Firstly, the Palaearctic chukar, *Alectoris chukar*, introduced to California, is a mediterranean bird in the western Palaearctic. It is interesting to observe that it has not succeeded as well as *Callipepla californica*, the other galliform species transplanted from one mediterranean region (California) to another (Chile). The reasons for the relative failure of *Alectoris* in mediterranean habitats, when compared to the success of *Callipepla*, are not clear. The number of potentially competing species (at least locally) would seem to be about the same in both regions: one other quail species in California (*Oreortyx pictus*) and one tinamou in Chile (*Nothoprocta perdicaria*). Other potential competitors (again, at least locally) might be species of doves (two species in California *Zenaida macroura* and *Columbina passerina*; and three species in Chile *Zenaida auriculata*, *Columbina picui* and *Metriopelia melanoptera*). (Note that in both regions *Zenaida* and *Columbina* occur and are very similar species pairs both taxonomically and ecologically.) A possible reason for the lack of success of *Alectoris chukar* could be that this species is the least mediterranean of the species or allospecies of the genus *Alectoris*. It occurs in very dry habitats, including desert, in its native western Asian range. Its Californian habitats may be likewise more arid than mediterranean.

My second concluding point is that the introduced *Molothrus bonariensis* in Chile has a native congeneric counterpart *Molothrus ater* in the mediterranean zone of California. The two species of cowbirds both being brood parasites, a comparative study of their impact on their respective avifaunas would be of interest.

Acknowledgements. I am very grateful to Francesco di Castri and Richard Groves for their invitation to write this chapter and to contribute to this book, and for their encouragement and patience. Although no field work was carried out specifically for the preparation of this chapter, I have drawn from the field experience gained during several trips to California and Chile. This field work had been supported by grants from the National Science Foundation, the Society of Sigma Xi, the National Geographic Society, and the Sanford Fund of the American Museum of Natural History.

Appendixes

Appendix 23.1. *Breeding species of land and water birds of California*

1. Podicipedidae
 1. 1. *Podilymbus podiceps*. 1. med.
 2. *Podiceps nigrocollis*. 2. med.
 3. *Aechmophorus occidentalis*. 3. med.
2. Pelecanidae
 4. *Pelecanus erythrorhynchos*. 4. med.
3. Phalacrocoriacidae
 5. *Phalacrocorax auritus*. 5. med.

Appendix 23.1. (*cont.*)

4. Ardeidae
   6. *Botaurus lentiginosus*. 6. med.
   7. *Ixobrychus exilis*. 7. med.
   8. *Ardea herodias*. 8. med.
   9. *Casmerodius albus*. 9. med.
   10. *Egretta thula*. 10. med.
   11. *Egretta caerulea*.
   12. *Bubulcus ibis*. 11. med. 1. INTRODUCED
   13. *Butorides straitus*. 12. med.
   14. *Nycticorax nycticorax*. 13. med.
5. Threskiornithidae
   15. *Plegadis chihi*. 14. med.
6. Anatidae
   16. *Dendrocygna bicolor*. 15. med.
   17. *Branta canadensis*. 16. med.
   18. *Aix sponsa*. 17. med.
   19. *Anas crecca*. 18. med.
   20. *Anas platyrrhynchos*. 19. med.
   21. *Anas acuta*. 20. med.
   22. *Anas discors*. 21. med.
   23. *Anas cyanoptera*. 22. med.
   24. *Anas clypeata*. 23. med.
   25. *Anas strepera*. 24. med.
   26. *Anas americana*.
   27. *Aythya valisineria*.
   28. *Aythya americana*. 25. med.
   29. *Aythya collaris*.
   30. *Histrionicus histrionicus*.
   31. *Bucephala islandica*.
   32. *Bucephala albeola*.
   33. *Mergus merganser*.
   34. *Oxyura jamaicensis*. 26. med.
7. Cathartidae
   35. *Cathartes aura*. 27. med.
   36. *Gymnogyps californianus*. 28. med.
8. Accipitridae
   37. *Pandion haliaetus*. 29. med.
   38. *Elanus caeruleus*. 30. med.
   39. *Haliaeetus leucocephalus*. 31. med.
   40. *Circus cyaneus*. 32. med.
   41. *Accipiter striatus*.
   42. *Accipiter cooperi*. 33. med.
   43. *Accipiter gentilis*.
   44. *Buteo lineatus*. 34. med.
   45. *Buteo swainsoni*. 35. med.

Appendix 23.1. (*cont.*)

---

46. *Buteo jamaicensis*. 36. med.
47. *Aquila chrysaetos*. 37. med.

9. Falconidae
    48. *Falco sparverius*. 38. med.
    49. *Falco peregrinus*. 39. med.
    50. *Falco mexicanus*. 40. med.
10. Phasianidae
    51. *Perdix perdix*. 2. INTRODUCED
    52. *Alectoris chukar*. 41. med. 3. INTRODUCED
    53. *Phasianus colchicus*. 42. med. 4. INTRODUCED
    54. *Pavo cristatus*. 43. med. 5. INTRODUCED
    55. *Dendragapus obscurus*.
    56. *Bonasa umbellus*.
    57. *Centrocercus urophasianus*.
    58. *Meleagris gallopavo*. 44. med. 6. INTRODUCED
    59. *Callipepla gambelii*.
    60. *Callipepla californica*. 45. med.
    61. *Oreortyx pictus*. 46. med.
11. Rallidae
    62. *Laterallus jamaicensis*.
    63. *Rallus longirostris*. 47. med.
    64. *Rallus limicola*. 48. med.
    65. *Porzana carolina*. 49. med.
    66. *Gallinula chloropus*. 50. med.
    67. *Fulica americana*. 51. med.
12. Gruidae
    68. *Grus canadensis*.
13. Charadriidae
    69. *Charadrius alexandrinus*. 52. med.
    70. *Charadrius vociferus*. 53. med.
14. Haematopodidae
    71. *Haematopus bachmani*. 54. med.
15. Recurvirostridae
    72. *Himantopus mexicanus*. 55. med.
    73. *Recurvirostra americana*. 56. med.
16. Scolopacidae
    74. *Catoptrophorus semipalmatus*.
    75. *Actitis macularia*.
    76. *Numenius americanus*.
    77. *Gallinago gallinago*.
    78. *Phalaropus tricolor*. 57. med.
17. Laridae
    79. *Larus delawarensis*.
    80. *Larus californicus*.
    81. *Larus occidentalis*. 58. med.

Appendix 23.1. (*cont.*)

82. *Sterna nilotica*. 59. med.
83. *Sterna caspia*. 60. med.
84. *Sterna elegans*. 61. med.
85. *Sterna forsteri*. 62. med.
86. *Sterna antillarum*. 63. med.
87. *Chlidonias niger*. 64. med.
88. *Rhynchops niger*. 65. med.

18. Columbidae
89. *Columba livia*. 66. med. 7. INTRODUCED
90. *Columba fasciata*. 67. med.
91. *Streptopelia risoria*. 68. med. 8. INTRODUCED
92. *Streptopelia chinensis*. 69. med. 9. INTRODUCED
93. *Zenaida asiatica*.
94. *Zenaida macroura*. 70. med.
95. *Columbina inca*.
96. *Columbina passerina*. 71. med.

19. Psittacidae
97. *Psittacula krameri*. 72. med. 10. INTRODUCED
98. *Nandayus nenday*. 73. med. 11. INTRODUCED
99. *Brotogeris versicolorus*. 74. med. 12. INTRODUCED
100. *Amazona viridigenalis*. 75. med. 13. INTRODUCED
101. *Amazona oratrix*. 76. med. 14. INTRODUCED

20. Cuculidae
102. *Coccyzus americanus*.
103. *Geococcyx californianus*. 77. med.

21. Tytonidae
104. *Tyto alba*. 78. med.

22. Strigidae
105. *Otus flammeolus*.
106. *Otus kennicottii*. 79. med.
107. *Bubo virginianus*. 80. med.
108. *Glaucidium gnoma*. 81. med.
109. *Micrathene whitneyi*. 82. med.
110. *Athene cunicularia*. 83. med.
111. *Strix occidentalis*. 84. med.
112. *Strix nebulosa*.
113. *Asio otus*. 85. med.
114. *Asio flammeus*. 86. med.
115. *Aegolius acadicus*.

23. Caprimulgidae
116. *Chordeiles acutipennis*.
117. *Chordeiles minor*. 87. med.
118. *Phalaenoptilus nuttallii*. 88. med.
119. *Caprimulgus vociferus*. 89. med.

Appendix 23.1 (*cont.*)

24. Apodidae
    120. *Cypseloides niger*. 90. med.
    121. *Chaetura vauxi*.
    122. *Aeronautes saxatalis*. 91. med.
25. Trochilidae
    123. *Archilochus alexandri*. 92. med.
    124. *Calypte anna*. 93. med.
    125. *Calypte costae*. 94. med.
    126. *Stellula calliope*.
    127. *Selasphorus platycercus*.
    128. *Selasphorus sasin*. 95. med.
26. Alcedinidae
    129. *Ceryle alcyon*. 96. med.
27. Picidae
    130. *Melanerpes lewis*. 97. med.
    131. *Melenerpes formicivorus*. 98. med.
    132. *Melanerpes uropygialis*.
    133. *Sphyrapicus varius*.
    134. *Sphyrapicus ruber*. 99. med.
    135. *Sphyrapicus thyroideus*.
    136. *Picoides scalaris*.
    137. *Picoides nuttallii*. 100. med.
    138. *Picoides pubescens*. 101. med.
    139. *Picoides villosus*. 102. med.
    140. *Picoides albolarvatus*. 103. med.
    141. *Picoides arcticus*.
    142. *Colaptes auratus*. 104. med.
    143. *Dryocopus lineatus*. 105. med.
28. Tyrannidae
    144. *Contopus borealis*.
    145. *Contopus sordidulus*. 106. med.
    146. *Empidonax traillii*. 107. med.
    147. *Empidonax hammondii*. 108. med.
    148. *Empidonax oberholseri*. 109. med.
    149. *Empidonax wrightii*.
    150. *Empidonax difficilis*. 110. med.
    151. *Sayornis nigricans*. 111. med.
    152. *Sayornis saya*. 112. med.
    153. *Pyrocephalus rubinus*. 113. med.
    154. *Myiarchus cinerascens*. 114. med.
    155. *Myiarchus tyrannulus*.
    156. *Tyrannus vociferans*. 115. med.
    157. *Tyrannus verticalis*. 116. med.
    158. *Tyrannus tyrannus*.

Appendix 23.1. (*cont.*)

29. Alaudidae
    159. *Eremophila alpestris*. 117. med.
30. Hirundinidae
    160. *Progne subis*. 118. med.
    161. *Tachycineta bicolor*. 119. med.
    162. *Tachycineta thalassina*. 120. med.
    163. *Stelgidopteryx serripennis*. 121. med.
    164. *Riparia riparia*. 122. med.
    165. *Hirundo pyrrhonota*. 123. med.
    166. *Hirundo rustica*. 124. med.
31. Corvidae
    167. *Perisoreus canadensis*.
    168. *Cyanocitta stelleri*. 125. med.
    169. *Aphelocoma coerulescens*. 126. med.
    170. *Gymnorhinus cyanocephalus*. 127. med.
    171. *Nucifraga columbiana*.
    172. *Pica pica*.
    173. *Pica nuttalli*. 128. med.
    174. *Corvus brachyrhynchos*. 129. med.
    175. *Corvus corax*. 130. med.
32. Paridae
    176. *Parus atricapillus*.
    177. *Parus gambeli*.
    178. *Parus rufescens*. 131. med.
    179. *Parus inornatus*. 132. med.
33. Remizidae
    180. *Auriparus flaviceps*.
34. Aegithalidae
    181. *Psaltriparus minimus*. 133. med.
35. Sittidae
    182. *Sitta canadensis*.
    183. *Sitta carolinensis*. 134. med.
    184. *Sitta pygmaea*. 135. med.
36. Certhiidae
    185. *Certhia americana*. 136. med.
37. Troglodytidae
    186. *Campylorhynchus brunneicapillus*.
    187. *Salpinctes obsoletus*. 137. med.
    188. *Catherpes mexicanus*. 138. med.
    189. *Thryomanes bewickii*. 139. med.
    190. *Troglodytes aedon*. 140. med.
    191. *Troglodytes troglodytes*. 141. med.
    192. *Cistothorus palustris*. 142. med.
38. Cinclidae
    193. *Cinclus mexicanus*.

Appendix 23.1. (*cont.*)

---

39. Muscicapidae, Sylviinae
    194. *Regulus satrapa.*
    195. *Regulus calendula.*
    196. *Polioptila caerulea.* 143. med.
    197. *Polioptila melanura.* 144. med.
    Muscicapidae, Turdinae
    198. *Sialia mexicana.* 145. med.
    199. *Sialia currucoides.* 146. med.
    200. *Myadestes townsendi.*
    201. *Catharus ustulatus.* 147. med.
    202. *Catharus guttatus.*
    203. *Turdus migratorius.* 148. med.
    204. *Ixoreus naevius.*
    Muscicapidae, Timaliinae
    205. *Chamaea fasciata.* 149. med.
40. Mimidae
    206. *Mimus polyglottos.* 150. med.
    207. *Oreoscoptes montanus.* 151. med.
    208. *Toxostoma bendirei.*
    209. *Toxostoma redivivum.* 152. med.
    210. *Toxostoma dorsale.*
    211. *Toxostoma lecontei.*
41. Motacillidae
    212. *Anthus spinoletta.*
42. Bombycillidae
    213. *Bombycilla cedrorum.*
43. Ptilogonatidae
    214. *Phainopepla nitens.* 153. med.
44. Laniidae
    215. *Lanius ludovicianus.* 154. med.
45. Sturnidae
    216. *Sturnus vulgaris.* 155. med. 15. INTRODUCED
46. Zosteropidae
    217. *Zosterops japonicus.* 156. med. 16. INTRODUCED
47. Vireonidae
    218. *Vireo bellii.*
    219. *Vireo vicinior.* 157. med.
    220. *Vireo solitarius.* 158. med.
    221. *Vireo huttoni.* 159. med.
    222. *Vireo gilvus.* 160. med.
48. Emberizidae, Parulinae
    223. *Vermivora celata.* 161. med.
    224. *Vermivora ruficapilla.*
    225. *Vermivora virginiae.*
    226. *Vermivora luciae.*

Appendix 23.1. (*cont.*)

227. *Dendroica petechia*. 162. med.
228. *Dendroica coronata*.
229. *Dendroica nigrescens*. 163. med.
230. *Dendroica occidentalis*.
231. *Oporornis tolmiei*.
232. *Geothlypis trichas*. 164. med.
233. *Wilsonia pusilla*. 165. med.
234. *Icteria virens*. 166. med.

Emberizidae, Thraupinae

235. *Piranga flava*.
236. *Piranga rubra*.
237. *Piranga ludoviciana*.

Emberizidae, Cardinalinae

238. *Cardinalis cardinalis*. 167. med. 17. INTRODUCED
239. *Pheucticus melanocephalus*.
240. *Guiraca caerulea*. 168. med.
241. *Passerina amoena*. 169. med.

Emberizidae, Emberizinae

242. *Pipilo chlorurus*.
243. *Pipilo erythrophthalmus*. 170. med.
244. *Pipilo fuscus*. 171. med.
245. *Pipilo aberti*.
246. *Aimophila ruficeps*. 172. med.
247. *Spizella passerina*. 173. med.
248. *Spizella breweri*. 174. med.
249. *Spizella atrogularis*. 175. med.
250. *Pooecetes gramineus*. 176. med.
251. *Chondestes grammacus*. 177. med.
252. *Amphispiza bilineata*. 178. med.
253. *Amphispiza belli*. 179. med.
254. *Passerculus sandwichensis*. 180. med.
255. *Ammodramus savannarum*. 181. med.
256. *Passerella iliaca*.
257. *Melospiza melodia*. 182. med.
258. *Melospiza lincolnii*.
259. *Zonotrichia leucophrys*. 183. med.
260. *Junco hyemalis*. 184. med.

Emberizidae, Icterinae

261. *Agelaius phoeniceus*. 185. med.
262. *Agelaius tricolor*. 186. med.
263. *Sturnella neglecta*. 187. med.
264. *Xanthocephalus xanthocephalus*. 188. med.
265. *Euphagus cyanocephalus*. 189. med.
266. *Quiscalus mexicanus*.
267. *Molothrus aeneus*.

Appendix 23.1 (*cont.*)

---

268. *Molothrus ater*. 190. med.
269. *Icterus cucullatus*. 191. med.
270. *Icterus galbula*. 192. med.
271. *Icterus parisorum*.

49. Fringillidae, Carduelinae
272. *Leucosticte arctoa*.
273. *Pinicola enucleator*.
274. *Carpodacus purpureus*. 193. med.
275. *Carpodacus cassinii*.
276. *Carpodacus mexicanus*. 194. med.
277. *Loxia curvirostra*.
278. *Carduelis pinus*.
279. *Carduelis psaltria*. 195. med.
280. *Carduelis lawrencei*. 196. med.
281. *Carduelis tristis*. 197. med.
282. *Coccothraustes vespertinus*.

50. Passeridae
283. *Passer domesticus*. 198. med. 18. INTRODUCED

51. Estrildidae
284. *Lagonosticta rubricata*. 199. med. 19. INTRODUCED

Compiled from Grinnell & Miller (1944), Miller (1951), Small (1974) and AOU (1983). Sequence and nomenclature follow AOU (1983).

Appendix 23.2. *Breeding species of land and water birds of Chile*

---

1. Tinamidae
1. *Nothoprocta ornata*.
2. *Nothoprocta perdicaria*. 1. med.
3. *Nothoprocta pentlandii*.
4. *Eudromia elegans*.
5. *Tinamotis pentlandii*.
6. *Tinamotis ingoufi*.

2. Rheidae
7. *Pterocnemia pennata*.

3. Podicipedidae
8. *Podilymbus podiceps*. 2. med.
9. *Podiceps occipitalis*. 3. med.
10. *Podiceps major*. 4. med.
11. *Rollandia rolland*. 5. med.

4. Phalacrocoracidae
12. *Phalacrocorax olivaceus*. 6. med.

Appendix 23.2. (*cont.*)

5. Ardeidae
    13. *Ixobrychus involucris*. 7. med.
    14. *Ardea cocoi*.
    15. *Casmerodius albus*. 8. med.
    16. *Egretta thula*. 9. med.
    17. *Bubulcus ibis*. 10. med. 1. INTRODUCED
    18. *Nycticorax nycticorax*. 11. med.
6. Threskiornithidae
    19. *Plegadis chihi*. 12. med.
    20. *Plegadis ridgwayi*.
    21. *Theristicus caudatus*. 13. med.
7. Phoenicopteridae
    22. *Phoenicopterus chilensis*.
    23. *Phoenicopterus andinus*.
    24. *Phoenicopterus jamesi*.
8. Anatidae
    25. *Coscoroba coscoroba*.
    26. *Cygnus melancoryphus*. 14. med.
    27. *Chloephaga melanoptera*.
    28. *Chloephaga poliocephala*.
    29. *Chloephaga rubidiceps*.
    30. *Chloephaga picta*.
    31. *Chloephaga hybrida*.
    32. *Lophonetta specularioides*.
    33. *Tachyeres patachonicus*.
    34. *Cairina moschata*. 15. med. 2. INTRODUCED
    35. *Anas specularis*. 16. med.
    36. *Anas flavirostris*. 17. med.
    37. *Anas sibilatrix*. 18. med.
    38. *Anas cyanoptera*. 19. med.
    39. *Anas platalea*. 20. med.
    40. *Anas georgica*. 21. med.
    41. *Anas versicolor*.
    42. *Anas puna*.
    43. *Netta peposaca*. 22. med.
    44. *Merganetta armata*. 23. med.
    45. *Oxyura jamaicensis*. 24. med.
    46. *Oxyura vittata*. 25. med.
    47. *Heteronetta atricapilla*. 26. med.
9. Cathartidae
    48. *Coragyps atratus*. 27. med.
    49. *Cathartes aura*. 28. med.
    50. *Vultur gryphus*.

Appendix 23.2. (*cont.*)

10. Accipitridae
    51. *Elanus leucurus*. 29. med.
    52. *Circus cinereus*. 30. med.
    53. *Accipiter bicolor*. 31. med.
    54. *Parabuteo unicinctus*. 32. med.
    55. *Geranoaetus melanoleucus*. 33. med.
    56. *Buteo polyosoma*. 34. med.
    57. *Buteo poecilochrous*.
    58. *Buteo albigula*.
    59. *Buteo ventralis*.
11. Falconidae
    60. *Polyborus plancus*. 35. med.
    61. *Phalcoboenus megalopterus*. 36. med.
    62. *Phalcoboenus albogularis*.
    63. *Phalcoboenus australis*.
    64. *Milvago chimango*. 37. med.
    65. *Falco sparverius*. 38. med.
    66. *Falco femoralis*. 39. med.
    67. *Falco peregrinus*. 40. med.
12. Phasianidae
    68. *Phasianus colchicus*. 41. med. 3. INTRODUCED
    69. *Callipepla californica*. 42. med. 4. INTRODUCED
13. Rallidae
    70. *Laterallus jamaicensis*. 43. med.
    71. *Rallus antarcticus*.
    72. *Porphyriops melanops*. 44. med.
    73. *Pardirallus sanguinolentus*. 45. med.
    74. *Gallinula chloropus*.
    75. *Fulica americana*.
    76. *Fulica armillata*. 46. med.
    77. *Fulica leucoptera*. 47. med.
    78. *Fulica rufifrons*. 48. med.
    79. *Fulica gigantea*.
    80. *Fulica cornuta*.
14. Charadriidae
    81. *Vanellus chilensis*. 49. med.
    82. *Vanellus resplendens*.
    83. *Charadrius collaris*.
    84. *Charadrius alexandrinus*. 50. med.
    85. *Charadrius alticola*.
    86. *Charadrius falklandicus*. 51. med.
    87. *Charadrius vociferus*.
    88. *Charadrius modestus*.
    89. *Oreopholus ruficollis*. 52. med.

Appendix 23.2. (*cont.*)

- 90. *Pluvianellus socialis.*
- 91. *Phegornis mitchellii.*

15. Haematopodidae
- 92. *Haematopus palliatus.* 53. med.
- 93. *Haematopus leucopodus.*
- 94. *Haematopus ater.*

16. Recurvirostridae
- 95. *Himantopus mexicanus.* 54. med.
- 96. *Recurvirostra andina.*

17. Rostratulidae
- 97. *Nycticryphes semicollaris.* 55. med.

18. Scolopacidae
- 98. *Gallinago gallinago.* 56. med.
- 99. *Gallinago andina.*
- 100. *Gallinago stricklandii.*

19. Thinocoridae
- 101. *Attagis gayi.*
- 102. *Attagis malouinus.*
- 103. *Thinocorus orbignyianus.*
- 104. *Thinocorus rumicivorus.* 57. med.

20. Laridae
- 105. *Larus scoresbii.*
- 106. *Larus modestus.*
- 107. *Larus dominicanus.* 58. med.
- 108. *Larus serranus.*
- 109. *Larus maculipennis.* 59. med.
- 110. *Sterna hirundinacea.* 60. med.
- 111. *Sterna trudeaui.* 61. med.
- 112. *Sterna lorata.*

21. Columbidae
- 113. *Columba livia.* 62. med. 5. INTRODUCED
- 114. *Columba araucana.* 63. med.
- 115. *Zenaida asiatica.*
- 116. *Zenaida auriculata.* 64. med.
- 117. *Columbina picui.* 65. med.
- 118. *Columbina cruziana.*
- 119. *Metriopelia ceciliae.*
- 120. *Metriopelia aymara.*
- 121. *Metriopelia melanoptera.* 66. med.

22. Psittacidae
- 122. *Cyanoliseus patagonus.* 67. med.
- 123. *Enicognathus ferrugineus.* 68. med.
- 124. *Enicognathus leptorhynchus.* 69. med.
- 125. *Bolborhynchus aurifrons.*

Appendix 23.2. (*cont.*)

23. Cuculidae
    126. *Crotophaga sulcirostris.*
24. Tytonidae
    127. *Tyto alba.* 79. med.
25. Strigidae
    128. *Bubo virginianus.* 71. med.
    129. *Glaucidium brasilianum.*
    130. *Glaucidium nanum.* 72. med.
    131. *Athene cunicularia.* 73. med.
    132. *Strix rufipes.* 74. med.
    133. *Asio flammeus.* 75. med.
26. Caprimulgidae
    134. *Chordeiles acutipennis.*
    135. *Caprimulgus longirostris.* 76. med.
27. Apodidae
    136. *Aeronautes andecolus.*
28. Trochilidae
    137. *Oreotrochilus estella.*
    138. *Oreotrochilus leucopleurus.*
    139. *Patagona gigas.* 77. med.
    140. *Sephanoides sephanoides.* 78. med.
    141. *Rhodopis vesper.*
    142. *Eulidia yarrellii.*
29. Alcedinidae
    143. *Ceryle torquata.*
    144. *Chloroceryle americana.*
30. Picidae
    145. *Picoides lignarius.* 79. med.
    146. *Colaptes rupicola.*
    147. *Colaptes pitius.* 80. med.
    148. *Campephilus magellanicus.*
31. Furnariidae
    149. *Geositta maritima.*
    150. *Geositta isabellina.*
    151. *Geositta rufipennis.*
    152. *Geositta cunicularia.* 81. med.
    153. *Geositta punensis.*
    154. *Geositta antarctica.*
    155. *Upucerthia dumetaria.* 82. med.
    156. *Upucerthia albigula.*
    157. *Upucerthia validirostris.*
    158. *Upucerthia ruficauda.*
    159. *Upucerthia andaecola.*
    160. *Chilia melanura.* 83. med.
    161. *Cinclodes patagonicus.* 84. med.

Appendix 23.2. (*cont.*)

162. *Cinclodes nigrofumosus*. 85. med.
163. *Cinclodes fuscus*. 86. med.
164. *Cinclodes oustaleti*.
165. *Cinclodes antarcticus*.
166. *Cinclodes atacamensis*.
167. *Sylviorthorhynchus desmursii*. 87. med.
168. *Aphrastura spinicauda*. 88. med.
169. *Phleocryptes melanops*. 89. med.
170. *Leptasthenura striata*.
171. *Leptasthenura aegithaloides*. 90. med.
172. *Asthenes humicola*. 91. med.
173. *Asthenes modesta*.
174. *Asthenes dorbignyi*.
175. *Asthenes pyrrholeuca*. 92. med.
176. *Asthenes anthoides*.
177. *Pygarrhichas albogularis*. 93. med.

32. Rhinocryptidae
178. *Pteroptochos castaneus*. 94. med.
179. *Pteroptochos tarnii*.
180. *Pteroptochos megapodius*. 95. med.
181. *Scelorchilus albicollis*. 96. med.
182. *Scelorchilus rubecula*. 97. med.
183. *Eugralla paradoxa*. 98. med.
184. *Scytalopus magellanicus*. 99. med.

33. Tyrannidae
185. *Agriornis livida*. 100. med.
186. *Agriornis microptera*.
187. *Agriornis montana*.
188. *Agriornis albicauda*.
189. *Neoxolmis rufiventris*.
190. *Pyrope pyrope*. 101. med.
191. *Muscisaxicola rufivertex*.
192. *Muscisaxicola albilora*.
193. *Muscisaxicola juninensis*.
194. *Muscisaxicola flavinucha*.
195. *Muscisaxicola capistrata*.
196. *Muscisaxicola frontalis*.
197. *Muscisaxicola albifrons*.
198. *Muscisaxicola alpina*.
199. *Muscisaxicola macloviana*.
200. *Muscisaxicola maculirostris*.
201. *Muscigralla brevicauda*.
202. *Lessonia rufa*. 102. med.
203. *Ochthoeca oenanthoides*.
204. *Ochthoeca leucophrys*.

Appendix 23.2. (*cont.*)

---

205. *Hymenops perspicillata*. 103. med.
206. *Elaenia albiceps*. 104. med.
207. *Pyrocephalus rubinus*.
208. *Pseudocolopteryx flaviventris*. 105. med.
209. *Tachuris rubrigastra*. 106. med.
210. *Anairetes parulus*. 107. med.
211. *Anairetes flavirostris*.
212. *Anairetes reguloides*.
213. *Coloramphus parvirostris*.

34. Phytotomidae
214. *Phytotoma rara*. 108. med.

35. Hirundinidae
215. *Tachycineta leucopyga*. 109. med.
216. *Pygochelidon cyanoleuca*. 110. med.
217. *Hirundo andecola*.

36. Troglodytidae
218. *Troglodytes aedon*. 111. med.
219. *Cistothorus platensis*. 112. med.

37. Muscicapidae, Turdinae
220. *Turdus chiguanco*.
221. *Turdus falcklandii*. 113. med.

38. Mimidae
222. *Mimus thenca*. 114. med.
223. *Mimus patagonicus*.

39. Motacillidae
224. *Anthus correndera*. 115. med.
225. *Anthus lutescens*.

40. Emberizidae, Thraupinae
226. *Conirostrum cinereum*.
227. *Conirostrum tamarugense*.
228. *Thraupis bonariensis*. 116. med.

Emberizidae, Emberizinae
229. *Volatinia jacarina*.
230. *Sporophila telasco*.
231. *Catamenia analis*.
232. *Catamenia inornata*.
233. *Sicalis uropygialis*.
234. *Sicalis auriventris*.
235. *Sicalis olivascens*.
236. *Sicalis lebruni*.
237. *Sicalis luteola*. 117. med.
238. *Phrygilus patagonicus*.
239. *Phrygilus gayi*. 118. med.
240. *Phrygilus atriceps*.
241. *Phrygilus fructiceti*. 119. med.

Appendix 23.2. (*cont.*)

242. *Phrygilus unicolor*.
243. *Phrygilus dorsalis*.
244. *Phrygilus erythronotus*.
245. *Phrygilus plebejus*.
246. *Phrygilus alaudinus*. 120. med.
247. *Diuca speculifera*.
248. *Diuca diuca*. 121. med.
249. *Melanodera melanodera*.
250. *Melanodera xanthogramma*.
251. *Xenospingus concolor*.
252. *Zonotrichia capensis*. 122. med.

Emberizidae, Icterinae

253. *Agelaius thilius*. 123. med.
254. *Curaeus curaeus*. 124. med.
255. *Sturnella loyca*. 125. med.
256. *Sturnella bellicosa*.
257. *Molothrus bonariensis*. 126. med. 6. INTRODUCED

41. Fringillidae, Carduelinae

258. *Carduelis crassirostris*.
259. *Carduelis magellanica*.
260. *Carduelis atrata*.
261. *Carduelis uropygialis*.
262. *Carduelis barbata*. 127. med.

42. Passeridae

263. *Passer domesticus*. 128. med. 7. INTRODUCED

Compiled from Hellmayr (1932), Philippi (1964), Johnson (1965, 1967, 1972) and Araya & Millie (1986). Sequence and nomenclature follow AOU (1983) when applicable, Araya & Millie (1986) otherwise, with a few minor exceptions.

References

AOU (American Ornithologists' Union) (1983). *Check-list of North American Birds*, 6th edn. Lawrence, Kansas: Allen Press.

Araya, B. M. & Millie, H. G. (1986). *Guia de Campo de las Aves de Chile*. Santiago: Editorial Universitaria.

Blondel, J., Vuilleumier, F., Marcus, L. F. & Terouanne, E. (1984*a*). Peuplements d'oiseaux sous bioclimat méditerranéen dans trois continents: convergences écologiques ou non? *Bulletin de la Société Botanique de France, Actualités Botaniques*, **131**, 345–63.

Blondel, J., Vuilleumier, F., Marcus, L. F. & Terouanne. E. (1984*b*). Is there ecomorphological convergence among bird communities of Chile, California, and France? In *Evolutionary Biology*, vol. 18, ed. M. K. Hecht, B. Wallace & G. T. Prance, pp. 141–213. New York: Plenum.

Cody, M. L. (1970). Chilean bird distribution. *Ecology*, **52**, 455–64.

Cody, M. L. (1974). *Competition and the Structure of Bird Communities*. Princeton, N.J.: Princeton University Press.

Erazo, L. S. & Valenzuela, L. A. (1985). Resultados preliminares de censos de aves en ambientes de estepa de espino (*Acacia caven*), V Regiôn, Chile. *Revista de Geografía de Valparaiso*, **16**, 25–30.

Grinnell, J. & Miller, A. H. (1944). *The Distribution of the Birds of California*. Berkeley: Pacific Coast Avifauna No. 27, Cooper Ornithological Club.

Handtke, K. & Mauersberger, G. (1977). Die Ausbreitung der Kuhreihers *Bubulcus ibis* (L.). *Mitteilungen aus dem Zoologischen Museum in Berlin*, **53** (Supplement), *Annalen Ornithologie*, **1**.

Hardy, J. W. (1973). Feral exotic birds in southern California. *Wilson Bulletin*, **85**, 506–12.

Hellmayr, C. E. (1932). The birds of Chile. *Field Museum of Natural History Publications 308, Zoological Series*, **19**, 1–472.

Inskipp, T. (1975). The importation of birds into Britain. *Bulletin of the International Council for Bird Preservation*, **12**, 98–102.

Johnson, A. W. (1965). *The Birds of Chile and Adjacent Regions of Argentina, Bolivia and Peru*, vol. 1. Buenos Aires: Platt Establecimientos Graficos S.A.

Johnson, A. W. (1967). *The Birds of Chile and Adjacent Regions of Argentina, Bolivia and Peru*, vol. 2. Buenos Aires: Platt Establecimientos Graficos S.A.

Johnson, A. W. (1972). *Supplement to the Birds of Chile and Adjacent Regions of Argentina, Bolivia and Peru*. Buenos Aires: Platt Establecimientos Graficos S.A.

Johnston, R. F. & Selander, R. K. (1973). Evolution in the house sparrow. III. Variation in size and sexual dimorphism in Europe and North and South America. *American Naturalist*, **107**, 373–90.

Mayr, E. (1965). The nature of colonization in birds. In *The Genetics of Colonizing Species*, ed. H. G. Baker & G. L. Stebbins, pp. 29–47. New York: Academic Press.

Meyer de Schauensee, R. (1966). *The Species of Birds of South America and Their Distribution*. Narberth, Pennsylvania: Livingston Publishing Company.

Meyer de Schauensee, R. (1982). *A Guide to the Birds of South America* (reprint). Philadelphia: Academy of Natural Sciences of Philadelphia, reprinted with additions by Intercollegiate Press Inc.

Miller, A. H. (1951). An analysis of the distribution of the birds of California. *University of California Publications in Zoology*, **50**, 531–644.

Mooney, H. A., Hamburg, S. P. & Drake, J. A. (1986). The invasions of plants and animals into California. In *Ecology of Biological Invasions of North America and Hawaii*, ed. H. A. Mooney & J. A. Drake, pp. 250–72. New York: Springer-Verlag.

Navas, J. (1971). Notas sobre aves del Parque Nacional Nahuel Huapi. II. La presencia de *Lophortyx californicus* en Neuquén y Río Negro. *Neotrópica*, **17**, 154–6.

Navas, J. (1987). Los vertebrados exóticos introducidos en la Argentina. *Revista del Museo Argentino de Ciencias Naturales 'Bernardino Rivadavia', Zoología*, **14**, 7–38.

Olrog, C. C. (1979). Nueva lista de la avifauna Argentina. *Opera Lilloana*, **27**, 1–324.

Owre, O. T. (1973). A consideration of the exotic avifauna of southeastern Florida. *Wilson Bulletin*, **85**, 491–500.

Philippi, B. R. A. (1964). Catálogo de la aves chilenas con su distribución geográfica. *Investigaciones Zoológicas Chilenas*, **11**, 1–179.

Sick, H. (1959). A invasão de América latina pelo Pardal, *Passer domesticus* Linnaeus 1758, com referencia especial o Brasil (Ploceidae, Aves). *Boletin do Museo Nacional, Nova Serie, Zoologia*, **207**, 1–31.

Sick, H. (1968). Ueber in Südamerika eingefuhrte Vogelarten. *Bonner Zoologische Beitrage*, **19**, 298–306.

Small, A. (1974). *The Birds of California*. New York: Winchester Press.

Vuilleumier, F. (1985). Forest birds of Patagonia: ecological geography, speciation, endemism, and faunal history. In *Neotropical Ornithology*, ed. P. A. Buckley, M. S. Foster, E. S. Morton, R. S. Ridgely & F. G. Buckley, pp. 255–304. Washington, D.C.: American Ornithologists' Union.

Wing, L. (1943). Spread of the starling and English sparrow. *Auk*, **60**, 74–87.

Wodzicki, K. (1965). The status of some exotic vertebrates in the ecology of New Zealand. In *The Genetics of Colonizing Species*, ed. H. G. Baker & G. L. Stebbins, pp. 425–60. New York: Academic Press.

24

Birds introduced to the fynbos biome of South Africa

R. K. BROOKE & W. R. SIEGFRIED

No bird species native to another continent occurs regularly or normally in intact fynbos vegetation, whether lowland or montane. Several introduced species (summarised in Table 24.1) occur in disturbed and human-modified environments in the fynbos biome. Introduced birds were reviewed recently for South Africa as a whole (Brooke *et al.*, 1986) and unreferenced statements made below are based on that earlier review: major references are cited in this review, however. This chapter covers much the same ground as Brooke *et al.* (1986) but only in respect of the fynbos biome because of its predominance in the mediterranean-climate area of South Africa. This chapter focuses on the impacts, if any, on fynbos vegetation. Escaped cage and aviary birds which have not bred in the wild are of little importance ecologically and are not normally observed in the fynbos biome.

The introduced bird with the greatest probable biomass in the fynbos biome is the helmeted guinea-fowl *Numida meleagris*, deliberately introduced to improve sport-shooting at the end of the last century (Skead, 1962). The source was *N. m. coronata* which is indigenous to the eastern Cape region of South Africa. Domestic guineafowl were, and to some extent still are, widely kept on farms and these are derived from West African *N. m. galeata*. Domestic and wild birds interbreed freely and a minority of wild birds show the pallid colouring derived from domestic populations. An introduction in the 1850s of domestic birds to Robben Island, Table Bay, has survived to the present (Brooke & Prins, 1986). Helmeted guineafowl freely enter the edges of fynbos vegetation but their activities and their impacts have not been investigated critically. It is possible, however, that they are significant consumers of seeds of fynbos plants and disperse some seeds to sites where they can germinate.

The most widespread and abundant introduced bird is the European starling *Sturnus vulgaris*. It was introduced deliberately to Cape Town in 1897 (Winter

Table 24.1. *Systematic summary (source, date of introduction, present status) of introduced bird species which have bred in the fynbos biome of South Africa*

Taxon	Source	Date of introduction	Present status or distribution
Ostrich (*Struthio camelus*)	Lebanon Nigeria	1870s 1880s	Some hybrid contamination
Mute swan (*Cygnus olor*)	Britain	1918	Extinct
Mallard (*Anas platyrhynchos*)	Britain	1940s	Cape Peninsula
Ring-necked pheasant (*Phasianus colchicus*)	Britain via Kimberley	1930s & 1940s	Extinct
Peacock (*Pavo cristatus*)	Unknown	1968	Robben Island
Chukar partridge (*Alectoris chukar*)	Unknown, but semi-domestic	1964	Robben Island[a]
Cape Francolin (*Francolinus capensis*)	Nearby mainland	1960s	Robben Island[b]
Helmeted guinea-fowl (*Numida meleagris*)	Eastern Cape Province	1890s	Throughout
	West Africa	1850s	Robben Island
Feral pigeon (*Columba livia*)	North-west Europe	1652+	Towns throughout
Red-eyed dove (*Streptopelia semitorquata*)	Beira	1933	Subspecific hybrids throughout
Ring-necked parakeet (*Psittacula krameri*)	Unknown	1850s	Extinct
Rosy-faced lovebird (*Agapornis roseicollis*)	Namibia (via cage-bird trade?)	1970s	Extinct
Song Thrush (*Turdus philomelos*)	Britain	1897	Extinct
Blackbird (*T. merula*)	Britain	1897	Extinct
European Starling (*Sturnus vulgaris*)	Scandinavia via Britain	1897	Throughout
Indian myna (*Acridotheres tristis*)	India or Burma via cage-bird trade	1986	Extinct
House sparrow (*Passer domesticus*)	India via Natal	1960s	Throughout

Table 24.1. (*cont.*)

Taxon	Source	Date of introduction	Present status or distribution
Red-crested cardinal (*Paroaria coronata*)	Southern South America via cage-bird trade	1950s	Extinct
Chaffinch (*Fringilla coelebs*)	Britain	1897	Cape Peninsula

[a] The Villiersdorp introduction about the same time is extinct.
[b] Replaced an 1850s introduction extinct before 1930.

bottom & Liversidge, 1954; Brooke *et al.*, 1986) by Cecil Rhodes as part of his program to improve the agriculture, horticulture and amenities of the Cape Colony. His birds were Scandinavian ones trapped on their wintering grounds in southern England. The European starling now occurs throughout the fynbos biome and has done so for the last 25 years. It is a disperser of the seeds of the introduced Australian tree *Acacia cyclops* (Glyphis *et al.*, 1981) and this assists the invasive spread of the plant into areas of fynbos vegetation.

Cecil Rhodes introduced a number of British species to his estate 'Groote Schuur', on the east side of Table Mountain, Cape Town, in addition to the European starling (Brooke *et al.*, 1986). The 200 or more rooks *Corvus frugilegus* and a smaller number of western nightingales *Luscinia megarhynchos* did not even breed. The blackbird *Turdus merula* and the song thrush *T. philomelos* bred for a while but died out, the former first, perhaps because of competition with the indigenous and similarly sized olive thrush *T. olivaceus*. The chaffinch *Fringilla coelebs* still breeds on the eastern side of the Table Mountain chain, usually in plantations of introduced *Pinus* spp., but its range is now slowly contracting after an initial expansion and its abundance is decreasing; it appears to be heading for local extinction.

The ostrich *Struthio camelus australis* originally occurred in all lowland fynbos vegetation types although probably sparsely. In the 1870s the farming of ostriches for their feathers was first attempted in the Cape Colony and a few males were imported from the northern hemisphere to improve feather quality by hybridisation (Smit, 1963). The first males thus introduced belonged to the small, now extinct race *S. c. syriacus* and came by way of Lebanon. Before the end of the century nominate race males had also been imported from Nigeria. Hybrid strains based on these imported birds were established. When the ostrich feather trade collapsed in 1914 many farm birds were allowed to run wild. Ostriches now occur throughout the lowlands of the fynbos biome but, at least in the western areas, few show obvious signs of carrying northern genes (Brooke *et al.*, 1986). Ostriches are almost entirely vegetarian and they freely consume flowers and other parts of many fynbos plants.

Red-eyed doves *Streptopelia semitorquata semitorquata* were imported from Beira, Mozambique. They bred freely in an aviary at Elgin, 65 km east of Cape Town, from where they were released in 1933. This happened at a time when the indigenous population *S. s. australis* was very sparse although the results of earlier tree planting stimulated by Cecil Rhodes were producing trees of a height and size favoured by red-eyed doves. A hybrid subspecific population now occurs throughout the western part of the fynbos biome wherever there are introduced trees more than about 4 or 5 m high (Brooke, 1984). The birds feed on seeds and fruit, usually of introduced trees and shrubs, found on the ground. They often feed to excess, spilling seeds when disturbed or drinking. They probably contribute to the dispersal of a number of plant species. They do not normally feed in intact fynbos.

The house sparrow *Passer domesticus* is as widespread as the European starling but not as abundant. The species was brought to Natal, South Africa, as pets by indentured sugar-cane labourers from India in the 1880s (Harwin & Irwin, 1966). The species spread outside Natal in 1948 and very rapidly covered southern Africa, reaching Cape Town by 1963 (Winterbottom, 1968). It is associated markedly with buildings, both domestic and for food storage, and is not known to have any impact on the structure or functioning of fynbos ecosystems, either natural or transformed.

Feral pigeons *Columba livia* occur in the centres of all large and most small towns of the fynbos biome, feeding in urban and suburban areas. The history of their introductions is not known (Brooke, 1981) although the first representatives of the species were imported as domestic stock in 1652 from Holland by Jan van Riebeeck. It is probable that all subsequent importations, including racing birds, have come from north-west Europe. In addition, many farmers and their labourers maintain dovecotes so that pigeons of this type may be seen anywhere in the farming areas. The species has no apparent impact on fynbos vegetation.

Ring-necked pheasants *Phasianus colchicus* were deliberately bred near Kimberley, South Africa, from birds imported from Britain. Resulting populations were liberated in at least five places in the fynbos biome in the 1930s and 1940s (Siegfried, 1962; Liversidge, 1985) to improve sport-shooting. No population persisted for more than a few years, probably owing to the large suites of indigenous predators and pathogens. By contrast, on Robben Island in Table Bay, Chukar partridges *Alectoris chukar* and peacocks *Pavo cristatus* (both of uncertain origin, although ultimately from India) live ferally in the absence of most continental predators (Brooke & Prins, 1986). Chukar partridges introduced to Villiersdorp, 80 km east of Cape Town, about the same time bred for a year or so but then died out (Liversidge, 1985). In addition, some farmers keep peacocks running wild in their gardens and they breed in the gardens of staff living in the Cape of Good Hope Nature Reserve (M. W. Fraser, personal communication). None of these species has any significant impact on fynbos vegetation.

A game and ornamental species from Britain that maintained a successful wild population for over 50 years was the mute swan *Cygnus olor* on several waterbodies between Cape Town and Port Elizabeth (Liversidge, 1985). In the late 1970s, however, they were apparently trapped and sold at high prices to wildfowl collectors and municipal parks until none were left in the wild (Brooke, 1986).

The mallard *Anas platyrhynchos* presents particular problems of interpretation owing to the difficulty in distinguishing some domestic strains for farms, ornamental varieties for parks and wild-type birds for sport-shooting. All three classes occur or have occurred in the fynbos biome and most farms carry domestic ducks. Occasional hybridisation with the indigenous yellow-billed duck *A. undulata* occurs (Brooke, 1986).

Parrots escape or are let loose in sufficient numbers to establish breeding populations but seldom do so (Brooke *et al.*, 1986). The species chiefly concerned are the Australian budgerigar *Melopsittacus undulatus* and cockatiel *Nymphicus hollandicus* and the Indian ring-necked parakeet *Psittacula krameri* which indeed bred ferally in Cape Town in the middle of the last century. An escaped or liberated population of Namibian rosy-faced lovebirds *Agapornis roseicollis* bred for a few years recently at Fish Hoek, 25 km south of Cape Town. A cage bird species that bred ferally for a short while was the South American red-crested cardinal *Paroaria coronata* at Hermanus, 100 km south-east of Cape Town. But none of these species appears to have survived or invaded fynbos vegetation. The Indian mynah *Acridotheres tristis*, a widespread introduction in Natal and the industrialised Witwatersrand region of the Transvaal (Brooke, 1976; Brooke *et al.*, 1986), has recently been noted breeding in the Cape Town suburb of Camps Bay. The species will probably not establish itself since the Cape Department of Nature & Environmental Conservation has decided to try to eradicate it. The source is unknown but presumed to be the cage-bird trade.

It is clear from the foregoing that introduced birds have had little impact on the ecology of intact or slightly human-modified fynbos. The two species, helmeted guinea-fowl and red-eyed dove, that have come closest to this are importations from other parts of southern Africa. Three species now have large populations in human-modified parts of the fynbos biome, the feral pigeon and European starling from north-western Europe, and the house sparrow from western India. What seems most significant is the number of species introduced to the biome which have failed to become established in the long term. There are three possible non-exclusive reasons for this: small size of initial propagules (though over 200 rooks were liberated), the side effects of the low levels of nutrients in sandstone-derived fynbos soils, and the large number of predators and pathogens indigenous to a mostly tropical continent.

References

Brooke, R. K. (1976). Morphological notes on *Acridotheres tristis* in Natal. *Bulletin of the British Ornithologists' Club*, **96**, 8–13.

Brooke, R. K. (1981). The feral pigeon a 'new' bird for the South African list. *Bokmakierie*, **33**, 37–40.

Brooke, R. K. (1984). A history of the redeyed dove in the south-western Cape Province, South Africa. *Ostrich*, **55**, 12–16.

Brooke, R. K. (1986). Bibliography of alien birds in southern and south-central Africa. *Foundation for Research Development – Ecosystem Programmes Occasional Report*, **14**, 1–66.

Brooke, R. K. & Prins, A. J. (1986). Review of alien species on South African offshore islands. *South African Journal of Antarctic Research*, **16**, 102–9.

Brooke, R. K., Lloyd, P. H. & de Villiers, A. L. (1986). Alien and translocated terrestrial vertebrates in South Africa. In *The Ecology and Management of Biological Invasions in Southern Africa*, eds. I. A. W. Macdonald, F. J. Kruger & A. A. Ferrar, pp. 63–74. Cape Town: Oxford University Press.

Glyphis, J. P., Milton, S. J. & Siegfried, W. R. (1981). Dispersal of *Acacia cyclops* by birds. *Oecologia*, **48**, 138–41.

Harwin, R. M. & Irwin, M. P. S. (1966). The spread of the house sparrow, *Passer domesticus*, in south-central Africa. *Arnoldia (Rhodesia)*, **2** (24), 1–17.

Liversidge, R. (1985). Alien bird species introduced into southern Africa. In *Proceedings of the Birds and Man Symposium, Johannesburg 1983*, ed. L. J. Bunning, pp. 31–44. Johannesburg: Witwatersrand Bird Club.

Siegfried, W. R. (1962). Introduced vertebrates in the Cape Province. *Cape Department of Nature Conservation Report*, **19**, 80–7.

Skead, C. J. (1962). A study of the crowned guinea fowl *Numida meleagris coronata* Gurney. *Ostrich*, **33** (2), 51–65.

Smit, D. J. v. Z. (1963). Ostrich farming in the Little Karoo. *Department of Agriculture Technical Services Bulletin*, **358**, 1–103.

Winterbottom, J. M. (1968). A check list of the land and fresh water birds of the western Cape Province. *Annals of the South African Museum*, **53**, 1–276.

Winterbottom, J. M. & Liversidge, R. (1954). The European starling in the south-west Cape. *Ostrich*, **25**, 89–96.

25

Species of introduced birds in mediterranean Australia

J. L. LONG & P. R. MAWSON

More than 70 and probably over 100 species of birds have been introduced to Australia (Long, 1981). At present, at least 22 species of birds of foreign origin are established (Table 25.1), at least half of which originated in the Mediterranean Basin. Four previously established species now appear to have become extinct and seven species are tenuously established in very restricted ranges. Two species are classed as 'feral' birds and at least 13 native species of birds have been translocated or reintroduced within Australia.

The majority of the species which were introduced and became established in Australia were deliberate releases by acclimatisation societies. Balmford (1978) discussed the events leading up to the establishment in 1861 of the first of these societies in Australia: namely, the Acclimatization Society of Victoria. Fortunately for Australia, the efforts made by these societies were less successful than their founders probably anticipated. Other successful species are the results of accidental arrivals, aviary escapes and releases from captivity and also from colonisation.

Generally, those taxa which have become at all widespread in Australia are those which have wide ecological niches in their region of origin. For birds, as for plants, there is a tendency for the more evolutionarily advanced taxa to be more successful than primitive ones (Sibley & Ahlquist, 1986). This tendency may be related to the fact that the more evolutionarily advanced taxa tend to inhabit more recent and human-made habitats. Consequently, many of the introductions to Australia have spread further only in more recent times in conjunction with such occurrences as changes in land use.

There have been a number of reviews and studies of the ecological and behavioural aspects of invasions of birds (Mayr, 1965; Long, 1981; Ehrlich, 1986; Moulton & Pimm, 1986), but only a few specific to Australia (Jones, 1986;

Table 25.1. *The origin and scientific names of various groups of species of birds introduced to mediterranean-climate Australia*

Region of origin	Scientific name
Established and widespread species	
Europe	*Sturnus vulgaris*
	Passer domesticus
	Alauda arvensis
	Turdus merula
	Acridotheres tristis
	Carduelis carduelis
	Anas platyrhynchos
	Columba livia
	Ardeola ibis
Asia	*Lonchura punctulata*
	Streptopelia chinensis
	Streptopelia senegalensis
Established but less widespread species	
Europe	*Passer montanus*
	Turdus philomelos
	Carduelis chloris
Asia	*Pycnonotus jocosus*
Established but very restricted species	
Europe	*Cygnus olor*
Asia	*Pavo cristatus*
	Numida meleagris
	Gallus gallus
	Phasianus colchicus
Africa	*Struthio camelus*
North America	*Lophortyx californica*
	Meleagris gallopavo
Previously established but now extinct species	
Europe	*Streptopelia decaocto*
Asia	*Pycnonotus cafer*
	Lonchura malacca
	Lophura nycthemera
Africa	*Euplectes albonotatus*
	Euplectes orix
Established translocated and/or reintroduced native species	
Australia	*Dromaius novaehollandiae*
	Cereopsis novaehollandiae
	Alectura lathami
	Geopelia placida
	Callocephalon fimbriatum
	Menura novaehollandiae

Table 25.1. *(cont.)*

Region of origin	Scientific name
	Neochmia ruficauda
	Lonchura castaneothorax
	Aegintha temporalis
	Poephila bichenovii
	Cacatua roseicapilla
	Cacatua galerita
	Dacelo gigas
	Atrichornis clamosus

Newsome & Noble, 1986) and none specific to the mediterranean-climate region of southern Australia.

Within Australia a significantly greater number of native species have been successfully introduced (or reintroduced) than foreign species (foreign in this context being defined as originating from outside Australia; Newsome & Noble, 1986). Most of the foreign introductions have expanded their range to some extent with continued expansion of agriculture, but this expansion appears to be decreasing. The interface between the communities of introduced species of low diversity and the communities of endemic taxa of higher diversity is no longer moving so rapidly. What will happen now? Will the introduced species become integrated with the remnants of the pre-European avifauna or will we see extinction and subsequent replacement of introduced and endemic species by species from each category?

Results of investigations by Green (1984) and Jones (1981) have shown that although introduced birds were often the most common species in an area they were rarely found in areas of natural vegetation. This finding suggests that there are particular characteristics of a site or habitat which affect the outcome of invasions. An invader in a new habitat must to some degree be pre-adapted. This is borne out by the higher success rate of species reintroduced into environments from which they have been extirpated (Ehrlich, 1986).

The ability to shift habitat preferences in part or whole is not unique to introduced species. For varying reasons several of the Australian endemics (Psittacidae and Columbidae) have undergone shifts in habitat preferences, primarily in relation to diet. Often these shifts in habitat and dietary preference are associated with particular species of introduced plants. An exceptional example of this is provided by *Calyptorhynchus magnificus* which now feeds almost entirely on *Emex australis*, a plant introduced from South Africa and now widespread in the central wheat-belt of Western Australia (Saunders, Rowley & Smith, 1985).

There appear to be few cases where two species of introduced birds have become competitors for the same resource. Neither is there much evidence that introduced species compete with native species for resources such as food or nesting sites. Green (1983) concluded that the decline in Tasmania of the eastern rosella *Platycercus eximius* coincided with the establishment and spread of *Sturnus vulgaris* and was a consequence of its nest-hole usurpation. Similar statements have been made for other Australian endemics, but the interaction between most introduced and endemic species remains largely unstudied.

The present dangers from introduced taxa in Australia would appear to come from birds presently held in captivity or from birds which are smuggled into Australia from South-east Asia and elsewhere.

Well-established species

The European starling *Sturnus vulgaris* and the house sparrow *Passer domesticus* are the two most widely distributed introduced birds in Australia. They were both released between the late 1850s and 1900 (Balmford, 1978; Long, 1981) and now occupy all of south-eastern Australia with a mediterranean-type climate northwards to the subtropical region of south-eastern Queensland. Both species spread rapidly, *S. vulgaris* at the rate of 20–25 km/year and *P. domesticus* through sparsely settled country in South Australia at 6.7 km/year and in Queensland through settled farmlands at a rate between 85.3 and 103.6 km/year (Blakers *et al.*, 1984). Both species have spread largely in association with settlement and disturbed habitats and both are still expanding their ranges. In 1976 *S. vulgaris* reached the border between Western Australia and South Australia and flocks began to invade south-east Western Australia. Pest control authorities in that state have been destroying colonising flocks now for more than ten years and so far have prevented any further spread westward.

Density measurements for *S. vulgaris* vary from that in partly cleared and grazed woodland of 0.31 birds/hectare (Ford & Bell, 1981) to that in irrigated orchard areas of 2.5 birds/hectare (Thomas, 1957). *Passer domesticus* reaches densities of 33.7 to 39.1 birds/hectare in some suburban areas (Jones, 1981, 1983). The communal roosting habits of *S. vulgaris* will ultimately cause it to become a far greater nuisance than it is at present. The main impact of both *S. vulgaris* and *P. domesticus* is in fouling in cities and towns and damage to agricultural crops (cherries and soft fruits).

The skylark *Alauda arvensis* is established widely in disturbed habitats in south-eastern Australia. From liberations of birds imported from Europe between 1854 and 1900 (Balmford, 1978; Long, 1981) it has spread slowly, although it has the potential to colonise much further. Its impact on agriculture has not been great, but recent damage to seedlings of oil poppy (Blakers *et al.*, 1984) suggests that further conflict with agriculture may be possible.

Now widespread from releases made between 1857 and 1872 (Balmford, 1978), the European blackbird, *Turdus merula*, is still expanding its range. Its impact appears to be greatest in areas such as Tasmania where soft fruits, particularly cherries, are grown extensively.

To control insect pests the common mynah *Acridotheres tristis* from Asia was released in Australia in 1862 and a number of times subsequently until 1872. Its present range is somewhat restricted to four large areas along the eastern coast, the boundaries of which are rapidly approaching one another because of further spread of the bird. This mynah is a potential destroyer of fruit crops, but is not yet numerous enough in Australia to have become much of a pest.

Two European cardueline finches (Fringillidae) – namely the goldfinch *Carduelis carduelis* and the greenfinch *C. chloris* – and one Asian estrilidine (Estrilidae) finch *Lonchura punctulata* are the only survivors of numerous finch and finch-like taxa which were released in the period 1860–1880 (Thomson, Long & Horton, 1988). Species such as the bullfinch *Pyrrhula pyrrhula* and chaffinch *Fringilla coelebs*, introduced from Europe, failed probably because of the adverse habitat and small numbers released. *Carduelis carduelis* survived, however, and now ranges over much of south-eastern Australia from Tasmania to Spencer's Gulf in South Australia to southern Queensland. A population in Western Australia suffered a sudden decline about 1970, more than likely because of destruction of habitat associated with a building boom, but also perhaps because an increased level of parasites (*Cnemidocoptes* spp.) affected its ability to utilise some available food resources.

Carduelis chloris has not spread as widely as *C. carduelis* but occupies the southern parts of south-eastern Australia and Tasmania. Less widespread still, the chestnut mannikin *L. punctulata* has become established along the eastern seaboard from Sydney north to Cairns, mainly in the warmer coastal areas of the subtropics and tropics. The present populations were derived from a series of escapes and releases of birds from captivity since 1930. Its spread has been rapid, especially in northern areas (Blakers *et al.*, 1984). Both *C. carduelis* and *C. chloris* as yet seem to be innocuous species in Australia although they have the potential to cause agricultural damage. *Lonchura punctulata* is capable of displacing a number of native taxa of finches by virtue of its wider breeding and feeding niches (Immelmann, 1960), but opinions differ as to whether this has actually happened (Frith, 1979; Immelmann, 1982).

Mallards *Anas platyrhynchos* were introduced from England to Australia in the 1860s or 1870s and are now well distributed on coastal and inland waters in southern Australia. The main expansion of range has occurred in eastern Australia since the 1950s (Blakers *et al.*, 1984), but how much has been because of natural increase rather than new introductions is not known. The small population in south-west Western Australia has not spread much, but this may be in part because of efforts by fauna authorities who remove them from ornamental waters whenever possible. *Anas platyrhynchos* will hybridise with the Pacific black duck *A. superciliosa* and produce fertile offspring. Hybrids have been obtained in Tasmania and Western Australia and the impact in this regard appears to be increasing.

The pigeon *Columba livia* has become feral in Australia as it has done in most countries of the world (Long, 1981). The original homing pigeons were imported from England and later from continental Europe. They escaped from the early European settlers and have continued to escape from successive generations of pigeon breeders. They are now present in all cities, larger towns, and often remote towns, throughout Australia. The largest concentrations occur in urban and suburban environments where there are often unlimited nesting places under the eaves and on the facades of older buildings. Generally, there are nearby food sources such as granaries, flour mills, railway lines, or parks, where they obtain 'hand-outs'. In a few areas pigeons have reverted completely to the wild form and roost and nest in cliffs or in trees away from human habitation. Feral pigeons have occasionally become agricultural pests, but their most serious impact may be the role they play in the dissemination of diseases both to domestic stock and to humans. In large cities, the excreta from large populations foul building structures and food depots where the food products may be destined for human consumption.

The two turtle doves *Streptopelia chinensis* and *S. senegalensis* are both southern Asian in origin and are now fairly widespread from liberations made in the late 1800s. *Streptopelia senegalensis* is widely distributed in south-western Australia and *S. chinensis* along the south-eastern and eastern coastline from Adelaide to Cairns, with a small population in metropolitan Perth, Western Australia. Both species have spread slowly, but without the aid of additional releases and escapes it appears that their present distribution might have been more restricted. Both species appear to prefer urban or suburban parklands and gardens to that of farmland and native bush. *Streptopelia senegalensis* appears likely to spread further north and south of its present range (Blakers *et al.*, 1984) and concern has been expressed that if it does so, then it may compete with some native columbid taxa. The impact of both doves has been slight. They occasionally cause some nuisance value by eating food set out for poultry and will eat vegetable and flower seedlings in seed nurseries.

The cattle egret *Ardeola ibis* was introduced into the north of Western Australia where it failed to become established, but subsequently it naturally colonised the area from southern Asia. From the tropical region of northern Australia it has spread southwards in the last 40 years into the regions of mediterranean climate and now occupies much of south-eastern and south-western Australia. It has not yet bred in many of the more southern areas or in Tasmania (Blakers *et al.*, 1984).

European tree sparrows *Passer montanus* were released in Australia at the same time as the house sparrow. The species has not had the same success, has spread at a slower rate, and has only a restricted range in south-eastern Australia. In some towns along the Murray River it has now become as numerous as the house sparrow. For instance, in the suburbs of Wagga Wagga, New South Wales, its density has been recorded as 29.2 birds/hectare (Jones, 1981). It has as yet had little impact in Australia, but if it continues to increase in numbers and spreads further it certainly will be as much a nuisance as the house sparrow.

The European song thrush *Turdus philomelos* has remained around Melbourne with little spread in the last 100 years. The reasons for its lack of success when compared with that of the allied *T. merula* is probably because it is less tolerant of hot dry conditions (Blakers *et al.*, 1984).

The South-east Asian bulbul *Pycnonotus jocosus* is well established with a restricted distribution but it is still spreading slowly. The first release was made in 1880 and there have been others subsequently, and probably some escapes from captivity. The species has been reported to damage fruit and vegetable crops and the potential to damage ripening fruits is considerable. At present, however, the species mainly inhabits gardens and parks in urban areas.

Established but very restricted species

The following taxa are established tenuously and may or may not continue to be successful. They have not become numerous enough to have had any impact on the fauna or flora of areas of mediterranean-climate Australia.

The African ostrich *Struthio camelus* was released in South Australia following attempts to establish feather and hide farms. A small population which sustains itself by natural breeding still survives in the Mount Lofty Ranges, near Adelaide.

The English mute swan *Cygnus olor* was released in Western Australia between 1897 and 1912 and still survives on the Avon River, 80 km east of Perth.

Both the peafowl *Pavo cristatus* from Asia and the guinea-fowl *Numida meleagris* from Africa are widely kept in a semi-feral state in Australia and occasionally become completely feral. A small colony of peafowl has survived on Rottnest Island, near Fremantle, Western Australia, for nearly 60 years. The domestic fowl *Gallus gallus* has often been released on islands off the eastern coast of Australia and still survives on Heron Island in subtropical Queensland.

The ring-necked pheasant *Phasianus colchicus* is established on Rottnest Island and possibly in some areas in south-eastern Australia and in Tasmania. The California quail *Lophortyx californica* has been introduced to two or three areas but survives only on King Island, between Tasmania and mainland Australia, where it has been established for about 40 years. The turkey *Meleagris gallopavo* is still present on Seal Island in the Furneaux Group of islands, Tasmania.

Established but now extinct species

The Asian bulbul *Pycnonotus cafer* may have been established in Australia by about 1917. The mannikin *Lonchura malacca* was established from escapees from aviaries in about 1929 and bred in the Sydney area until about 1940. Likewise, the African wydah *Euplectes albonotatus* was established in the Sydney area for about 30 years, but has not been recorded since the early 1970s. The African weaver *Euplectes orix* became established near Adelaide in 1926 but has been considered extinct since 1976. The silver pheasant *Lophura nycthemera* became established in the Porongorup Range, Western Australia, when a free-ranging group was abandoned by the owner. These birds remained in the area where they survived and bred until removed by the park authorities about five years later (Long, 1988).

Although all these species are potential pests of grain and fruit crops they each failed to become numerous enough or widespread enough to cause any impact in Australia.

Established native species which have been translocated or reintroduced

At least 13 native species have been either established in areas outside their native ranges or have been reintroduced into their previous ranges in Australia. The majority were released deliberately but some have become established as a result of escapes from captivity. In most cases little is known of any impact from these translocations and reintroductions.

The emu *Dromaius novaehollandiae* was introduced to Maria Island, Tasmania in 1968 where it is now established, and on Kangaroo Island, South Australia, where it became extinct in 1927; it was reintroduced to Kangaroo Island in 1957 and now breeds there. Reintroductions of emus to areas formerly inhabited by them have also occurred locally in parts of Queensland, New South Wales and Victoria (Blakers *et al.*, 1984). Cape Barren geese *Cereopsis novaehollandiae* have been released successfully at Tidbinbilla, Australian Capital

Territory, on Maria, Three Hummock and Hunter Islands, Tasmania, and on Kangaroo Island, South Australia. The brush turkey *Alectura lathami*, the dove *Geopelia placida*, and the cockatoo *Callocephalon fimbriatum* were also successfully established on Kangaroo Island. The latter two species may not now be present there.

The lyrebird *Menura novaehollandiae* is established and breeding on a small scale in two areas in Tasmania, from introductions made in 1934. The range of the species is slowly expanding.

A small population of finches *Neochmia ruficauda* was breeding in the Sydney area in 1980. Both the mannikin *Lonchura castaneothorax* and the finch *Aegintha temporalis* are established and breeding, the former in the metropolitan area of Perth and the latter in the Darling Range, east of Perth. A third finch species *Poephila bichenovii* may also be established in the Perth metropolitan area. Most of these taxa escaped from aviaries to become established. They may or may not be successful in the long term, but *A. temporalis* has been established for nearly 30 years (Long, 1988).

Changes in land use such as clearing for agriculture and the removal of natural forests, as well as the release and escape of aviary birds, appear to be responsible for the southwards colonisation in southern Australia of the galah *Cacatua roseicapilla*. The cockatoo *Cacatua galerita* from eastern Australia was established and breeding in agricultural and suburban areas of Perth between the 1960s and 1970s. The species has now been largely eliminated by the efforts of pest control authorities.

The most widespread of the introduced native species in Australia is the kookaburra *Dacelo gigas* which now occupies south-west Western Australia, Tasmania, Kangaroo Island and Flinders Island as a result of the deliberate release of birds. In Western Australia its impact on native birds by occasionally preying on small species is held by some to be detrimental, but there is no conclusive evidence for this claim.

One of the most widely publicised reintroductions recently has been that of the noisy scrub-bird *Atrichornis clamosus*. This species, at one time thought to be extinct, was rediscovered on the south coast of Western Australia in 1961 (Smith, 1987). To ensure its survival two small groups have been translocated, one to Mount Manypeaks and one to Walpole, where they appear to be well established.

Concluding remarks

Closer examination of the species by species summaries given above shows that of the species considered widespread and common, two thirds originated from areas in which a mediterranean climate influenced part or whole of their native range. In addition, of those 12 introduced species, more than 90 per cent have very large native distributions. Further to this, two-thirds of those species either live in flocks (although flock size varies considerably) or roost communally, and with the possible exception of *Lonchura punctulata*, all are polyphagous.

Few if any of the less common and more restricted introduced species exhibit such characteristics. The success of those species now widespread in Australia was greatly assisted by the creation of environments which provided 'European-type' plants as a food source and provided areas with a greatly reduced species diversity. Lower diversity means less competition for food, shelter and nesting resources and a much smaller likelihood of competition between congeners or morphologically similar species. These characteristics have been considered necessary for successful introduction into new environments (see Ehrlich, 1986).

The continued survival of rarer or less widespread species is not necessarily contentious, since the mean survival time of such species should not be any less than that of the widespread species. Limiting factors such as the amount of suitable habitat available and the number of initial introductions have played major roles in determining the status of rarer species.

In fact, many of the rarer species were introduced into Australia at much later dates than species such as *Sturnus vulgaris*, *Passer domesticus* and *Columba livia*; further they were released into habitats which although suitable for colonisation had effectively been made into islands by agricultural and urban development. Once populations had expanded to saturation levels, opportunities for migration to other suitable areas were not forthcoming. The extinction of rarer and less widespread species will result if large-scale habitat modification occurs. Such habitat modification (or destruction) appears to have led to the extinction of those species mentioned previously, although very small founding numbers may have contributed equally.

References

Balmford, R. A. (1978). Early introductions of birds to Victoria. *Australian Bird Watcher*, **7**, 237–48, 262–5.

Blakers, M., Davies, S. J. J. F. & Reilly, P. N. (ed.) (1984). *The Atlas of Australian Birds*. Melbourne: Royal Australasian Ornithologists Union & Melbourne University Press.

Ehrlich, P. R. (1986). Which animals will invade? In *Ecology of Biological Invasions of North America and Hawaii*, ed. H. A. Mooney & J. A. Drake, pp. 79–95. New York: Springer-Verlag.

Ford, H. A. & Bell, H. (1981). Density of birds in eucalypt woodland affecting the varying degrees of dieback. *Emu*, **81**, 202–8.

Frith, H. J. (1979). *Wildlife Conservation*. Sydney: Angus & Robertson.

Green, R. H. (1983). The decline of the eastern rosella and other Psittaciformes in

Tasmania concomitant with the establishment of the introduced European starling. *Records of the Queen Victoria Museum*, **82**, 1–5.

Green, R. J. (1984). Native and exotic birds in a suburban habitat. *Australian Wildlife Research*, **11**, 181–90.

Immelmann, K. (1960). The spread of introduced birds in northern Queensland. *Australian Journal of Science*, **23**, 130–1.

Immelmann, K. (1982). *Australian finches*. Sydney: Angus & Robertson.

Jones, D. (1986). Exotic birds – selected examples. In *The Ecology of Exotic Animals & Plants: Some Australian Case Histories*, ed. R. L. Kitching, pp. 93–107. Brisbane: Wiley.

Jones, D. N. (1981). Temporal changes in the suburban avifauna of an inland city. *Australian Wildlife Research*, **8**, 109–19.

Jones, D. N. (1983). The suburban bird community of Townsville, a tropical city. *Emu*, **83**, 12–18.

Long, J. L. (1981). *Introduced Birds of the World*. Sydney: A. H. & A. W. Reed.

Long, J. L. (1988). *Introduced Birds and Mammals in Western Australia*. Perth: Access Press.

Mayr, E. (1965). The nature of colonization in birds. In *The Genetics of Colonizing Species*, ed. H. G. Baker & G. L. Stebbins, pp. 29–43. New York: Academic Press.

Moulton, M. P. & Pimm, S. L. (1986). Species introductions to Hawaii. In *Ecology of Biological Invasions of North America and Hawaii*, ed. H. A. Mooney & J. A. Drake, pp. 231–49. New York: Springer-Verlag.

Newsome, A. E. & Noble, I. R. (1986). Ecological and physiological characters of invading species. In *Ecology of Biological Invasions: An Australian Perspective*, ed. R. H. Groves & J. J. Burdon, pp. 1–20. Canberra: Australian Academy of Science.

Saunders, D. A., Rowley, I. & Smith, G. T. (1985). The effect of clearing for agriculture on the distribution of cockatoos in the south-west of Western Australia. In *Birds of Eucalypt Forests and Woodlands: Ecology, Conservation and Management*, ed. A. Keast, H. F. Recher, H. Ford & D. Saunders, pp. 309–21. Mooney Ponds, Victoria & Chipping Norton, N.S.W.: Royal Australasian Ornithologists Union & Surrey Beatty & Sons Ltd.

Sibley, C. G. & Ahlquist, J. E. (1986). Reconstructing bird phylogeny by comparing DNA's. *Scientific American, Feb. 1986*, 68–78.

Smith, G. (1987). A bird in the bush – the noisy scrub-bird story. *Australian Natural History*, **22**, 189–92.

Thomas, H. F. (1957). The starling in the Sunraysia district, Victoria. *Emu*, **57**, 31–48, 131–44, 151–60, 269–84.

Thomson, J. M., Long, J. L. & Horton, D. R. (1988). Human exploitation of and introductions to the Australian fauna. In *Fauna of Australia*, Vol. 1A, ed. G. R. Dyne & D. W. Walton, pp. 227–49. Canberra: Australian Government Publishing Service.

Part IV Applied aspects of mediterranean invasions

The biogeography of mediterranean invasions is of fundamental interest, as the preceding sections set out to show, and it is a major theme of this volume that studies of biological invasions have considerable application to the solution of major problems of land use in regions of mediterranean climate. As people, plants and animals have moved between the several regions of mediterranean climate problems in land use have arisen. Some plants have spread and invaded land used as range or for cropping or for human amenity. The ecology of these plants has been discussed in several preceding chapters but in this section they are discussed again in a more applied sense in relation to their control.

One potentially powerful way to control invasive organisms is to deliberately introduce highly selective natural enemies in a program of biological control. There have been several cases where such programs for invasive organisms in mediterranean regions have been successful, and sometimes spectacularly so, e.g. the introduction of the myxoma virus for rabbit control in southern Australia and the introduction of a chrysomelid beetle for control of St John's wort in California. These and other examples have been extensively reviewed elsewhere and will not be repeated here. In this section we include, however, a chapter on the introduction of dung beetles to southern Australia from France in a novel program to limit the breeding of native flies in the dung of introduced herbivores. Although there are dung beetles native to southern Australia they are relatively ineffective at reducing the rate of decomposition of dung pads from the domesticated herbivores introduced only 200 years ago.

26

Weed invasion in agricultural areas

J. L. GUILLERM

Plant populations have always been subject to species replacements and demographic shifts in density, both of which may lead to invasion. As agriculture developed over millennia, so the chances for plant invasion increased, especially among those plants we now call weeds. Weeds are colonising species capable of occupying otherwise unoccupied space. Only those species with several reproductive strategies and those with life cycles adapted both to summer-dry, winter-wet conditions and to cultivation have become major weeds in regions of mediterranean climate. The selection forces affecting size and spread of populations of weeds have been described by Sagar (1982). After introduction, some plants may spread immediately; alternatively, the spread of some other plants may be delayed until a sizeable population of propagules has built up and only then will spread occur.

The weeds of mediterranean areas

In mediterranean-climate areas weediness comes from the biological ability of a plant species to survive a mediterranean-type climate and a particular cultivation regime associated with the development of agriculture in the Mediterranean Basin. Human effects on weed invasion have increased, not always gradually, but sometimes episodically (Holzner & Glauniger, 1982; Pignatti, 1983; Zohary, 1983; Pons & Quézel, 1985). The long and complex history of agricultural disturbance, especially in the Mediterranean Basin, has led to a large and diverse weed flora, which is still evolving.

About 11 000 years BP a Neolithic culture change occurred in south-western Asia, and over the Mediterranean Basin, which was to have an irreversible effect upon human history. This very antiquity of agriculture results in a complex pattern of establishment and invasion of weed species in both time and space in the Mediterranean Basin. Shifts within the weed flora, although largely unrecorded, have probably been considerable, and may even exceed the increases recorded for regions such as California (Frenkel, 1979; Rejmanek, this volume).

From prehistoric times, the processes of deforestation and conversion of land for pasture and cultivation have resulted in the evolution of rural landscapes which are a complex of different plant communities. These plant communities result from diverse land uses (see, for instance, Thellung, 1912; Kamishev, 1959; Kornas, 1968; Greuter, 1971; Guillerm, 1980; Holzner, 1982). Kuhnholtz-Lordat (1964) identified three main assemblages of weed species: the 'ager' (cultivated fields), the 'saltus' (fallows, grasslands, pastures) and the 'silva' (shrublands, forests), each of which provide for large flows of weed species within the same land-use unit, or between adjoining land-use units.

The diversity of a weed community will be influenced by the introduction of species and the proximity to cultivated fields, as a source of colonising species. Species-rich neighbourhoods offer a variety of potential invaders, whilst depauperate ones do not. From this diverse source of weed propagules in space and time, and as a result of different lengths of propagule viability, most agricultural soils contain a lot of weed seeds that are well adapted to the mediterranean climate and cultivation type.

The size and shapes of fields and of their surrounding landscape units, determined by past and present agricultural land-use patterns (Long *et al.*, 1971), influence the richness of the soil propagule bank; they are very important factors to understand weed composition and shifts within the weed communities, in both space and time. In the last few decades, these aspects have also changed within perennial cultivations with the intensive use of chemicals for weed control (Brullo, 1979; Fort, 1979; Guillerm, 1980; Le Maignan, 1981; Maillet, 1981; Muracciole, 1981; Guillerm & Maillet, 1984; Ribeiro, 1984; Loudyi, 1985; Protopapadakis, 1985; Espirito-Santo, 1986; Moreira, 1986).

The antiquity of the weed flora

In Egypt, agriculture began some seven to eight thousand years ago (Kosinova, 1974). The most important sources of information are plant remains from the ancient tombs. About 50 weeds have been recorded and listed chronologically from the Neolithic to the Coptic periods (Boulos & El Hadidi, 1984). Most of them are, today, distributed commonly all over the Mediterranean region.

Present biogeographical spectra of weed communities of the Mediterranean Basin show the predominance of Mediterranean species originating from other climatic zones (Table 26.1). This number is higher in Morocco, southern Italy, Sicily, Djerba Island (Tunisia), northern Cyrenaica and to the west of Syria, and decreases in France, Spain and Italy (except southern Italy), because of the

Table 26.1. *Percentages of Mediterranean species within the weed communities of different regions of the Mediterranean Basin*

Geographical region	Medit. species (%)	Reference
Cordoba, Spain	55	Pujadas & Hernandez-Bermejo, 1984
Languedoc, France	32	Guillerm & Maillet, 1982
Italy	40	Franzini, 1982
Southern Italy	70	Franzini, 1982
Sicily, Italy	65	Maugeri, 1979
Palermo, Italy (cereals)	81	Raimondo *et al.*, 1979
Palermo, Italy (oranges)	61	Raimondo *et al.*, 1979
Lattaquie, Syria	89	Soufi, 1987
Egypt	41	Kosinova, 1974
Northern Cyrenaica, Libya	77	Brullo, 1979
Djerba, Tunisia	84	van den Bergen, 1980
Meknes plateau, Morocco	77	Loudyi, 1985

presence of species originating in northern Europe and Eurasia. In Egypt, the lower number of Mediterranean species results from the presence of Saharo-Sindian and tropical species.

The antiquity of agricultural disturbance results in a homeostasis in Mediterranean weed communities when they are faced with the entry of invasive plants. A new species may need a lag period to build sizeable founder populations, firstly in ruderal places and there await a favourable event in order to invade nearby fields, as, for instance, when a herbicide is applied to a weed community. *Senecio inaequidens* originated in the Cape province of South Africa, and was introduced to France as a wool contaminant (Jovet & Bosserdet, 1968). From 1935 to 1970 *S. inaequidens* occurred only in ruderal places, but recently it has increased in vineyards in southern France where it seems to occupy vacant spaces created by the disappearance of species susceptible to recurrent herbicide applications (Guillerm *et al.*, 1990).

Modern cultivation methods (e.g. levelling, uniformity of fertilizer application, water regulation, intensive use of herbicides) favour those weeds with requirements close to those of crop species (Holzner, 1978).

In those mediterranean-climate countries settled more recently and now having a high level of agrotechnology, such as parts of Australia, South Africa and California, massive changes in agricultural land use have occurred. Such a drastic change has favoured the establishment of weeds introduced from European, Mediterranean and tropical regions. The low number of native species adapted to this recent agricultural disturbance, the lack of time for selection of populations adapted to this drastic level of disturbance or to build a sizeable founder population able to compete with newcomers originating from countries with an older agricultural pattern, together with the lack of local predators, ensures success of these introduced weeds. In South Africa the weed flora contains 57 per cent introduced species (Wells & Stirton, 1982). In Israel since 1890, 73 weeds have been introduced of which about 20 are noxious (Dafni & Heller, 1982). These species have been favoured by the recent and rapid agricultural development happening in Israel.

Species reactions to the mediterranean climate and to cultivation

One characteristic of mediterranean weeds, whether native or introduced, is their adaptation to the mediterranean climate, with its winter-cold, summer-dry conditions and irregular timing of rainfall events in autumn. According to this climatic regime, spread of weed species depends on a number of factors including genetic and somatic variability, sexual or clonal reproduction (morphology and physiology of reproductive organs), architectural organisation, life form, life cycle and plant habit.

The climatic regime of the Mediterranean Basin and the timing of germination are very important for weed establishment and spread. This ability of species to adapt their germination date, as shown by winter and spring cycles for northern European or Eurasian species, or summer cycles for tropical species, may determine the success of weed invasion. For perennials, the development of strong and efficient regenerative systems ensures their establishment. Holzner & Immomen (1982) have shown, along a gradient of weed distribution from Finland and Austria to Italy, the decreasing differences between the weeds of cereals and row crops from the warmer parts of southern Europe towards the cooler and more humid areas of northern Europe.

The other particular main characteristic of mediterranean weeds is the way their life cycle is adapted to types of cultivation. The percentages of annuals and perennials differ with the crop cultivar and the crop rotation, which results in different spread patterns and different weed communities, according to the different agronomic practices, their timing and the type of weed control practised. For instance, concerning the latter factor, the selection of resistant biotypes to particular herbicides has occurred recently in populations of both annual and perennial weeds.

The biological responses of species to the mediterranean climate and agronomic practices influence their germination ecology, their life cycles, competitive abilities, nutrient requirements, reproductive strategies and their dispersion rates. I shall now comment on just a few examples of the diverse reactions of weeds to these factors.

The species level

Seed dormancy is a general adaptation of many plants to survive harsh environments as a result of different seed morphology and/or vegetative strategies.

Many plant species are able to produce different types of seeds (Harper, 1977) that will germinate immediately or at a precise time, or else show erratic germination over one or several years. These species are mainly short-lived fugitive species of the families Asteraceae, Chenopodiaceae, Cruciferae and Gramineae. Some species are adapted for dispersal, whilst others have no obvious dispersal properties. Seed polymorphism is often associated with physiological polymorphism, with regard to dormancy and to growth characteristics.

Dispersed seeds may have no dormancy whilst non-dispersed ones may be dormant. Plants from dormant seeds may be more vigorous and produce more seeds; they have a higher probability of being outbred. Plants from non-dormant seeds are usually intolerant of stress but are able to produce some seeds under extreme conditions; they seem to be more inbred.

Some species produce heavy basal seeds and light upper ones. For *Avena fatua*, populations coming from heavy seeds are more competitive than populations from light seeds (Peters, 1985). The heavy seeds are less dormant than light seeds and give rise to the first wave of emergence, making these populations highly competitive. The light seeds arising later give rise to much less vigorous plants.

Other species have different types of seeds, localised in the inflorescence. For *Carduus pycnocephalus* and *C. tenuiflorus*, seeds situated in the centre of the capitulum are not dormant and are readily dispersed, whilst outer seeds can be dormant and have no apparent means of dispersal (Olivieri *et al.*, 1983; Olivieri & Berger, 1985; Olivieri *et al.*, this volume). The easily dispersed inner seeds are able to germinate within 2 days given sufficient moisture, thereby allowing for rapid colonisation. The non-dispersed outer seeds show variable dormancy from 2 days to 3 years, thereby allowing for long-term persistence in the seed bank.

Hard-seededness is a common dormancy mechanism in families Leguminosae, Convolvulaceae, Chenopodiaceae and Malvaceae. Many weed species have hard seeds that enable their long persistence in soil seed banks and gradual germination over time (Rolston, 1978).

The association between sexual reproduction and vegetative spread occurs differently in perennial weeds, according to the agroecosystem in which they occur. For *Paspalum distichum* as many as 100000 seeds may be produced (Okuma & Chikura, 1984). This species spreads rapidly both by seeds, creeping rhizomes, and aerial stems. Under drought conditions, however, the aerial shoots are more susceptible to desiccation and high temperature than are the underground rhizomes (Huang *et al.*, 1987). Thus in areas with a distinct dry summer period, over-wintered seeds and rhizomes are more likely to be sources of infestation. Other perennial species, originating from tropical zones, are able to spread under mediterranean and colder climates either by seeding or vegetatively. For *Sorghum halepense*, for instance, the existence of biotypes enables this species as a whole to survive different climatic stresses. Some biotypes overwinter from rhizomes, other ones do not, but re-establish exclusively from seeds (Warwick *et al.*, 1986). Expansion of *S. halepense* at the northern limits of its range, where rhizomes are susceptible to freezing temperatures, comes only from seed production.

Some weed species are able to compete with crops because they possess higher seedling emergence and establishment rates. *Avena sativa* has a competitive ability of its roots greater than that of its shoots. Prolific development of crown roots favours absorption of nutrients, thereby producing greater shoot growth, more tillers, a larger root system and higher seed production (Martin & Field, 1987).

Plasticity in life form within some species enables them to adapt to different stages of a crop rotation system. For *Solanum eleagnifolium*, at present a spreading and noxious introduction to Morocco, different reproductive strategies occur that are correlated with the type and timing of cultivation (for more details see Yannitsaros & Economidou, 1974; Vigna *et al.*, 1981; Tanji *et al.*, 1984; Guillerm *et al.*, 1990). This species survives and spreads from seeds (a so-called 'therophytic' strategy), from buds close to the soil surface ('hemicryptophytic' strategy), from buds situated on the hypocotyl or by regeneration from root pieces, whether vertical or horizontal ('geophytic' strategy).

Many perennial weeds spread mainly vegetatively, e.g. *Cynodon dactylon* and species of *Oxalis* and *Equisetum*. *Oxalis pes-caprae* is currently spreading through all the Mediterranean areas; it varies in ploidy level and style length and reproduces mainly from bulbs. In Australia the weedy genotype is a pentaploid, short-lived form that reproduces only vegetatively (Michael, 1964). Within the Mediterranean Basin, Chabrolin (1934) showed that in Tunisia the occurrence of sexual reproduction is exceptional, but that the late flowers (in February and March) are able to produce seeds, although they are sterile. Spread of this species thus comes from bulb production. Michael (1965) showed that there was a critical stage in the plant's life cycle for control at which the old bulbs were becoming exhausted of reserves in favour of shoot growth, and before the new bulbs were sufficiently well formed to be self-perpetuating. Viable bulbs may sprout in the same year and behave as annuals (Lane, 1984).

Adaptation to climate

In annual grasses, the existence of non-dormant and dormant seeds enable species to adapt their life cycle to periods of climatic stress. Species such as *Poa annua* germinate throughout the year and may produce several germinations per year. For species such as *Avena fatua* germination occurs mainly in spring, although some germination occurs in autumn. For other species such as *Avena sterilis* ssp. *sterilis* germination occurs after autumn rains. Some seedlings arise in summer but they do not give rise to fertile populations. Grass weeds originating from tropical zones, such as *Digitaria sanguinalis* and *Paspalum* spp., show only summer germination.

Populations of wild oats, a species common and widespread in all regions of mediterranean climate, occur within different zones according to their temperature requirements. Along a decreasing temperature gradient in western Europe, *Avena sterilis* ssp. *sterilis* is a southern taxon, *A. sterilis* ssp. *ludoviciana* is more Atlantic, and *A. fatua* more northern. The latter is present in Spain and Portugal, but only on plateaux, and in the northern areas of Greece and Yugoslavia.

Datura stramonium and *D. ferox* are noxious weeds in South African summer crops, although they may germinate throughout the year. The production of viable seeds over several years and their ability to germinate at any suitable time of the year enable *D. stramonium* to evade standard weed control measures and thereby spread (Malan *et al.*, 1982).

Ammi majus is a xerophilous species (Montegut, 1984) which requires mild winters to flower. In North Africa and southern Spain, its occurrence is not controlled by soil pH, whereas in southern France is occurs on clayey calcareous soils and in the area around Paris it occurs only on warm calcareous soils.

Adaptation to cultivation

Several factors, such as types of cultivation, type and timing of agricultural practices, and crop rotation systems, all interfere with weed invasion. The cropping schedule may influence weed spread. For *Euphorbia nutans*, originally from North America and a recent invader in several western European countries, mainly in Spain (Hernandez Bermejo *et al.*, 1984) and Italy (Pignatti, 1982), it completes its life cycle in irrigated cotton fields. Its development is poor and its life cycle shortened in wheat in Andalusia, because the wheat in that region is harvested early. Moreover, the species is resistant to most herbicides. The biology of this species is possibly similar to that of *Euphorbia geniculata*, a summer annual of neo-tropical origin, which has extended its distribution into Israel, also in irrigated cotton fields (Dafni & Hellner, 1980). Its seedlings have a strong potential to regenerate after injury from adventitious buds produced on the hypocotyl. Seed germination, viability and dispersal are different for populations with dormant and non-dormant seeds (Kigel *et al*., 1984).

The type of ploughing may determine the size of the population of viable seeds. For *Lolium rigidum* in Australia, subsequent ryegrass densities were reduced more by mouldboard-ploughing (Reeves & Smith, 1975). Timing of mechanical or chemical weed control may influence weed survival. For *Oxalis pes-caprae* (see earlier), the critical period for control is before the production of new bulbs.

Crop rotation systems may also interfere with weed spread. For *Lolium rigidum* in Australia, its biology is adapted to a wheat-pasture rotation. Weed control in the pasture phase in between the cropping phases can reduce the size of the ryegrass seed bank (Monaghan, 1980) and hence increase yield of the subsequent crop.

The ability to employ different life form strategies favours the adaptation of some species to different crop rotations. For instance, the diverse life forms of *Solanum eleaegnifolium* in Morocco (Tanji *et al*., 1984) enable this species to invade a range of agricultural systems including autumn crops, sugar beet, cotton, maize, vegetables and orchards, as well as ruderal places (see earlier).

Space and time scales and weed invasions

Weeds exhibiting high rates of increase are likely to become invasive. Weeds with low rates of spread do not have wide-ranging distributions in general, but they may pose serious problems within restricted areas and they may increase with time. Analysis of rates of spread (into abundant, frequent, invading or casual), from the global to the local scale, enables present and potential threats to be detected. Forcella (1985) suggested that rates of spread of species in a regional weed flora that have not yet reached their final distributions may be used as indices to identify future weed invasions. Macworter & Chandler (1982) also emphasised the importance of knowledge of rates and patterns of spread in successful control programs for weeds.

From the global to the local, there is an infinite range of scales which result in different patterns of weed distributions. The so-called cosmopolitan weeds are related to a world scale and result from previous introductions. Patterns on a local scale point to a potential extension or else provide evidence for a potential invasion. Real cosmopolitan or ubiquitous species do not exist in fact. All species show ecological preferences, and even among those which are widely distributed, they may be absent from certain specialised sites. Moreover, a large area of distribution usually implies infra-specific variation. With a knowledge of the ecology and present spread of *Solanum eleaegnifolium*, it should be possible to predict the potential of the closely related *S. karsense*. The latter species is native to Australia and is presently invading summer crops in South Africa; it has a similar ecology and habit to that of *S. eleaegnifolium* in Australia (Monaghan & Brownlee, 1981). Potentially it is invasive in other regions of mediterranean climate were it ever introduced.

The very diversity and fluctuations in weed flows within an agroecosystem provide evidence for considering the system as an information channel whose efficiency can be estimated from the species entropy of weeds within the communities sampled (see, for more information on this concept, Godron, 1968; Guillerm, 1971; Morris & Guillerm, 1974). From a large set of samples relative to species distribution in space and time, a species that is either present or absent will have an entropy value equal to zero, the minimum value, and a species present in half the samples will have an entropy value equal to one, the maximum value. From a local to a global scale, it is possible to detect spread of local, regional or cosmopolitan weeds. Over a definite time scale, shifts in entropy value indicate maintenance, decrease or increase in weed spread.

Biogeographical and ecological data will provide information on present and potential weediness of the sort shown by Forcella *et al.* (1986) for *Echium italicum*, *E. vulgare* and *E. plantagineum*, which are respectively rare, uncommon and common in Australia.

Discussion

As changes in agriculture introduce new disturbances to the agroecosystem, so the weed community will continue to change and new chances for plant invasion be created.

The earliest weed communities result from the ability of pre-existing species, adapted to open and perturbed spaces, to colonise new areas created for cultivation. After emergence and the progressive differentiation and specialisation of the weed flora, weed invasion has been favoured, and will be favoured increasingly by human traffic, reclamation of new land for cultivation, by shifts in agricultural practices and weed control methods, as well as by the introduction and extension of the range of new crops.

Weeds are both specialist species, co-evolved with cultivation, as well as being generalist species, in terms of invasion. These two adaptive strategies have been emphasised by Holzner (1982) and by Oka & Morishima (1982). The balance between local species (natives, archaeophytes or naturalised species) and aliens (neophytes and present introduced newcomers) differs from country to country, in all regions of mediterranean climate, according to the type of weed control and the level of agricultural technology, such as extensive or intensive cultivation, irrigation, glasshouse culture, traditional or new crops. Other factors that may influence weed invasion in agricultural systems include the level of heterogeneity in the areas surrounding cultivated fields, the level of transportation of propagules to the place of cultivation or to a new area.

Inconsistently, it appears that an invasive plant may or may not be originally a weed, a native or alien species, an annual or perennial species, a generalist or a specialist. The fluctuating status of an invasive plant, in space and time, results in 'decreasing' or 'increasing' phases, with very extensive distribution (cosmopolitanism) or with only localised distribution. Unfortunately for human activities associated with agriculture there will always be at least one population of an opportunistic weed species to respond to the human interference and to co-evolve with humans.

From the large amount of research results on weed biology reviewed in this and other chapters, answers are still required to the following questions:

1. Why do some species require a lag period in which to build a founder population of sufficient numbers before they can invade, whilst populations of some other species are able to quickly adapt to new and sometimes major agricultural disturbances, such as repeated herbicide applications?
2. What initial population sizes and rates of spread are required for invasion success of different weed species?
3. What biological attributes influence the responses of a species to agricultural disturbances?
4. What are the major pressures in agroecosystems that act on weed populations as they spread and coexist with crop species?
5. What are the main biological and environmental factors that affect the balance in a community of weed species between reproduction by seeds and reproduction by vegetative means?

In attempting to answer these questions, future research will need to incorporate studies of the morphology of different weed species, especially their architecture and patterns of bud initiation.

References

Boulos, L. & El Hadidi, M. N. (1984). *The Weed Flora of Egypt*. Cairo: The American University of Cairo Press.

Brullo, S. (1979). La vegetazione infestante messicola della Cirenaica settentrionale. In *La Vegetazione Infestante e Sinantropica*, ed. G. G. Lorenzoni & C. Ferrari, pp. 171–88. Pavia: Societa Italiano di Fitosociologica.

Chabrolin, Ch. (1934). Les graines d'*Oxalis cernua* Thung. en Tunisie. *Bulletin de la Societé d'Histoire Naturelle de l'Afrique du Nord*, **25**, 396–8.

Dafni, A. & Heller, D. (1980). The threat posed by alien weeds in Israel. *Weed Research*, **20**, 277–83.

Dafni, A. & Heller, D. (1982). Adventive flora of Israel. Phytogeographical, ecological and agricultural aspects. *Plant Systematics & Evolution*, **149**, 1–18.

Espirito-Santo, D. (1986). Evoluçao de flora infestante des vinhas do Bombarral. *Revue Ciencia e Tecnica Vitivinicola, Lisboa*, **5**, 31–52.

Forcella, F. (1985). Final distribution is related to rate of spread in alien weeds. *Weed Research*, **25**, 181–91.

Forcella, F., Wood, J. T. & Dillon, S. P. (1986). Characteristics distinguishing invasive weeds within *Echium* (Bugloss). *Weed Research*, **26**, 351–64.

Fort, G. (1979). La non culture en vignoble et les évolutions de la flore. X. Journées d'études sur le désherbage. *Comptes Rendus du Troisième Colloque International sur l'Ecologie et la Biologie des Mauvaises Herbes*, pp. 967–71.

Franzini, E. (1982). Weeds of Italy. In *Biology and Ecology of Weeds*, ed. W. Holzner & N. Numata, pp. 245–56. The Hague: Junk.

Frenkel, R. E. (1979). Ruderal vegetation along some California roadsides. *University of California Publications in Geography*, **20**, 1–163.

Godron, M. (1968). Quelques applications de la notion de fréquence en écologie végétale. *Oecologia Plantarum*, **3**, 185–212.

Greuter, W. (1971). L'apport de l'homme à la flore spontanée de Crète. *Boissiera*, **19**, 329–37.

Guillerm, J. L. (1971). Profils écologiques et information mutuelle entre espèces et facteurs écologiques. *Oecologia Plantarum*, **6**, 209–26.

Guillerm, J. L. (1980). Stratégies dans les phytocenoses post-culturales, émergences et liaisons entre stades évolutifs. In *Recherches d'Ecologie Theorique. Les Stratégies Adaptives*, ed. R. Barbault, P. Blandin & J. A. Meyer, pp. 237–50. Paris: Maloine.

Guillerm, J. L. & Maillet, J. (1982). Weeds of western Mediterranean countries of Europe. In *Biology and Ecology of Weeds*, ed. W. Holzner & N. Numata, pp. 227–43. The Hague: Junk.

Guillerm, J. L. & Maillet, J. (1984). Influence de l'environnement sur la flore des vignes désherbées chimiquement. *Proceedings 3rd International Symposium of E.W.R.S. on Weed Problems in the Mediterranean Area, Oeiras, Portugal*, **1**, 49–56.

Guillerm, J. L., Le Floc'h, E., Maillet, J. & Boulet, C. (1990). The invading weeds within the western Mediterranean Basin. In *Biological Invasions in Europe and the Mediterranean Basin*, ed. F. di Castri, A. J. Hansen & M. Debussche, pp. 61–84. Dordrecht: Kluwer.

Harper, J. L. (1977). *Population Biology of Plants*. London: Academic Press.

Hernandez-Bermejo, J. E., Saavedra, M., Hidalgo, B., Montero, J. M. & Garcia-Torres, L. (1984). Weed flora in the irrigated crops of the Guadalquivir River valley. *Euphorbia nutans* Lag., a new weed species. *Proceedings 3rd International Symposium of E.W.R.S. on Weed Problems in the Mediterranean Area, Oeiras, Portugal*, **3**, 621–8.

Holzner, W. (1978). Weed species and weed communities. *Vegetatio*, **38**, 13–20.

Holzner, W. (1982). Concepts, categories and characteristics of weeds. In *Biology and Ecology of Weeds*, ed. W. Holzner & N. Numata, pp. 3–20. The Hague: Junk.

Holzner, W. & Glauniger, H. (1982). Weed shifts. In *Improving Agriculture*, pp. 150–2. Rome: FAO.

Holzner, W. & Immomen, H. (1982). European overview. In *Biology and Ecology of Weeds*, ed. W. Holzner & N. Numata, pp. 203–6. The Hague: Junk.

Huang, W. Z., Hsiao, A. I. & Jordan, L. (1987). Effect of temperature, light and certain growth regulating substances on sprouting, rooting and growth of single node rhizome and shoot segments of *Paspalum distichum* L. *Weed Research*, **27**, 57–67.

Jovet, P. & Bosserdet, P. (1968). *Senecio harveianus* MacOwan. Rélévé chronologique des observations en France. *Bulletin Centre d'Etudes Recherches Scientifiques*, **7**, 417–20.

Kamishev, N. S. (1959). On the classification of anthropochores. *Botanicheskii Zhurnal SSSR*, **44**, 1613–16.

Kigel, J., Lior, E. & Rubin, B. (1984). Seed germination and viability in the weedy summer annual *Euphorbia geniculata* Ortega. *Proceedings 7th International Symposium of E.W.R.S. on Weed Ecology, Biology and Systematics, Paris*, pp. 43–52.

Kornas, J. (1968). Geograficzno-historycna klasyfikacja roslin synantropijnych. *Materialy Zakladu Fitosocjologii Stosowanej (Uniwersytet Warszawski)*, **25**, 33–41.

Kosinova, J. (1974). Studies on the weed flora of cultivated lands in Egypt. IV. Mediterranean and tropical elements. *Candollea*, **29**, 281–95.

Kuhnholtz-Lordat, G. (1964). La silva, le saltus et l'ager de garrigue. *Annales de l'Ecole Nationale Agriculture de Montpellier*, **261**, 1–82.

Lane, D. (1984). Factors affecting the development of populations of *Oxalis pes-caprae* L. *Weed Research*, **24**, 219–25.

Le Maignan, I. (1981). Contribution à l'étude des groupements de mauvaises herbes des cultures en France. Aspects synsystematiques et biologiques. Thèse 3ème cycle, Université de Paris-Sud, Orsay.

Long, G., Rimbault, G. & Guillerm, J. L. (1971). Remote sensing of man-made ecosystems which result from variability of historical and human pressure. *Proceedings 7th International Symposium on Remote Sensing of the Environment, Ann Arbor, Michigan*, pp. 395–407.

Loudyi, M. C. (1985). Etude botanique et écologique de la végétation spontanée des terres cultivées du plateau de Meknes (Maroc). Thèse 3ème cycle, Université des Sciences et Techniques du Languedoc, Montpellier.

Macworter, C. G. & Chandler, J. M. (1982). Conventional weed control technology. In *Biological Control of Weeds with Plant Pathogens*, ed. R. Charudattan & H. L. Walker, pp. 5–28. New York: Wiley.

Maillet, J. (1981). Evolution de la flore adventice dans le Montpellierais sous la pression des techniques culturales. Thèse Dr. Ing., Université Sciences et Techniques du Languedoc, Montpellier.

Malan, C., Visser, J. H. & Grobbelaar, N. (1982). Control of problem weeds of maize on the Transvaal high veld (South Africa). II. *Datura stramonium* L. *Weed Research*, **22**, 101–4.

Martin, M. P. L. D. & Field, R. J. (1987). Competition between vegetative parts of wild oat (*Avena fatua* L.) and wheat (*Triticum aestivum* L.). *Weed Research*, **27**, 119–24.

Maugeri, G. (1979). *Flora e Vegetazione Infestante le Colture della Sicilia*. Poitiers: Faculté de Sciences, Université de Poitiers.

Michael, P. W. (1964). The identity and origin of varieties of *Oxalis pes-caprae* L. naturalised in Australia. *Transactions of the Royal Society of South Australia*, **88**, 167–74.

Michael, P. W. (1965). Studies on *Oxalis pes-caprae* L. in Australia. II. The control of the pentaploid variety. *Weed Research*, **5**, 133–40.

Monaghan, N. M. (1980). The biology and control of *Lolium rigidum* as a weed of wheat. *Weed Research*, **20**, 111–21.

Monaghan, N. M. & Brownlee, H. (1981). Control of *Solanum karsensis* in grain of *Sorghum*. *Weed Research*, **21**, 43–6.

Montegut, J. (1984). Causalité de la répartition des mauvaises herbes, espèces indicatrices du biotype cultural. *La Recherche Agronomique Suisse*, **23**, 15–46.

Moreira, I. (1986). Le point sur les techniques d'entretien des sols viticoles au Portugal. *Proceedings 2nd International Symposium of the Institut Techniques Vin sur la Non Culture de la Vigne, Montpellier*, pp. 65–78.

Morris, J. W. & Guillerm, J. L. (1974). The ecological profiles technique applied to data from Lichtenburg, South Africa. *Bothalia*, **11**, 355–64.

Muracciole, M. (1981). *Etude de la Flore Adventice des Cultures Pérennes en Corse Orientale*. Versailles: Mémoires de l'Ecole Nationale Superieure d'Horticulture.

Oka, H. I. & Morishima, H. (1982). Ecological genetics and the evolution of weeds. In *Biology and Ecology of Weeds*, ed. W. Holzner & N. Numata, pp. 73–89. The Hague: Junk.

Okuma, M. & Chikura, S. (1984). Ecology and control of a subspecies of *Paspalum distichum* L. 'Chikugo suzumenohie' growing in a creek in the paddy area on the lower reaches of Chijugo River in Kyushu. IV. Possibility of reproduction by seeds. *Weed Research, Japan*, **28**, 31–4.

Olivieri, I. & Berger, A. (1985). Seed dimorphism for dispersal: physiological, genetic and demographical aspects. In *Genetic Differentiation and Dispersal in Plants*, ed. P. Jacquard, G. Heim & J. Antonovics, pp. 413–29. Berlin: Springer-Verlag.

Olivieri, I., Swan, M. & Gouyon, P. H. (1983). Reproductive system and colonizing strategy of two species of *Carduus* (Compositae). *Oecologia*, **60**, 114–17.

Peters, N. C. B. (1985). Competitive effects of *Avena fatua* L. plants derived from seeds of different weights. *Weed Research*, **25**, 67–77.

Pignatti, S. (1982). *Flora d'Italia*. 3 vols. Bologna: Edagricole.

Pignatti, S. (1983). Human impact on the vegetation of the Mediterranean Basin. In *Man's Impact on Vegetation*, ed. W. Holzner, M. J. A. Werger & I. Ikusima, pp. 151–61. The Hague: Junk.

Pons, A. & Quézel, P. (1985). The history of the flora and vegetation and past and present human disturbance in the Mediterranean region. In *Plant Conservation in the Mediterranean Region*, ed. C. Gomez-Campo, pp. 25–43. The Hague: Junk.

Protopapdakis, E. (1985). Changement de la flore adventice des vergers d'agrumes en Crète sous la pression du désherbage chimique. *Agronomie*, **5**, 833–9.

Pujadas, A. & Hernandez-Bermejo, J. E. (1984). Taxonomical, ecological and phytogeographical interpretation of weed flora in Cordoba (Spain). *Proceedings 3rd International Symposium of E.W.R.S. on Weed Problems in the Mediterranean Area, Oeiras, Portugal*, **1**, pp. 33–9.

Raimondo, F., Ottonello, D. & Castiglia, G. (1979). Aspetti stagionali e caratteri biocorologici della vegetazione infestante gli agrumeti del palermitano. In *La Vegetazione Infestante e Sinantropica*, ed. G. G. Lorenzoini & G. Ferrari, pp. 159–70. Pavia: Societa Italiano di Fitosociologica.

Reeves, T. G. & Smith, I. S. (1975). Pasture management and cultural methods for the control of annual ryegrass (*Lolium rigidum*) in wheat. *Australian Journal of Experimental Agriculture & Animal Husbandry*, **15**, 527–30.

Ribeiro, J. A. (1984). A preliminary ecological weed survey of the north-eastern region of Portugal. *Proceedings of the 3rd International Symposium on Weed Problems in the Mediterranean Area, Oeiras, Portugal*, **1**, 41–8.

Rolston, M. P. (1978). Water impermeable seed dormancy. *Botanical Review*, **44**, 365–96.

Sagar, G. R. (1982). An introduction to the population dynamics of weeds. In *Biology and Ecology of Weeds*, ed. W. Holzner & N. Numata, pp. 161–8. The Hague: Junk.

Soufi, Z. (1987). Mauvaises herbes de la région maritime de Syrie. Thèse, Université Sciences et Techniques du Languedoc, Montpellier.

Tanji, A., Boulet, C. & Hammoumi, M. (1984). Contribution à l'étude de la biologie de *Solanum eleaegnifolium* Cav. (Solanacees), adventice des cultures dans le périmètre irriqué du Tadla (Maroc). *Weed Research*, **24**, 401–9.

Thellung, A. (1912). La flore adventice de Montpellier. *Memoire de la Societé Nationale des Sciences Naturelles et Mathematiques du Cherbourg*, **37**, 57–728.

Tuganaev, V. V. (1978). Classification of vegetal weeds in relation to modern agricultural cultivation. *Ekologiya*, **3**, 87–8.

Van den Bergen, C. (1980). Observations sur la végétation adventice des moissons. *Bulletin de la Societé Royale Botanique du Belgique*, **113**, 33–44.

Vigna, M. R., Fernandez, O. A. & Brevedan, R. E. (1981). Biologia y control de *Solanum eleaegnifolium* Cav. *Revue Faculdad Agronomia Buenos Aries*, **2**, 79–89.

Warwick, S. I., Phillips, D. & Andrews, C. (1986). Rhizome depth: the critical factor in winter survival of *Sorghum halepense* (L.) Pers. (Johnson grass). *Weed Research*, **26**, 381–7.

Wells, M. J. & Stirton, C. H. (1982). Weed problems of South African pastures. In *Biology and Ecology of Weeds*, ed. W. Holzner & N. Numata, pp. 429–34. The Hague: Junk.

Yannitsaros, A. & Economidou, E. (1974). Studies on the adventive flora of Greece. I. General remarks on some recently introduced taxa. *Candollea*, **29**, 111–19.

Zohary, M. (1983). Man and vegetation in the Middle-East. In *Man's Impact on Vegetation*, ed. W. Holzner, M. J. A. Werger & I. Ikusima, pp. 277–85. The Hague: Junk.

27

Plant invasions in the rangelands of the isoclimatic mediterranean zone

H. N. LE HOUÉROU

This chapter is concerned with plant invasions in the rangelands of the isoclimatic mediterranean zone of the Mediterranean Basin, California and Chile. Australian and South African examples will not be included, although the general points to be made may apply equally well to these other two regions of mediterranean climate.

Plant invasions in rangelands usually result from disturbances of various kinds, either natural or human-induced. The former include erosion, sedimentation, flooding, wildfires, climatic variation and outbreaks of insects and/or rodents. Anthropogenic disturbances include deforestation, drainage, overstocking, human-induced fires, water and grazing management and changes in livestock husbandry. Rangelands themselves are often the result of either natural or human-induced processes. More often, however, they arise from a mixture of both kinds of process, such as deforestation and wildfires, which jointly influence vegetation dynamics and which in turn may favour the formation of grazing disclimaxes. Most of the grasslands of the world are the result of wildfires as, for instance, in tropical savannas, mediterranean shrublands, most of the American prairies and perhaps also in the origin of large tracts of the Asian steppes. In such cases the suppression of wildfires has often led to bush encroachment, as in the expanding mesquite bushlands in the rangelands of the southern United States and of various other shrubs in tropical savannas. On the other hand, the cessation of grazing in mediterranean shrublands usually leads to fuel accumulation and increased fire hazard (Le Houérou, 1974, 1977*a*, 1981, 1987), which may also have undesirable consequences.

The nature and causes of rangeland invasion

Invasion by native species should be distinguished from that caused by introduced species (Le Houérou & Le Floc'h, 1987; Le Floc'h *et al.*, 1990). Broadly speaking, invasion in the Mediterranean Basin rangelands comes chiefly from native species, whilst in Californian and Chilean rangelands invasion is the result of colonisation by annual plants introduced from the Mediterranean Basin, but occasionally also from South Africa. There are, of course, a number of exceptions to this generalisation.

The causes of invasion are many; among them are changes in patterns of land-use and in grazing systems arising from development of watering points, fencing, control of stock numbers, etc. Changes in stocking rates and in livestock husbandry are other causes of invasion, e.g. a shift from one species of livestock to another or from monospecific husbandry to the raising of several species. All these general causes may bring about new ecological balances and altered conditions as a result of changed physical, chemical and biological processes. These processes include trampling of the soil surface, leading to compaction, sealing of the soil surface, formation of a crust, increased run-off, decreased water infiltration and storage, the formation of bare patches, erosion and ultimately desertization of the site. Nutrient enrichment of the soil through accumulation of faeces and urine in densely stocked areas such as around watering points, tracks, sheepyards, pens, etc., may result in invasion by nitrophilous plants. Overgrazing, which ensures a strong selective advantage to species with a lower preference to stock, may result in drastic changes to botanical composition, ground cover, biomass, productivity and animal production and/or invasion by either introduced or native species which may be better adapted to the changed conditions. Disturbance of the soil surface with respect to soil or water conservation (e.g. furrowing, pitting, benching, etc.) may change the competitive balance between species.

Plants are classified, in the terminology of range science (Dyksterhuis, 1949), as either decreasers (those preferred by stock), increasers (those species of lesser preference) or as invaders (those ignored by stock and not representative of the pristine site condition; the so-called 'climax'). This classification, increasingly challenged, is somewhat ambiguous, since a given species may 'decrease' in the number of its individuals, level of ground cover, biomass, productivity or animal production which may be sustained under continuous heavy grazing; it may 'increase' under light stocking rates or in exclosures. Similarly, an invasive plant can usually be 'decreased' or even eliminated by appropriate management techniques.

Most invasive plants show an *r* type of selection strategy (McArthur, 1962) with heavy inputs of energy to seed production, a fact that provides them with an aggressive colonisation potential, particularly in areas where *K*-selected species may have been weakened by the processes mentioned above. Among the *r*-selected species are many members of the family Asteraceae and specifically various kinds of thistles and representatives of the tribes Anthemidae, Astereae, Calenduleae, Cichoreae, Cynareae, Heliantheae, Inuleae and Senecioneae. Many annual grasses are also *r*-selected, as for instance in the tribes Aveneae, Bromeae, Festuceae and Hordeae.

Competition for water

Invasion often occurs as a result of competition favouring the expansion of a more opportunistic introduced species over native species, irrespective of range condition and management practices. A well-documented case of such a mechanism is the invasion of cheat grass (*Bromus tectorum*) into the arid and semi-arid ranges of the northern intermountain region (Utah, Nevada, Idaho, Oregon and Washington) of the U.S.A. (Harris, 1967). The region is subject to a mediterranean-type climate with autumn, winter and spring rains and a dry summer.

Cheat grass appeared in the region before the middle of the nineteenth century and quickly became dominant over very large areas of range which used to be dominated by the perennial bluebunch wheat grass (*Agropyron spicatum*). Most of the rangeland condition was close to pristine at the time of initial invasion. Wheat grass was progressively eliminated from its native habitat over very large areas. Wheat grass is a deep-rooted perennial, whereas cheat grass is an equally deep-rooted annual originating in the Mediterranean Basin. The phenology of the two species is similar, insofar as germination (in the case of *Bromus*) and the breaking of dormancy (*Agropyron*) start with the first rains in autumn, they both grow slowly during winter and very rapidly in spring, setting flowers in June and seeds in July; plants either die (*Bromus*) or enter dormancy (*Agropyron*) in summer, as is customary for Mediterranean species.

Harris (1967) observed that the presence of cheat grass greatly reduced the vigour and survival of wheat grass. In fact, wheat grass could not establish wherever the density of cheat grass was 1000 individuals per square metre or higher. Harris (1977) further showed that the roots of cheat grass grew much faster than those of its competitor, although to a similar final depth (90–100 cm). In spring, for instance, the roots of cheat grass reached a depth of 90 cm where the soil water potential was −0.1 MPa (i.e. close to field capacity), whilst those of wheat grass reached no further than 20 cm, a depth at which the soil water potential was −1.5 MPa (i.e. close to the wilting point of mesophytes). Cheat grass thus had exhausted the moisture at various depths before wheat grass roots could reach these depths. Wheat grass was thus weakened and finally eliminated, even without any grazing, as cheat grass, because of its rapid root growth in winter, had a strong selective advantage in spring and early summer over wheat grass.

Competition for nutrients

Competition for nutrients is a well-established factor in range invasion, notably in the 'piosphere' (Lange, 1969), where the abundance of dung enriches the soil in organic matter and nutrients, particularly nitrogen, which favours nitrophilous plants such as *Aizoon hispanicum*, *Asphodelus* spp., *Carthamus lanatus*, *Malva* spp., *Marrubium alysson*, *Mesembryanthemum cristallinum*, *Peganum harmala*, *Scolymus hispanicus*, *Silybum marianum*, *Sisymbrium irio* and *Urginea* spp. Piospheres may develop as concentric circles around water points (the sacrificed areas), along livestock tracks to water or along transhumance tracks.

Competition due to overgrazing

Continuous overgrazing weakens the highly palatable and preferred species by defoliation. Such species therefore have not a chance to accumulate reserves of carbohydrate in the stem bases and roots. Regrowth becomes progressively weaker with each defoliation. A spiral of elimination/invasion is thereby initiated, the preferred species finally being eliminated by species that are less sought after by stock and therefore more vigorous and stronger competitors. The latter will, in turn, undergo the same process if stock pressure remains heavy and continuous, until the botanical composition of the range includes only those species that are ignored by livestock.

Such a scenario is common in the Mediterranean Basin where range condition is almost invariably poor, because less palatable species have had a selective advantage over preferred species for more than 3000 years! In the Mediterranean Basin, invasion involves mainly local native species of low palatability, often spiny or more or less toxic. Lethal toxicity is, however, extremely rare and only found in a relatively few species such as *Nerium oleander* and some species of *Datura* and *Hyoscyamus*.

The most common invasive plants in these various categories belong to some 45 families and include about 260 genera (see Appendix 27.1; see also Le Floc'h, this volume; Le Houérou, 1977*a*, 1981; Le Houérou & Le Floc'h, 1987; Le Floc'h *et al.*, 1990). These native invasive species may become locally dominant or even constitute almost pure stands with very low grazing value over large areas of the Mediterranean Basin.

Among the relatively few introduced species having a significant impact in Mediterranean rangelands the following taxa are the most important (see also Le Floc'h, this volume): *Agave americana*, *Aloe vera*, *Aster squamatus*, *Atriplex semibaccata*, *Blakiella inflata*, *Calotropis procera*, *Conyza bonariensis*, *C. canadensis*, *Datura metel*, *D. stramonium*, *Heliotropium curassavicum*, *Kochia indica*, *Nicotiana glauca*, *Nigella sativa*, *Opuntia ficus-indica* var.

amyclaea, *O. dillenni*, *Oxalis pes-caprae*, *Phalaris canariensis*, *Pennisetum villosum*, *Oenothera biennis*, *Phytolacca americana*, *Salsola longifolia*, *S. kali*, *Withania somnifera*, *Xanthium brasilicum*, *X. spinosum*, *X. strumarium*.

Invasion of mediterranean species into California

Invasion of plants from the Mediterranean Basin into California started with the first Spanish settlement at San Diego in 1769 (Raven, 1977). These plants now number 480 species belonging to 250 genera (see Appendix 27.2 and Rejmanek, this volume). Monocotyledons comprise 120 species of this total, 80 per cent of which are annual grasses, i.e. nearly 100 species. Among the 360 species of dicotyledons are included 1 tree, 2 vines (*Lonicera*), 20 shrubs, 120 biennial or perennial herbs and 210 annuals. Annuals comprise 57 per cent of the naturalised flora (Naveh, 1966; Raven, 1977; Raven & Axelrod, 1978). Species originating in the Mediterranean Basin thus represent a little over 11 per cent of the Californian floristic province. The entire naturalised flora comprises 654 species, i.e. 16 per cent of the flora west of the Sierra Nevada–Cascade divide. Many of these invasive Mediterranean species are found in the California prairie of the San Joaquin valley, in the sagebrush steppe, the oak parkland and the juniper savanna (Heady, 1977; Heady *et al.*, 1977).

The invasion of Californian rangeland probably resulted from the conjunction of overgrazing, burning and cultivation, followed by abandonment by the early settlers of the eighteenth and nineteenth centuries (Robbins, 1940). Part of the invasion came from Chile through the importation of cereals and straw in the early nineteenth century. The peak in livestock numbers was reached between 1850 and 1900.

The pristine native vegetation seems to have been a prairie dominated by the following perennial grasses: *Aristida hamulosa*, *A. oligantha*, *Deschampsia danthonoides*, *Festuca idahoensis*, *F. pacifica*, *Koeleria cristata*, *Melica californica*, *M. imperfecta*, *Poa scabrella*, *Sitanion hystrix*, *Stipa cernua* and *S. pulchra* (Heady, 1977). The California prairie was closely related to the Palouse prairie, which occurs in similar climates in the drier parts of Oregon and Washington (Heady, 1977).

The pristine prairie was used by wild ungulates, lagomorphs and rodents and specifically by pronghorn antelopes (*Antilocapra americana*), mule deer (*Odocoilus hemionus*), elk (*Cervus elaphus nannodes*), jack rabbit (*Lepus californicus*), kangaroo rat (*Dipodomys hermanni*) and pocket gopher (*Thomomys bottae*). Antelopes occurred in herds of 2000–3000 and elks by the hundreds (Heady, 1977). Owing to the heavy stocking during the second half of the nineteenth century, ungulates diminished, perennial grasses were eliminated from most of the rangelands and replaced by annual native and introduced grasses and herbs, mainly from the Mediterranean Basin (Heady, 1977; Heady *et al.*, 1977).

The major invasive plants introduced from the Mediterranean Basin are listed in Appendix 27.2; see also Rejmanek, this volume). Many of them are weeds of cultivation in the Mediterranean Basin, a fact that reinforces the probable strong impact of cultivation followed by abandonment on the invasion of the California prairie. The impact of cultivation has been still stronger – and undisputed – in the rangelands of central Chile.

Invasion of mediterranean species in the rangelands of central Chile

The process of range invasion in central Chile is similar to that of California, although invasion probably began some 200 years earlier because the Spanish settled there between 1540 and 1560, compared with a date of 1769 in Alta California. The same processes took place: the conjunction of cultivation, overgrazing and burning. Cultivation, either continuous or episodic, seems to have played a major role (M. Etienne, personal communication).

Introduction of plants from Spain took place by the importation of hay (to feed the imported livestock during the oceanic journey from Spain to Chile). It also probably occurred to some extent by transport of propagules in fleeces as well as in imported cereal grains.

The most common invasive plants are virtually the same as in California with large proportions of annual grasses and species belonging to the families Asteraceae and Geraniaceae (see Appendix 27.3, as well as Mooney *et al.*, 1970; Gulmon, 1977; Mooney, 1977; Solbrig *et al.*, 1977; Montenegro *et al.*, this volume). In terms of ground cover and biomass, annual grasses and Geraniaceae each represent about 40 per cent (M. Etienne, personal communication).

A number of species moved from Chile to California by way of exported cereal grain which occurred from the end of the eighteenth century to the early twentieth century (M. Etienne, personal communication). These species were both earlier introductions from the Mediterranean Basin (see above) as well as native Chilean species (such as *Bromus trinii*, *Hordeum hystrix*, *Madia sativa*, *Micropsis nana*, *Sisyrynchium arenarium* and *Vulpia megalura* and species of the genera *Adesmia*, *Amsinckia*, *Calandrinia*, *Chaetanthera*, *Cryptantha*, *Helenium*, *Pectocarya*, *Plantago*, *Orthocarpus*, *Trifolium* and *Trisetobromus*). Conversely, some native Californian species moved to Chile, e.g. *Eremocarpus setigerus*, *Eschscholtzia californica* and *Oenothera dentata*.

The most common species introduced from the Mediterranean Basin and now naturalised in Chile are listed in Mooney *et al.* (1970), Gasto & Contreras (1971), Olivares & Gasto (1971), Gulmon (1977), Solbrig *et al.* (1977), Gasto (1979),

Etienne & Ovalle (1981), Etienne *et al.* (1982*a*, *b*), Ovalle (1986) and Montenegro *et al.* (this volume).

From an analysis of a series of quadrats studied in California (at Palo Alto, Hopland and San Diego) and in Chile (neighbourhood of Santiago), Gulmon (1977) found 94 genera of annuals in California, 34 of which (36 per cent) were in common with the region of Santiago. As in California, annual grasses in Chile represented about 50 per cent of the genera in the sampled communities, the main other families being Asteraceae (15 per cent of the species present), Geraniaceae (12 per cent), Fabaceae (10 per cent) and Plantaginaceae (8 per cent); the remaining 5 per cent included, in decreasing order, Convolvulaceae, Scrophulariaceae, Rubiaceae and Caryophyllaceae. The frequencies of biological types in the quadrats studied were approximately 70–90 per cent annual grasses, 30–50 per cent annual broad-leaved rosettes, 30–50 per cent annual legumes, 10–20 per cent annual forbs and 10–20 per cent perennials. But, in terms of plant cover and biomass, annual grasses and Geraniaceae represent about 40 per cent each in the mediterranean zone of Chile (M. Etienne, personal communication).

Comparison between the Mediterranean Basin, California and Chile

In the Mediterranean Basin, invasion of rangelands is mostly by native species; introduced species play only a minor role in terms of number of species (less than 5 per cent) and an even lesser role when frequency, ground cover and biomass are considered (although they may play a greater role locally in limited areas). In California and Chile annual species represent 80 to 90 per cent of species numbers, ground cover and biomass. Species numbers are about evenly shared between naturalised Mediterranean species and native annuals, whereas the former represent most of the ground cover and biomass and have a higher frequency.

In spite of the high degree of similarity between California and Chile, there are, however, two major differences in the causes of invasion and the dynamic situation of mediterranean-climate rangelands. In California, the main cause of invasion seems to be the conjunction of overgrazing and burning (either deliberate or natural). Cultivation seems to have played a relatively minor role in California. Conversely, in Chile, cultivation followed by abandonment played the major role. Californian annual ranges do not seem able to revert to pristine perennial grassland, even after decades of total stock exclusion (Heady, 1977), whereas in Chile exclosures lead to colonisation by perennial grasses representative of the pristine situation, such as species of the genera *Stipa*, *Piptochaetium* and *Nassella* (M. Etienne, personal communication). Invasion in California also included a number of annual species native to central Chile, whereas the opposite occurred for a few species only.

Conclusion

Invasion of rangelands in the Mediterranean Basin is fundamentally different in nature from that in California and Chile. In the Mediterranean Basin invasion comes essentially from native local species shared about equally between annuals and perennials in terms of species numbers, ground cover and biomass. The invasive plants are species of lesser preference to livestock or are those ignored by domestic herbivores. The latter species have acquired a strong selective advantage through various mechanisms such as fire, overstocking (or most often a combination of both), trampling, changes in land use and in grazing patterns, development of watering points, etc.

In California and Chile the same causes, in conjunction with episodic cultivation, resulted in the invasion by annual plants, both native and naturalised from the Mediterranean Basin in the case of Chile, and from the Mediterranean Basin *and* Chile in the case of California. These plants include a large proportion of annual grasses and Asteraceae in terms of species numbers, and of annual grasses and Geraniaceae in terms of ground cover, frequency and biomass.

Experience shows that Californian rangelands invaded by annuals cannot return to the pristine condition of perennial grassland, whereas this is not so for the rangelands of central Chile.

Appendixes

Appendix 27.1. *The most common plants invasive in the rangelands of the Mediterranean Basin.*
Genera are arranged alphabetically within families

Aizoaceae: *Aizoon, Gisekia, Glinus, Mesembryanthemum, Mollugo, Trianthema*
Apiaceae: *Ammi, Apium, Bunium, Bupleurum, Daucus, Eryngium, Ferula, Foeniculum, Pituranthos, Ridolphia, Smyrnium, Thapsia, Turgenia*
Asteraceae: *Achillea, Amberboa, Ambrosia, Anthemis, Artemisia, Asteriscus, Atractylis, Bubonium, Calendula, Carduncellus, Carduus, Carlina, Carthamus, Centaurea, Chrysanthemum, Cirsium, Cnicus, Conyza, Cotula, Crithmum, Cynara, Dittrichia, Erigeron. Filago, Galactites, Hertia, Matricaria, Notobasis, Onopordum, Ormenis, Pallenis, Pulicaria, Scolymus, Xanthium*
Asclepiadaceae/Apocynaceae: *Caralluma, Cyanchum, Gomphocarpus, Nerium, Pergularia*
Amaranthaceae: *Amaranthus*
Berberidaceae: *Berberis, Leontice*
Boraginaceae: *Borago, Cerinthe, Echium, Heliotropium, Lappula, Lithospermum*

Appendix 27.1. (*cont.*)

Brassicaceae: *Alyssum, Brassica, Cakile, Capsella, Conringia, Coronopus, Diplotaxis, Erucaria, Hirschfeldia, Matthiola, Moricandia, Muricaria, Nasturtiopsis, Notoceras, Oudneya, Raphanus, Sinapis, Sisymbrium, Vella*
Capparidaceae: *Cleome*
Caryophyllaceae: *Agrostemma, Alsine, Arenaria, Cerastium, Dianthus, Gypsophila, Melandryum, Minuartia, Polycarpon, Pteranthus, Silene, Tunica, Vaccaria*
Cistaceae: *Cistus, Fumana, Halimium, Helianthemum*
Convolvulaceae: *Calystegia, Convolvulus, Cressa, Cuscuta, Ipomoea*
Crassulaceae: *Cotyledon, Crassula, Sedum*
Dipsacaceae: *Dipsacus, Knautia, Scabiosa, Succisa*
Euphorbiaceae: *Andrachne, Chrozophora, Euphorbia, Mercurialis, Ricinus*
Fabaceae: *Anagyris, Anthyllis, Astragalus, Genista, Lathyrus, Lotus, Lupinus, Melilotus, Ononis, Psoralea, Spartium. Trigonella*
Frankeniaceae: *Frankenia*
Geraniaceae: *Erodium, Geranium*
Gramineae: *Aristida, Brachypodium, Bromus, Corynephorus, Ctenopsis, Cymbopogon, Cynosurus, Echinaria, Elymus, Festuca, Gastridium, Hordeum, Hyparrhenia, Imperata, Koeleria, Lagurus, Lamarkia, Lygeum, Molinia, Nardus, Nardurus, Panicum, Phragmites, Poa, Polypogon, Scleropoa, Setaria, Sorghum, Sphenopus, Sporobolus, Stipa, Vulpia, Vulpiella, Wangenheima*
Hypericaceae: *Hypericum*
Juncaceae: *Juncus*
Lamiaceae: *Ajuga, Ballota, Lamium, Marrubium, Mentha, Phlomis, Prasium, Salvia, Satureja, Teucrium, Thymbra, Thymus*
Liliaceae, Iridaceae, Amaryllidaceae, Orchidaceae & Araceae: *Allium, Aphyllanthes, Arisarium, Asphodeline, Asphodelus, Bellevallia, Biarum, Colchicum, Dipcadi, Gagea, Gladiolus, Iris, Muscari, Ophris, Orchis, Pancratium, Scilla, Urginea*
Lythraceae: *Lythrum, Peplis*
Onagraceae: *Circea, Epilobium, Oenothera*
Malvaceae: *Althaea, Hibiscus, Lavatera, Malope, Malva*
Papaveraceae: *Glaucium, Papaver, Roemeria*
Polygonaceae: *Emex, Polygonum, Rumex*
Phytolaccaceae: *Phytolacca*
Plumbaginaceae: *Armeria, Limoniastrum, Limonium*
Resedaceae: *Astrocarpus, Caylusea, Oligomeris, Randonia, Reseda*
Rosaceae: *Agrimonia, Alchemilla, Coris, Geum, Potentilla, Rosa, Rubus, Sarcopoterium*
Ranunculaceae: *Adonis, Anemone, Ceratocephalus, Delphinium, Ficaria, Ranunculus*
Rubiaceae: *Crucianella, Galium, Putoria, Rubia, Sherardia*
Rutaceae: *Haplophyllum, Ruta*
Salsolaceae: *Agathophora, Anabasis, Atriplex, Bassia, Beta, Camphorosma, Chenopodium, Halocnemon, Halogeton, Halopeplus, Haloxylon, Hammada, Noaea, Salsola, Suaeda*
Scrophulariaceae: *Anarrhinum, Antirrhinum, Bellardia, Celsia, Euphragia, Linaria, Odontites, Onosma, Scrophularia, Verbascum*
Solanaceae: *Datura, Hyoscyamus, Mandragora, Solanum, Withania*

Appendix 27.1. (*cont.*)

Thymeleaceae: *Daphne, Thymelea*
Verbenaceae: *Lippia, Verbena, Vitex*
Zygophyllaceae: *Fagonia, Nitraria, Peganum, Tetradiclis, Tribulus, Zygophyllum*

Appendix 27.2. *The most common plants of Mediterranean origin now invasive in Californian rangelands*

Asteraceae: *Anthemis cotula, Centaurea melitensis, C. solstitialis, Crepis radicata, Filago gallica, Hedypnois cretica, Hypochoeris glabra, H. radicata, Lactuca serriola, Silybum marianum*

Fabaceae: *Lotus angustissimus, L. micranthus, Medicago hispida, Melilotus indica, Trifolium dubium, T. filiforme, T. glomeratum, T. pratense, Vicia benghalensis*

Geraniaceae: *Erodium botrys, E. cicutarium, E. malachoides, E. moschatum, Geranium dissectum, G. molle*

Gramineae: *Aira caryophyllea, A. praecox, Anthoxanthum odoratum, Avena barbata, A. fatua, Briza minor, Bromus diandrus, B. madritensis, B. marginatus, B. mollis, B. rubens, B. sterilis, B. tectorum, Cynosurus cristatus, C. echinatus, Gastridium ventricosum, Holcus lanatus, Lagurus ovatus, Lamarkia aurea, Lolium rigidum, Taeniatherum asperum (Elymus caput-medusae), Trachynia distachya (Brachypodium distachyon), Vulpia dertonensis, V. myuros*

Various other families: *Anagallis arvensis, Brassica nigra, Cerastium viscosum, Convolvulus arvensis, Daucus carota, Galium aparine, G. murale, Heliotropium curassavicum, Juncus bufonius, Lythrum hyssopifolia, Plantago lanceolata, P. ovata, Raphanus sativus, Rumex acetosella, R. crispus, Sherardia arvensis, Silene gallica.*

Appendix 27.3. *The most common plants of Mediterranean origin now invasive in Chilean rangelands*

All families: *Aira caryophyllea, Anthemis cotula, Arenaria* spp., *Avena barbata, A. fatua, Briza media, B. minor, Bromus mollis, Centaurea melitensis, Cerastium viscosum, Chondrilla juncea, Convolvulus arvensis, Cynara cardunculus, Erodium botrys, E. cicutarium, E. malachoides, E. moschatum, Filago gallica, Galium murale, Gastridium ventricosum, Hypochoeris chrysantha, H. glabra, H. radicata, Juncus bufonius, Lolium multiflorum, Lythrum hyssopifolia, Medicago hispida, M. littoralis, M. truncatula, Parentucellia latifolia, Plantago lanceolata, Raphanus raphanistrum, Rapistrum rugosum, Rumex acetosella, R. crispus, Sherardia arvensis, Silene gallica, Sisymbrium officinale, Spergula arvensis, Tolpis barbata, Torilis nodosa, Trifolium campestre, T. filiforme, T. glomeratum, T. hirtum, Veronica peregrina, Vulpia bromoides, V. dertonensis.*

References

Dyksterhuis, E. J. (1949). Condition and management of rangeland based on quantitative ecology. *Journal of Range Management*, **2**, 104–15.

Etienne, M. & Ovalle, C. (1981). Carta de potencialidades pastorales de un sector de la provincia de Cauquenes. In *Cartografia de la Vegetacion y sus Applicaciones en Chile*, ed. M. Etienne & D. Contreras. *Facultad di Ciencias e Agronomia, Universidad de Chile, Boletin Tecnio*, **46**, 1–27.

Etienne, M., Caviedes, E., Gonzalez, C. & Prado, C. (1982*a*). Cartografia de la vegetacion de la zona arida mediterranea de Chile. Transecto 1: Puerto Oscuro, Combarbalà, Monte Patria, Ovalle, La Serena. *Terrae Aridae*, **1**, 1–73.

Etienne, M., Gonzalez, C. & Prado, C. (1982*b*). Cartografia de la vegetacion de la zona arida mediterranea de Chile. Transecto 2: Los Vilos, Illapel, Combarbalà. *Terra Aridae*, **1**, 81–126.

Gasto, J. (1979). *Ecologia: El Hombre y La Transformacion de la Naturaleza*. Santiago: Ediciones Universitariz, Universidad de Chile.

Gasto, J. & Contreras, D. (1971). Bioma pratense de la region mediterranea de pluviometria limitada. *Facultad de Agronomia, Universidad de Chile, Boletin Tecnico*, **35**, 3–29.

Gulmon, S. L. (1977). A comparative study of the grassland of California and Chile. *Flora*, **166**, 261–78.

Harris, G. A. (1967). Some competitive relationships between *Agropyron spicatum* and *Bromus tectorum*. *Ecological Monographs*, **37**, 89–111.

Harris, G. A. (1977). Root phenology as a factor of competition among grass seedlings. *Journal of Range Management*, **30**, 172–7.

Heady, H. F. (1977). Valley grasslands. In *Terrestrial Vegetation of California*, ed. M. G. Barbour & J. Major, pp. 491–514. New York: Wiley.

Heady, H. F., Foin, T. C., Hektner, M. M., Taylor, D. W., Barbour, M. G. & Barry, W. J. (1977). Coastal prairie and northern coastal scrub. In *Terrestrial Vegetation of California*, ed. M. G. Barbour & J. Major, pp. 733–57. New York: Wiley.

Lange, R. T. (1969). The 'piosphere': sheep tracks and dung patterns. *Journal of Range Management*, **22**, 396–400.

Le Floc'h, E., Le Houérou, H. N. & Mathez, J. (1990). History and patterns of plant invasion in northern Africa. In *Biological Invasions in Europe and the Mediterranean Basin*, ed. F. di Castri, A. J. Hansen & M. Debussche, pp. 105–33. Dordrecht: Kluwer.

Le Houérou, H. N. (1974). Fire and vegetation in the Mediterranean Basin. *Proceedings of the Annual Tall Timbers Fire Ecology Conference*, **13**, 237–77.

Le Houérou, H. N. (1977*a*). Plant sociology and ecology applied to grazing lands research, survey and management in the Mediterranean Basin. In *Handbook of Vegetation Science*, vol. 13: *Application of Vegetation Science to Grassland Husbandry*, ed. W. Krause, pp. 213–74. The Hague: Junk.

Le Houérou, H. N. (1977*b*). Fire and vegetation in North Africa. In *Proceedings of the Symposium on the Environmental Consequences of Fire and Fuel Management in Mediterranean Ecosystems*, ed. H. A. Mooney & C. E. Conrad, pp. 334–41. USDA Forest Service General Technical Report WO3, Washington, D.C.

Le Houérou, H. N. (1981). Impact of man and his animals on mediterranean vegetation. In *Ecosystems of the World*, vol. 11, *Mediterranean-type Shrublands*, ed. F. di Castri, D. W. Goodall & R. L. Specht, pp. 479–521. Amsterdam: Elsevier.

Le Houérou, H. N. (1987). Vegetation wildfires in the Mediterranean Basin: evolution and trends. *Ecologia Mediterranea*, **13**, 13–24.

Le Houérou, H. N. & Le Floc'h, E. (1987). Flore et dynamique à long terme de la végétation en Tunisie centrale et méridionale. In *La Végétation Tunisienne: Essai de Synthèse*, vol. 2, ed. M. Nabli, in press. Tunis: Faculté des Sciences, Université de Tunis.

McArthur, R. H. (1962). Some generalized theorems of natural selection. *Proceedings of the National Academy of Sciences, USA*, **48**, 1893–7.

Mooney, H. A. (1977). Introduction. In *Convergent Evolution in Chile and California: Mediterranean Climate Ecosystems*, ed. H. A. Mooney, pp. 1–12. Stroudberg, Pennsylvania: Dowden, Hutchinson & Ross.

Mooney, H. A., Dunn, E. L., Shropshire, F. & Song, L. (1970). Vegetation comparisons between the mediterranean climate areas of California and Chile. *Flora*, **159**, 480–96.

Naveh, Z. (1966). Mediterranean ecosystems and vegetation types in California and Israel. *Ecology*, **48**, 445–59.

Olivares, A. & Gasto, J. (1971). Comunidades de terofitas en subseres postasaduras y en exclusion en la estepa de *Acacia caven* (Mol.) Hook. & Arn. *Facultad de Agronomia, Universidad de Chile, Boletin Tecnico*, **34**, 3–24.

Ovalle, C. (1986). Etude de système écologique sylvopastoral à *Acacia caven* (Mol.) Hook. & Arn. Application à la gestion des ressources renouvelables dans l'aire climatique méditerranéenne humide et subhumide du Chili. Thèse, Université Sciences et Techniques du Languedoc, Montpellier.

Raven, P. H. (1977). The California flora. In *Terrestrial Vegetation of California*, ed. M. G. Barbour & J. Major, pp. 109–38. New York: Wiley.

Raven, P. H. & Axelrod, D. I. (1978). Origin and relationships of the California flora. *University of California Publications in Botany*, **72**, 1–134.

Robbins, W. W. (1940). Alien plants growing without cultivation in California. *California Experiment Station Bulletin*, **637**, 1–128.

Solbrig, O. T., Cody, M. L., Fuentes, E. R., Glanz, W., Hunt, J. H. & Moldenke, A. R. (1977). The origin of the biota. In *Convergent Evolution in Chile and California: Mediterranean Climate Ecosystems*, ed. H. A. Mooney, pp. 13–26. Stroudburg, Pennsylvania: Dowden, Hutchinson & Ross.

28

Forest plantations and invasions in the mediterranean zones of Australia and South Africa

L. D. PRYOR

Forest plantations tend to be the source of propagules if tree species become invasive. Such plantations in the part of southern Australia with a mediterranean-type climate are largely of pines, especially *Pinus radiata*. There is a smaller component of *Pinus pinaster*, particularly on soils of low fertility in south-west Western Australia. Small trial plots have been established in arboreta of many other species in the genus, especially *P. laricio*, *P. ponderosa*, *P. contorta*, *P. muricata*, *P. halepensis*, and some other coniferous genera such as *Cupressus*.

Forest plantations of these species have been established principally on sites which formerly carried tall open-forest or open-woodland (Specht, 1970), in which *Eucalyptus* species were the main trees, and species of *Acacia* (either shrubs or small trees) were associated species. After establishment it has usually been necessary to remove regrowth of the indigenous species, either as coppice shoots from stumps, as regenerated seedlings, or as root suckers in the case of some *Acacia* species, to ensure establishment of the pines. Such persistent plants are often referred to as 'woody weeds', although in no way do they possess invasive characteristics.

At a rotation age of about 30 to 40 years, the pine stand is typically without associated species, either indigenous or introduced, except for a few low-growing herbs on the forest floor and some bryophytes on wetter sites. Occasionally, in moist localities, European blackberry (*Rubus* spp.) may enter the stand and persist as an undesirable species in the management system. In recent years, blackberry has been removed where necessary by application of herbicides but on most sites blackberry thins out and becomes innocuous towards rotation age.

In the context of this chapter, it is of greater interest, however, to consider the extent to which the plantation tree species becomes invasive in vegetation surrounding the forest plantation. In southern Australia there are very extensive interfaces between pines and native eucalypt stands which present an opportunity for seed dispersal into the native vegetation.

In *Pinus radiata* the seed is shed in hot summer weather when the ripe cones open and the winged seeds fall spinning to the ground. According to wind speed, seed may be carried some distance from the source tree. Most seeds released in this way travel a distance not more than 2 or 3 times the height of the source tree but occasionally seedlings establish themselves at considerable distances from the only possible source, thereby indicating that seed is carried occasionally in strong winds or by other means for a few kilometres. From the known quantity of seed fall, it is apparent that most which falls on adjoining sites does not lead to successful seedling establishment; after germination the resultant seedlings fail to survive.

The most obvious exception to the above pattern is the colonisation by *Pinus radiata* seedlings of sites where bare mineral earth is exposed, such as on road cuttings, borrow pits or other disturbed areas. On such sites vegetation has been largely or completely removed earlier. Such colonisation is frequently observed adjoining forest stands and seems to be a general property of the various pine species, whether extensively planted or established merely in small plots. Can *Pinus radiata* invade indigenous eucalypt vegetation, especially if such is in an undisturbed state? In attempting to answer this question it is important to recognise what is meant by the term 'undisturbed' in the human context, since apart from direct clearing, there are varying degrees of human influence such as domestic or feral grazing and the previous fire regime.

In various localities in southern Australia, *Pinus radiata* seedlings establish themselves within a zone commonly of about 500 m from a seed-producing stand (Burdon & Chilvers, 1977; Chilvers & Burdon, 1983). The seedlings usually establish on the downwind side in what is ostensibly undisturbed natural tall open-forest, dominated by indigenous eucalypts. Burdon & Chilvers studied the recruitment of pines into a eucalypt stand over time, taking account of the cone-bearing level of the parent pine stand with age; they related this level to eucalypt regeneration in the same stand over a similar period. It was concluded that *Pinus radiata* is capable of colonising the rather open type of tall open-forest they studied. Burdon & Chilvers also suggested that the question of whether the pines were filling an otherwise unoccupied niche in the eucalypt forest or were displacing the eucalypts remains to be resolved.

Taken on its own, the study of Burdon & Chilvers (1977) could imply that ultimately, in the conditions of their test, a pine/eucalypt community could result in a solely pine stand; they did not, however, attempt to extrapolate this process in the long term to arrive at probable end results.

The most obvious factor which could affect the long-term outcome of pine invasion in a eucalypt forest is fire. Eucalypt forests are fire-prone vegetation and adaptation to fire is a prominent feature of them (Gill, 1975). By comparison, *Pinus radiata* is fire-sensitive. If the foliage is killed by fire the tree dies and the same is true for most other species of *Pinus* planted or tested in the areas of mediterranean-climate in southern Australia, *P. canariensis* being an exception. In terms of ecological balance, therefore, it seems likely that in the normal Australian environment, pines will continue to invade the sclerophyll eucalypt forest but will be turned back periodically by fire (if the seed source itself is not eliminated). When regulation of such an invasion is sought the planned use of fire is an inexpensive and effective way to achieve control, but this idea has so far been scarcely integrated into current management plans. In one case, in the Kosciusko National Park in New South Wales, the complete removal of a conifer plantation, principally of *P. laricio* (= *P. nigra*), *P. contorta* and *P. ponderosa*, has been deemed necessary to eliminate the seed sources for such a potential invasion.

Invasion of pines into substantially disturbed areas of natural vegetation, or even agricultural or pastoral areas, has not so far been recognised as a problem. Again, with the exception only of *Pinus canariensis*, not only are they fire-susceptible but they do not coppice when cut to ground level; elimination is simple and ordinarily follows from current land use practices.

Broad-leaved forest species

Forest plantations of broad-leaved species are few in the mediterranean-climate zone of southern Australia. In recent years some planting of poplar has been carried out, mainly with *Populus deltoides* or of clones with this species as one of the parents. Whilst female clones have been included and seed production has been prolific, no invasive establishment of seedlings has yet been observed. Some species of *Populus* develop suckers from the roots and this characteristic may constitute an unwanted invasion of very local proportions only; such species are not yet used in forest plantations. The species concerned are mainly *Populus alba* or related species of the Section Leuce in the genus.

Limited plantings of willow (*Salix* spp.) have also been undertaken but with no evidence of any invasive characteristics, except by fragmentation in waterways with subsequent establishment of broken branches as cuttings downstream from the plantings.

Species with potential for future invasion

The practice of ornamental, amenity and crop planting in southern Australia has two aspects. Firstly, it has led to the introduction to Australia of a range of tree species for these purposes alone. Secondly, it has been associated with the transfer of indigenous Australian species widely within the country.

In the first category, there are a few tree species which, like *Populus alba*, sucker vigorously from the roots and may cause local nuisance. The most notable species in this category are *Robinia pseudoacacia*, *Ulmus procera* and related clones in horticultural use (mainly of European origin) and also *Ailanthus altissima*. Another group of species within this first category is those which grow readily from branch fragments, of which the most frequent are willows, mainly clones of hybrids between *Salix fragilis* and *S. alba*, as well as *Salix babylonica*. The only species which might be considered a tree, albeit a small one, in this category which has become invasive following seed distribution appears to be the olive *Olea europaea*. Active establishment in degraded woodlands and grazing areas by olive has been a feature in the region of Adelaide, South Australia during this century. Olive invasion is also now becoming a problem in one area to the west of Sydney (J. Dellow, personal communication). The source of seed for the Adelaide invasion has no doubt been originally the oil-producing plantations in the region. Control has been tried by digging out the plants,including the root system, or more recently by the application of herbicides to cut stumps.

In the second category, in spite of the transfer of many indigenous tree species from their zone of natural habitat to one in which they did not occur spontaneously, there seem to be no significant examples of extensive invasion by them in their new surroundings. This is certainly true of the eucalypts where seedling growth on exposed mineral soil limits spread from the sites to which the transfer has been made. There has been a greater tendency for acacias to spread when transferred to a new Australian habitat. If *Acacia baileyana* is considered a tree (cf. a tall shrub), then it does regenerate spontaneously in disturbed areas and in gardens. *Acacia mearnsii*, a somewhat larger small tree, has the capacity to sucker from the roots, as well as to produce some seedling regeneration. The botanically related legume *Albizia lophantha*, native to Western Australia, has shown a tendency to spontaneously regenerate as seedlings in South Australia. The general situation at present, however, is that the indigenous tree species of Australia do not tend to become invasive when introduced to other localities in Australia in which they do not occur naturally.

Tree species as weeds in South Africa

An outstanding example of tree invasions in a mediterranean-type climate is provided by Australian species introduced to South Africa (Roux, 1964; Wells, this volume). South Africa also provides striking examples of non-Australian species introduced from elsewhere with a similar propensity to invade (Taylor *et al.*, 1985). Notable examples of the first group are species of the genus *Acacia* in which both trees and shrubby species are involved. The tree species are *Acacia melanoxylon*, *A. mearnsii* and *A. dealbata* and the shrubby species *Acacia saligna*, *A. longifolia* and *A. cyclops* (Stirton, 1980).

Acacia melanoxylon was deliberately introduced to suppress weed growth in forest openings following clearance operations in the Knysna area of Cape Province. The species performed this task effectively, but at the same time it inhibited the regeneration of indigenous species which it was hoped would be favoured by the weed suppression. The species exhibited a tendency to expand its range as an invasive species on disturbed areas, so that the use of *A. melanoxylon* for its original purpose was discontinued some time ago (Phillips, 1928).

Acacia mearnsii has been widely planted for tan bark production in the non-mediterranean-climate regions of South Africa and parts of eastern Africa, so that many sources exist from which it might spread. Although it is not clear to what extent disturbance is necessary, the species may compete vigorously with indigenous vegetation and form thickets. Survival of seed in the soil, in common with many other legumes, presents problems for its long-term control. Present control methods rely on cutting the stems back to their root junction below the soil.

The position with other Australian acacias is similar, although *A. cyclops* and *A. longifolia* do not coppice and are more readily eliminated by cutting close to the ground. The different *Acacia* species have somewhat different habitat preferences even when invasive. *Acacia cyclops* favours low elevations near the coast, as do *A. longifolia* and *A. saligna*: *A. mearnsii* spreads in areas of higher rainfall; *A. melanoxylon* favours moist cool areas (Stirton, 1980). *Albizia lophantha* behaves in a manner similar to that of the acacias in the coastal zone.

Many species of the genus *Eucalyptus* are widely used in forest plantations in South Africa in the zone of mediterranean climate. In common with other areas into which eucalypts have been introduced from Australia, there is little sign of aggressive invasion by these species. Colonisation of sites where mineral soil has been exposed is the most common location of spontaneously regenerating seedlings. An occurrence of dense seedling growth of *E. lehmannii* in the Cape of Good Hope Nature Reserve is recorded by Taylor *et al.* (1985) on an area following a fire in the reserve. These authors also noted some spread by water-borne seed of the same species and its establishment on stream banks, but in general the occurrences are of low intensity and the species is not as invasive as some *Acacia* species.

According to one unconfirmed report, the only species of eucalypt for which discontinuation of planting has been considered in South Africa because of potential invasiveness is *Eucalyptus sieberi*. This species regenerates very vigorously from seed in its natural habitat in Australia – a characteristic which perhaps makes it somewhat exceptional amongst those species otherwise commonly used in forest plantations in South Africa.

The lack of aggressive spread by eucalypts in the area with which they were dealing led Taylor *et al.* (1985) to suggest that eucalypts pose little threat as invasive species; this seems also to accord with the general situation on a world basis, although at times claims have been made otherwise (Burley, 1984). There is however little published to claim the genus as providing invasive trees in countries to which they have been introduced. It is reported (Poore *et al.*, 1988) that eucalypts in California are proving invasive, but this is unreferenced and probably concerns spread from trees introduced for horticulture rather than for plantation forestry (but see Rejmanek, this volume; Montenegro, this volume).

In South Africa, the principal offender in the group of non-Australian trees is *Pinus pinaster*, which was an early nineteenth century introduction to South Africa from Europe. The species thrives in areas where the winter rainfall exceeds 500 mm a year. The spread of *P. pinaster* into the region of fynbos shrubland is aided by fire but, paradoxically, fire can be used in conjunction with mechanical methods to secure its removal (Stirton, 1980; Guillarmod, 1983).

Tree invasion in other zones of mediterranean climate

Published information does not disclose evidence that tree species are invasive in the Mediterranean Basin itself or the mediterranean-climate areas of California or Chile, although in the south of France there is some sign that *Acacia dealbata* has such a tendency. There has been a suggestion that, although several species of Australian *Acacia* (such as *A. cyclops* and *A. saligna*) are widely cultivated and flower and fruit in abundance in California, they do not appear to be as invasive in California as they are in the Cape of Good Hope region (F. J. Kruger, personal communication).

The susceptibility of the South African mediterranean environment to invasion by Australian tree species seems to be a unique situation and perhaps reflects the paucity of trees in the South African sclerophyll vegetation zone. In terms of ecological exchange, the reverse movement of species which are shrubby (e.g. *Chrysanthemoides monilifera*) or herbaceous (e.g. *Emex australis*, *Oxalis pes-caprae*) from South Africa to Australia and elsewhere has been associated with aggressive invasion of sites with a mediterranean environment. Broadly, it appears that trees are not frequently weedy and invasive in the world zones with a mediterranean climate.

Control of invasive trees

The most commonly reported methods for control of invasive trees are removal by hand or by mechanical means with attention given to the characteristics of the species involved and especially their capacity to coppice from the stump or sucker from the roots. Economic considerations lead often to the substitution of mechanical methods with the application of chemicals for control, some herbicides being very successful. In common with general experience in forestry and agriculture, however, long-term risks to users, either apparent or real, often emerge and these impose constraints on the more widespread use of herbicides.

Since invasive trees are usually introduced to the areas in which they spread vigorously and are detached from their natural pests and enemies, much thought has been given to developing methods of biological control. In at least the case of one shrubby Australian species, viz. *Hakea sericea*, this control method offers promise.

The use of different fire regimes to control some invasive species is perhaps less well recognised than it merits. In the case of invasive coniferous tree species, such as *Pinus pinaster* in South Africa and *P. radiata* in Australia, the conifers are more fire-sensitive than the indigenous vegetation in either case. Whilst fire will kill established coniferous trees, it also favours seedling regeneration, so that not just a single fire of appropriate intensity is needed but a program of several fires in sequence is required to achieve control or elimination of the unwanted invasive tree (Taylor *et al.*, 1985). On the other hand, many *Acacia* species produce prolific seed crops and seeds stored in the ground will remain viable for long periods. Seeds may be stimulated to germinate by a fire of sufficiently high intensity. This characteristic clearly would involve a different program of fire management from that appropriate to conifer control (see Pieterse & Cairns, 1988).

Ecological management of invasive species, especially in the use of fire, whilst a relatively well-understood tool in some instances, appears to offer further opportunities for control programs for invasive tree and shrub species beyond those used commonly at present. More refined programs are needed for the different invasive species based on their regenerative characteristics.

References

Burdon, J. J. & Chilvers, G. A. (1977). Preliminary studies on a native Australian eucalypt forest invaded by exotic pines. *Oecologia*, **31**, 1–12.

Burley, J. (1984). Global needs and problems of collection, storage and distribution of multipurpose tree germplasm. In *Multipurpose Tree Germplasms*, ed. J. Burley & P. von Carlowitz, p. 67. Nairobi: International Council for Research in Agroforestry.

Chilvers, G. A. & Burdon, J. J. (1983). Further studies on a native Australian eucalypt forest invaded by exotic pines. *Oecologia*, **59**, 239–45.

Gill, A. M. (1975). Fire and the Australian flora: a review. *Australian Forestry*, **38**, 4–25.

Guillarmod, A. F. M. G. Jacot (1983). Recovery of eastern Cape heathland after fire. *Bothalia*, **14**, 701–4.

Phillips, J. F. V. (1928). The behaviour of *Acacia melanoxylon* R.Br. (Tasmanian blackwood) in the Knysna forests: an ecological study. *Transactions of the Royal Society of South Africa*, **16**, 31–43.

Pieterse, P. J. & Cairns, A. L. P. (1988). The population dynamics of the weed *Acacia longifolia* Fabaceae in the absence and presence of fire. *South African Forestry Journal*, **45**, 25–7.

Poore, D., Adlard, P. G. & Arnold, M. (1988). Environmental implications of eucalypts as replacements for the slower growing indigenous species of the world. In *Proceedings of the International Forestry Conference for the Australian Bicentenary*, Part 1, 11 pp.

Roux, E. R. (1964). The Australian acacias in South Africa. In *Ecological Studies in Southern Africa*, ed. D. H. S. Davis, pp. 137–42. The Hague: Junk.

Specht, R. L. (1970). Vegetation. In *The Australian Environment*, 4th edn (rev.), ed. G. W. Leeper, pp. 44–67. Melbourne: CSIRO Australia & Melbourne University Press.

Stirton, C. H. (ed.) (1980). *Plant Invaders. Beautiful but Dangerous*, 2nd edn. Cape Town: Department of Nature & Environmental Conservation, Cape Provincial Administration.

Taylor, H. C., Macdonald, S. A. & Macdonald, I. A. W. (1985). Invasive alien woody plants in the Cape of Good Hope Nature Reserve. II. Results of a second survey from 1976 to 1980. *South African Journal of Botany*, **51**, 21–9.

29

The importation of mediterranean-adapted dung beetles (Coleoptera: Scarabaeidae) from the northern hemisphere to other parts of the world

A. A. KIRK & J.-P. LUMARET

Dung beetles fly to fresh dung and disperse or bury the dung whilst feeding or providing brood masses for their larvae. The recycling of dung in this way improves soil texture and returns nutrients and water to the soil (Bornemissza, 1970), as well as destroying fly eggs and nematodes which breed in dung (Fincher, 1973, 1975; Waterhouse, 1974). It has therefore been considered possible that the introduction of dung beetles from overseas could bring about improved breakdown of dung and also control of flies and parasites in areas such as Australia and the USA where native dung beetle activity is low (Bornemissza, 1960; Waterhouse, 1974; Fincher, 1981; Kirk, 1983; Ridsdill-Smith & Kirk, 1985).

Dung beetles have been introduced into areas of mediterranean climate in Australia and the USA from similar climatic areas in Europe, North Africa and western Asia as biocontrol agents for dung-breeding flies and dung (Ridsdill-Smith & Kirk, 1982, 1985; Kirk, 1983; Kirk & Ridsdill-Smith, 1986). They have been introduced into south-western Australia to improve the breakdown of cattle dung and thereby reduce populations of the dung-breeding bush fly *Musca vetustissima*. South-western Australia has a climate with cool moist winters and warm dry summers. As part of this program, beetles were selected from south-western Spain, which has a similar climate and where about 50 Mediterranean species of dung beetles (Coleoptera: Scarabaeidae) have been recorded (Baraud, 1977). The beetles selected for introduction were adapted to feed on and bury dung in the spring (Ridsdill-Smith & Kirk, 1985) at a time when bush fly populations in south-western Australia increase rapidly (Matthiessen, 1983) and when there is little activity of native dung beetles (Ridsdill-Smith & Hall, 1984).

Importation of adult dung beetles into Australia is prohibited by strict quarantine measures designed to prevent the introduction of disease. Only surface- sterilised eggs of dung beetles may be introduced into Australia (Waterhouse, 1974), a ruling which also applies to dung beetle eggs imported into the USA (Fincher, 1986). As part of the overall program, eggs were air-freighted in sterile peat and boxes sent previously from Australia.

Until 1982 very large numbers of Mediterranean beetles were sent from Europe to the CSIRO quarantine station at Pretoria, South Africa, for rearing and subsequent introduction to Australia. Three Mediterranean-adapted species (*Euoniticellus pallipes*, *Onthophagus taurus* and *Geotrupes spiniger*) were released and recovered in Australia between 1976 and 1979. Because of the difficulties encountered in rearing, however, it was decided to rear beetles collected in Spain in a CSIRO laboratory at Montpellier, France. The main problems encountered in Europe were during rearing (Kirk & Feehan, 1984), surface sterilisation and handling of the extremely fragile eggs. In Australia, the main problems encountered were high larval mortality and retarded adult emergence of some species. Overall mortality was high with a 20 per cent chance of survival to emergence. Twenty thousand eggs of six species of Mediterranean dung beetles were shipped to Australia directly from Europe. Three species (*Copris hispanus*, *Onthophagus vacca* and *Bubas bison*) were released in south-western Australia in 1984.

Comparable introductions of African dung beetle species have been made via Australia to New Caledonia and Vanuatu and the following species are now established there: *Onthophagus gazella*, *Euoniticellus intermedius*, *Sisyphus spinipes* and *Liatongus militaris* (Gutierrez *et al.*, 1988).

No specifically Mediterranean-adapted beetle has been introduced into the USA. *Onthophagus taurus*, a climatically unrestricted beetle which is abundant in mediterranean-climate areas (Lumaret, 1978; Kirk & Ridsdill-Smith, 1986), appeared, however, in Florida and Georgia after an unknown release (Fincher, 1986). It was introduced subsequently from south-eastern USA to the mediterranean-climate areas of California (Fincher, 1986).

During the last century soil was used as ballast in cargo ships coming from Europe to pick up timber from Canada and the USA. Accidental introductions of nine species of *Aphodius* (*A. distinctus*, *A. erraticus*, *A. fimetarius*, *A. fossor*, *A. granarius*, *A. haemorrhoidalis*, *A. prodromus*) and also *Onthophagus nuchicornis* and *Geotrupes stercorarius* were made probably because of the dumping of ballast containing the beetles along the east coast of North America (Brown, 1940). More European species of Coleoptera occur there than in any other part of North America (Brown, 1950).

The account which follows is based mainly on the importation of Mediterranean dung beetles from Europe to Australia.

Methods

The selection of beetles

Surveys for dung beetles on the Iberian peninsula were made between 1976 and 1983, generally in spring when dung beetles were highly active (Ridsdill-Smith & Kirk, 1985), but also at other times. Nearly all provinces in Spain and Portugal were covered. Droppings were examined and dung beetles, if present, were collected from them or from the soil directly under and around them. Altitude, dropping type, soil type and bioclimatic group were noted at each site.

Dung beetle collection sites in both south-western Spain and their equivalents in Australia were categorised using the UNESCO/FAO (1963) bioclimatic classification. These bioclimatic maps are based on a study that recognises temperature, precipitation, and relative humidity as important climatic variables. The principal climatic categories are defined by the mean temperature of the coldest month, 't', so that when 't' is greater than 0 °C, the zone is considered temperate, warm-temperate or warm. Within the temperate or warm-temperate categories, a mediterranean climate is defined (UNESCO/FAO, 1963) as one with a definite dry period of from 1 to 8 months coincident with the period of longest daylight (summer). Precipitation is the total monthly precipitation 'P', a 'dry' month being defined as one in which P is equal to or less than twice the mean temperature of the month. A dry season consists of consecutive dry months. To recognise the influence of atmospheric humidity in the absence of measurable precipitation, e.g. mist or dew, an index of hot weather drought (the so-called xerothermic index, 'x') was evolved (UNESCO/FAO, 1963). This index describes the degree of drought as defined by the number of days in a month which are dry from a biological point of view: a physiologically 'dry day' is one when the relative humidity is less than 40 per cent.

The xerothermic index, xm, may be calculated as:

$$\text{xerothermic index} = \{\text{no. of days in month} - (\text{no. days with rain} + \text{no. days with mist or dew}/2)\} \times \text{humidity factor} \quad (1)$$

where days with mist or dew count as half a dry day.

The humidity factor, h, is an arbitrary value based on the mean monthly relative humidity (RH), such that h = 0.9 when RH is 40–60 per cent (equivalent to 0.9 of a dry day), 0.8 when RH is 61–80 per cent, 0.7 when RH is 81–90 per cent, and 0.6 when RH is 91–100 per cent. For example, to calculate the xerothermic index for July when there are 4 days of rain, 8 days of mist or dew, and a mean monthly humidity of 65 per cent relative humidity (h = 0.8):

$$\begin{aligned} xm &= \{31 - (4 + 8/2)\} \times 0.8 \\ &= 18.4 \end{aligned} \quad (2)$$

The xerothermic index is obtained by summing the values of x for those months which constitute the dry season: x = 0–40 being considered transitional between the temperate and mediterranean climates; x = 40–200, mediterranean; and x = 200–250, subdesertic.

Phenology and numbers of beetles

During 1980, the spring of 1981, and in 1982 and 1983, beetles were collected in pitfall traps, identified, counted and the dry weight of each species determined.

Rearing of beetles

Onthophagus vacca, *Bubas bison*, *Onitis belial* and *O. ion* all dig tunnels and bury masses of dung into which eggs are oviposited. The buried dung serves as a provision for the developing larvae and is termed a brood mass. These species were reared in gauze-covered boxes or buckets of soil.

Copris hispanus, a species from the Mediterranean region, is a brood-caring beetle which in nature produces only 4 to 5 eggs per year. It was difficult therefore to obtain large numbers of its eggs for shipment to Australia. Fabre (1911) and Tyndale-Biscoe (1983) extended the oviposition period of *Copris* species by removing the original brood balls and providing the beetle with a fresh supply of dung. *Copris hispanus* was reared following the method of Kirk & Feehan (1984) (see Figure 29.1) from collections initially made in Spain in March 1982 and 1983.

Eggs of all species were air-freighted to Canberra directly from Montpellier once a week. Co-workers in Canberra picked up the boxes and unpacked them in quarantine. The eggs were transferred into pre-formed brood balls and kept under controlled conditions in an insectary. The brood balls were surveyed until adults emerged. Adults were kept in the insectary until the relevant activity period when they were sent to Western Australia for field release.

Results

Distribution and bioclimatic classification

Forty-two species (2 Coprini, 2 Oniticellini, 19 Onthophagini, 5 Onitini, 6 Scarabaeini and 8 Geotrupidae) were identified from the surveys and trapping of dung beetles on the Iberian peninsula (Kirk & Ridsdill-Smith, 1986) (see Appendix 29.1, this volume). A group of mediterranean-adapted species was delimited by the use of canonical variate analysis from this fauna which represented a pool of 16 species restricted to a mediterranean-type climate together with 8 species unrestricted to a particular climatic type but existing in mediterranean-type climates. No dung beetle restrictions to soil or dung type were apparent from this analysis (Kirk & Ridsdill-Smith, 1986).

Phenology and importance

Copris hispanus and *B. bison*, both nocturnal fliers, were active in spring and were the most important elements of the fauna. Of the dry weight of the fauna trapped at dawn-set and dusk-set dung pads, they represented 73 per cent and 86 per cent respectively and were responsible for burying more than 40 per cent of fresh dung pads in 7 days (Ridsdill-Smith & Kirk, 1985). In addition, the species restricted to the mediterranean-climate area, viz. *Onitis belial*, *O. ion* and *Gymnopleurus flagellatus*, were active in spring and diurnal. The most important non-restricted species was the diurnal *Onthophagus vacca* which has a very high activity in spring in the Mediterranean areas studied (Lumaret, 1978/9; Ridsdill-Smith & Kirk, 1985; Lumaret & Kirk, 1987).

Rearing

Twenty thousand eggs were produced at Montpellier by beetles adapted to the mediterranean climate (Table 29.1). Table 29.1 also gives information on the origin and establishment of mediterranean dung beetle species in Australia and the USA. The new technique of inducing *C. hispanus* to oviposit more than the four eggs produced in nature, resulted in 18 eggs/female (Kirk & Feehan, 1984).

Shipment, reception and liberation of beetles

Mortality levels were high, with only about 20 per cent of all eggs sent to Australia via South Africa, or directly from France, developing into adults. The

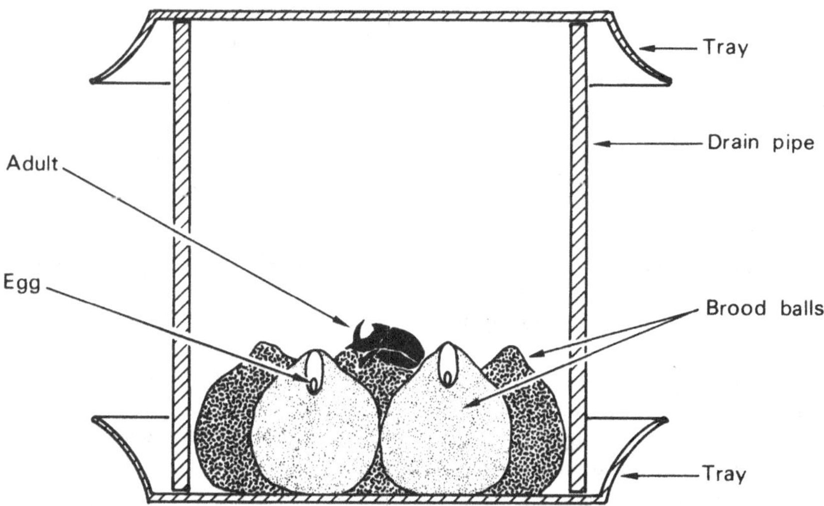

Figure 29.1. Rearing chamber in section (from Kirk & Feehan, 1984).

Table 29.1. *Introduction of mediterranean dung beetles to Australia and the USA from Europe and South Africa*

Species	Country of origin	Egg number	Status		
			Insectary	Released	Recovered
Into Australia via South Africa (Pretoria)					
Euoniticellus fulvus	Turkey		—	1978	—
E. pallipes	Turkey		—	1978	1979
Onitis belial	Morocco & Spain		F	—	—
Bubas bison	Morocco		F	—	—
Onthophagus andalusicus	Morocco		F	—	—
O. opacicollis	Greece		—	1982	—
O. taurus	Spain/Italy		—	1975	1979
	Turkey		—	1977	1979
O. vacca	Morocco		F	—	—
	France		—	1980	—
Copris hispanus	Morocco, Spain & Italy		F	—	—
C. lunaris	Italy		F	—	—
Geotrupes spiniger	France		—	1979	1984
	Greece		C (1982)	—	—
Into Australia directly from France (Montpellier)					
C. hispanus	France/Spain	5000	—	1983	—
B. bison	France/Spain	7000	—	1983	1986
O. vacca	France/Spain	6000	—	1983	—
Onitis ion	Spain	1000	C (1984)	—	—
O. belial	Spain	1000	C (1984)	—	—
Gymnopleurus flagellatus	Spain	20	C (1984)	—	—
Into Australia directly from winter-rainfall area of South Africa[a]					
Onthophagus binodis	S. Africa		—	1972	1975
Onitis aygulus	S. Africa		—	1977	1980
O. caffer	S. Africa		—	1983	—
Into the USA					
Onthophagus taurus	Europe via Florida & Georgia		1982	—	In most SE states & California

F, failed in Canberra insectary; C, in insectary on date indicated.

[a] These species proved, however, to have a bimodal activity pattern in south-western Australia (summer- and autumn-active) (Davies, 1987).

first instar larvae of a number of species failed to develop and in some cases adults remained in their dung masses and died (M. Tyndale-Biscoe, personal communication). Early problems of contamination of dung masses in Canberra by fungi and sciarids (Diptera: Sciaridae) were resolved (J. E. Feehan, personal communication). Of the species sent to Australia directly from France, three (*C. hispanus*, *B. bison* and *O. vacca*) were released in south-western Australia in 1984.

Discussion

The mediterranean-climate areas are challenging environments for dung beetles. In order to adapt to the often extreme conditions (Lumaret, 1991; Lumaret & Kirk, 1987), beetles have developed life histories which enable them to exploit dung, especially during the moister cooler period of the year. Some workers have recorded particular associations between dung beetles and soil texture (Nealis, 1977; Lumaret, 1978/9; Doube, 1983), between dung beetles and dung type (Lumaret, 1978/9; Oppenheimer & Begum, 1978; Walter, 1978; Ricou, 1981; Martin-Piera, 1982; Hanski, 1983), and between dung beetles and temperature and rainfall (Matthews, 1972, 1974; Lumaret, 1978; Tyndale-Biscoe *et al.*, 1981). Associations of this sort were not evident generally for the beetles studied on the Iberian peninsula (Kirk & Ridsdill-Smith, 1986). Adaptation and restriction to particular climates determined the distribution of beetles on the Iberian peninsula (Kirk & Ridsdill-Smith, 1986). The bioclimatic classification of UNESCO/FAO (1963) enabled Spradbery & Kirk (1978) to separate the distribution of the Siricid wood wasp species (Hymenoptera: Siricidae) present in Europe and to predict the likely spread of *Sirex noctilio* in Australian coniferous forests (Kirk, 1974). The use of the classification to determine the distribution of Mediterranean beetles was an important tool in the selection of beetles for Australia (Kirk, 1983; Ridsdill-Smith & Kirk, 1985; Kirk & Ridsdill-Smith, 1986). *Copris hispanus* and *B. bison* were selected for introduction as they were the most important nocturnal, spring-active beetles encountered in the survey. Additional support for this decision came from southern France where *B. bison* was the only large dung beetle (216 mg dry weight) active in spring on humid soils where it buried 70 per cent of dung (Kirk, 1983). In addition, *O. belial*, *O. ion* and *G. flagellatus* were selected on the basis of being the most important diurnal species in south-western Spain.

The South African species imported into Australia from winter-rainfall areas (Table 29.1) had a summer activity period in south-western Australia which did not fit in with the need for spring-active beetles (Ridsdill-Smith & Kirk, 1985). Davies (1987) reported on South African species endemic to the narrow band of mediterranean climate in the southern Cape and pointed out that there may be vegetational constraints on the activity and presence of those species; these constraints make them unsuitable for introduction to Australia where they may compete with the species endemic to similar vegetation (heath) in Australia (Ridsdill-Smith *et al.*, 1983).

Precise selection of beetles based on phenology, activity and climatic classification should result in more successful introductions. The problems encountered during rearing and shipment were, however, the limiting factors in beetle establishment. Dung beetle eggs are naturally cocooned in dung up to 30 cm under the ground. The chamber in which they are put experiences little or only very slow changes in temperature, humidity and air content. Dung beetle eggs have thin walls and are prone to burst on drying out at the surface or when pressure changes are very abrupt (Kirk & Feehan, unpublished results). Not surprisingly, considering the relatively rough treatment (compared to natural conditions) the eggs received in manipulations plus the unknown effects of 3 per cent formalin on hatching and development, mortality rates were very high. The use of new techniques to increase egg production (Kirk & Feehan, 1984) may also mean that infertile eggs are produced. It is not known whether females of *C. hispanus* are able to store enough sperm from one mating to repeatedly fertilise the five-fold increase over natural egg production. More work is required on this topic and on the possibility that diapause (as an adaptation perhaps to mediterranean-climatic conditions) is the cause of the delayed third instar larval development in *B. bison* in the insectary, as seen by Edwards (1986) in the African species *Onitis caffer* or delayed adult emergence in *B. bison* and *C. hispanus*.

If dung beetles could be shown not to harbour introduced livestock diseases, then the best method would be to import adult beetles directly into quarantine in south-western Australia and allow beetles to oviposit naturally in order to increase numbers for release.

Appendix

Appendix 29.1. *List of species of dung beetles collected on the Iberian peninsula (*Aphodius *spp. not included)*

	Species	Bioclimatic status
Scarabaeidae		
Scarabaeinae		
Coprini (2 species)	*Copris hispanus*	M
	C. lunaris	T
Oniticellini (2 species)	*Euoniticellus pallipes*	M
	E. fulvus	U

Appendix 29.1. (*cont.*)

	Species	Bioclimatic status
Onthophagini (19 species)	*Euonthophagus amyntas*	U
	Onthophagus lemur	U
	O. maki	M
	O. merdarius	M
	O. opacicollis	M
	O. punctatus	M
	O. ruficapillus	M
	O. furcatus	U
	O. taurus	U
	O. vacca	U
	O. fracticornis	T
	O. grossepunctatus	U
	O. ovatus	T
	O. similis	T
	O. andalusicus	M
	O. hirtus	M
	O. hispanicus	M
	O. melitaeus	M
	Caccobius schreberi	U
Onitini (5 species)	*Onitis belial*	M
	O. ion	M
	Bubas bison	M
	B. bubalus	M
	Chironitis hungaricus	M
Scarabaeini (6 species)	*Scarabaeus cicatricosus*	M
	S. laticollis	M
	S. sacer	M
	S. semipunctatus	M
	Gymnopleurus flagellatus	M
	G. sturmi	M
Geotrupidae (8 species)	*Geotrupes ibericus*	M
	G. spiniger	T
	G. stercorarius	T
	G. mutator	T
	G. niger	M
	Ceratophyus hoffmanseggi	M
	Typhaeus typhaeus	U
	T. momus	M

M, Mediterranean; U, ubiquitous; T, temperate.

References

Baraud, J. (1977). *Coléoptères Scarabaeoidea.* Nouvelle Revue d'Entomologie, (Supplément) **7**, 1–352.

Bornemissza, G. F. (1960). Could dung-eating insects improve our pasture? *Journal of the Australian Institute of Agricultural Science*, **26**, 54–6.

Bornemissza, G. F. (1970). Insectary studies on the control of dung-breeding flies by the activity of the dung beetle *Onthophagus gazella* F. (Coleoptera: Scarabaeidae). *Journal of the Australian Entomological Society*, **9**, 31–41.

Brown, W. J. (1940). Notes on the American distribution of some species of Coleoptera common to the European and North American continents. *The Canadian Entomologist*, **72**, 65–78.

Brown, W. J. (1950). The extralimital distribution of some species of Coleoptera. *The Canadian Entomologist*, **82**, 197–205.

Davies, A. L. V. (1987). Geographical distribution of dung beetles (Coleoptera: Scarabaeidae) and their seasonal activity in south-western Cape Province. *Journal of the Entomological Society of Southern Africa*, **50**, 275–85.

Doube, B. M. (1983). The habitat preference of some bovine dung beetles (Coleoptera: Scarabaeidae) in Hluhluwe Game Reserve, South Africa. *Bulletin of Entomological Research*, **73**, 357–71.

Edwards, P. B. (1986). Development and larval diapause in the southern African dung beetle *Onitis caffer* Boheman (Coleoptera: Scarabaeidae). *Bulletin of Entomological Research*, **76**, 109–17.

Fabre, J. H. (1911). The Spanish Copris. In *The Life and Love of the Insect*, ed. J. H. Fabre, pp. 63–75 (trans. A. T. de Mattos). London: A. & C. Black.

Fincher, G. T. (1973). Nidification and reproduction of *Phanaeus* spp. in three textural classes of soil (Coleoptera: Scarabaeidae). *Coleopterists Bulletin*, **27**, 33–8.

Fincher, G. T. (1975). Effect of dung beetle activity on the number of nematode parasites acquired by grazing cattle. *Journal of Parasitology*, **56**, 378–83.

Fincher, G. T. (1981). The potential value of dung beetles in pasture ecosystems. *Journal of the Georgia Entomological Society*, (Supplement), **16**, 316–33.

Fincher, G. T. (1986). Importation, colonization and release of dung-burying scarabs. In *Biological Control of Muscoid Flies*, ed. R. S. Patterson & D. A. Rutz, pp. 61–76. Hyattsville, Maryland: Entomological Society of America, Miscellaneous Publication No. 61.

Gutierrez, J., Macqueen, A. & Brun, L. O. (1988). Essais d'introduction de quatre espèces de bousiers Scarabaeinae en Nouvelle Calédonie et au Vanuatu. *Oecologia Applicata*, **9**, 39–53.

Hanski, I. (1983). Distributional ecology and abundance of dung and carrion-feeding beetles (Scarabaeidae) in tropical rain forests in Sarawak, Borneo. *Acta Zoologica Fennica*, **167**, 1–45.

Kirk, A. A. (1974). Bioclimates of Australian *Pinus radiata* areas and *Sirex noctilio* localities in the northern hemisphere. *Australian Forestry*, **37**, 126–31.

Kirk, A. A. (1983). The biology of *Bubas bison* (L.) (Coleoptera: Scarabaeidae) in southern France and its potential for recycling dung in Australia. *Bulletin of Entomological Research*, **73**, 129–36.

Kirk, A. A. & Feehan, J. E. (1984). Method for increased production of eggs of *Copris hispanus* L. and *Copris lunaris* L. (Coleoptera: Scarabaeidae). *Journal of the Australian Entomological Society*, **23**, 293–4.

Kirk, A. A. & Ridsdill-Smith, T. J. (1986). Dung beetle distribution patterns in the Iberian Peninsula. *Entomophaga*, **31**, 183–90.

Lumaret, J. P. (1978). Biogéographie et écologie des Scarabéides coprophages du Sud de la France. Thèse Doctorat, Université des Sciences et Techniques du Languedoc, Montpellier.

Lumaret, J. P. (1978/9). Biogéographie et écologie des Scarabéides coprophages du Sud de la France. I. Méthodologie et modèles de répartition. *Vie et Milieu*, **28–9** (**C**), 1–34.

Lumaret, J. P. (1991). Use of excrement by dung beetles in drought affected areas. In *Time Scales of Biological Responses to Water Constraints*, ed. J. Roy, J. Aronson & F. di Castri, Ecological Studies, Heidelberg: Springer-Verlag (in press).

Lumaret, J. P. & Kirk, A. A. (1987). Ecology of dung beetles in the French mediterranean region (Coleoptera: Scarabaeidae). *Acta Zoologica Mexicana (n.s.)*, **24**, 1–55.

Martin-Piera, F. (1982). Los Scarabaeinae (Coleoptera: Scarabaeoidea) de la Peninsula Iberica e Islas Baleares. Tésis Doctoral, Universidad Complutense de Madrid.

Matthews, E. G. (1972). A revision of the scarabaeine dung beetles of Australia. I. Tribe Onthophagini. *Australian Journal of Zoology, Supplementary Series*, **9**, 1–330.

Matthews, E. G. (1974). A revision of the scarabaeine dung beetles of Australia. II. Tribe Scarabaeini. *Australian Journal of Zoology, Supplementary Series*, **24**, 1–211.

Matthiessen, J. N. (1983). The seasonal distribution and characteristics of bush fly *Musca vetustissima* (Walker) populations in south-western Australia. *Australian Journal of Ecology*, **8**, 383–95.

Nealis, V. G. (1977). Habitat associations and community analysis of south Texas dung beetles (Coleoptera: Scarabaeinae). *Canadian Journal of Zoology*, **55**, 138–47.

Oppenheimer, J. R. & Begum, J. (1978). Ecology of some dung beetles (Scarabaeidae and Aphodiinae) in Dacca district. *Bangladesh Journal of Biology*, **6**, 23–9.

Ricou, G. (1981). Contribution à l'étude de la dynamique des populations coprophiles. Biocoenoses des fèces en Margeride Lozérienne. Mémoire Diplôme Etudes Supérieures Sciences Naturelles, Université de Rouen.

Ridsdill-Smith, T. J. & Hall, G. P. (1984). Beetles and mites attracted to fresh cattle dung in south-western Australian pastures. CSIRO Australia Division of Entomology Report No. 34, pp. 1–29.

Ridsdill-Smith, T. J. & Kirk, A. A. (1982). Dung beetles and dispersal of cattle dung. In *Proceedings 3rd Australasian Conference on Grassland Invertebrate Ecology, Adelaide, 1981*, ed. K. E. Lee, pp. 215–20. Adelaide: Government Printer.

Ridsdill-Smith, T. J. & Kirk, A. A. (1985). Selecting dung beetles (Scarabaeinae) from Spain for bushfly control in south-western Australia. *Entomophaga*, **30**, 217–24.

Ridsdill-Smith, T. J., Weir, T. A. & Peck, S. B. (1983). Dung beetles (Scarabaeidae: Scarabaeinae and Aphodiinae) active in forest habitats in south-western Australia during winter. *Journal of the Australian Entomological Society*, **22**, 307–9.

Spradbery, J. P. & Kirk, A. A. (1978). Aspects of the ecology of siricid woodwasps (Hymenoptera: Siricidae) in Europe, North Africa and Turkey with special reference to the biological control of *Sirex noctilio* F. in Australia. *Bulletin of Entomological Research*, **68**, 341–59.

Tyndale-Biscoe, M. (1983). Effects of ovarian condition on nesting behaviour in a brood-caring dung beetle, *Copris diversus* Waterhouse (Coleoptera: Scarabaeidae). *Bulletin of Entomological Research*, **73**, 45–52.

Tyndale-Biscoe, M., Wallace, M. M. H. & Walker, J. M. (1981). An ecological study of an Australian dung beetle, *Onthophagus granulatus* Boheman (Coleoptera: Scarabaeidae), using physiological age-grading techniques. *Bulletin of Entomological Research*, **71**, 137–52.

UNESCO/FAO (1963). *Ecological Study of the Mediterranean Zone. Bioclimatic Map of the Mediterranean Zone*. Paris: UNESCO Press.

Walter, P. (1978). Recherches écologiques et biologiques sur les Scarabéides coprophages d'une savane du Zaire. Thèse Doctorat, Université des Sciences et Techniques du Languedoc, Montpellier.

Waterhouse, D. F. (1974). The biological control of dung. *Scientific American*, **230**, 100–9.

Part V Overview

30

The biogeography of mediterranean plant invasions

R. H. GROVES

The propagules of some plant species are dispersed naturally over long distances by agents such as birds, water or wind. We usually refer to these species collectively as the so-called 'cosmopolitan flora' of a region (Kloot, this volume). Such species often occur along coasts or around lakes of more than one region. Kloot estimates that from 50 to 60 plant species – about 5 per cent – occurring presently in the flora of South Australia may have arrived naturally by the process of long-distance dispersal. Of the 134 species that comprise the major weeds of the western Mediterranean Basin, Guillerm *et al.* (1990) estimate 24 – 18 per cent – to be cosmopolitan, of which 12 (9 per cent) species are grasses. But most species which migrate from their native region to other parts of the world appear to have done so by human means.

From the earliest times of documented records, plants have been moved from one region to another for human purposes. Records of human settlements in the Mediterranean Basin provide evidence for the deliberate movement of plants in cases where trees or cereals were of direct benefit to humans. For instance, Darius (530–522 BC) wrote to his steward Gadatar, '. . . you are taking trouble over my estates in that you are transferring trees and plants from beyond the Euphrates to Asia Minor' (as quoted in Thirgood, 1981, p. 45). Trees such as species of *Pinus* (especially *P. pinea* with its edible seeds) were planted widely by the Etruscans, Greeks, Romans and Arabs, so that their original areas of distribution are now obscure (see, e.g., Mirov, 1967). This age-old redistribution of tree species continues; an instance of this is the recent widespread planting of the Australian tree genera *Eucalyptus* and *Casuarina* in the Mediterranean Basin and other regions of mediterranean climate (Pryor, this volume).

In a similar way and for similar purposes, herbaceous crop plants such as the cereals and pulses were also moved about by humans (see Naveh & Vernet, this volume). Deacon (this volume) suggests that several food crops (e.g. *Sorghum*) moved southwards in Africa with waves of human migrations over the last 2000 years. Deacon records that sorghum had not reached the winter-rainfall region of southern Africa by the time of first European contact 500 years ago.

Concomitant with the recorded movements of beneficial plants were the unplanned and largely unrecorded movements of those mostly non-beneficial plants which in this volume have been termed 'invasive'. In some cases the deliberate establishment of invasive plants was initially beneficial. The Australian acacias in the western Cape or the southern African *Chrysanthemoides monilifera* in eastern Australia were beneficial in that they stabilised sand dunes prone to erosion. Additionally, the Australian acacias provided a valuable source of wood for fuel, as indeed they still do in some parts of Africa. Increasingly, however, many introduced plants became too successful in their region of introduction and became nuisances, firstly to human well-being, and subsequently to nature conservation values. In this volume these latter 'nuisance' aspects have been highlighted. Let us not forget, however, the economic and social benefits which some so-called invasive plants have brought and continue to bring to humans living in regions of mediterranean climate.

This overview chapter concentrates on the approximately 1000 to 1500 plant species which are currently invasive in the five regions of mediterranean climate. I shall describe the course of one of the best-documented invasions of plants in the mediterranean-climate region of France, in the vicinity of Montpellier. I then will consider some biogeographical features of the introduced flora of the regions of mediterranean climate. Aspects of invasive plant management in terms of their control or eradication will be discussed. Finally, I shall address the question of whether plant invasions in regions of mediterranean climate show characteristics different from those in regions with other climatic regimes.

The Flora Juvenalis

Le Floc'h (this volume) discusses one well-known documentation of a plant invasion which up to now seems not to have been readily available to English-speaking readers. The particular invasion concerns the so-called *Flora Juvenalis* around Montpellier, southern France. Over the last 200 years an invasion of introduced plants occurred which is very well documented botanically because of the long tradition of Montpellier for excellence in plant collection and identification.

The story starts in 1686 with the construction of the Lez canal from the Mediterranean Sea to Montpellier to enable wool stores to operate at Port Juvenal from 1750. At this site, wool was received from all over the world, washed in hot water and spread over adjoining fields to dry. Thus many plant propagules arrived at the site. Some species established and became the *Flora Juvenalis*.

With time the number of species increased until by 1859 there were 458 adventive species established in the area (Figure 30.1). But with subsequent changes in wool technology, the storing and washing of wool at Port Juvenal was discontinued in 1880. By 1905–10 (Thellung, 1911–12) ten species remained and by 1950, only six remained (Rioux & Quézel, 1950).

This plant invasion is of interest because it shows a long lag or 'latent' period for the invasion. The lag period was followed by a very rapid increase that may occur in an introduced flora within about 40 years (in this case, between 1813 and

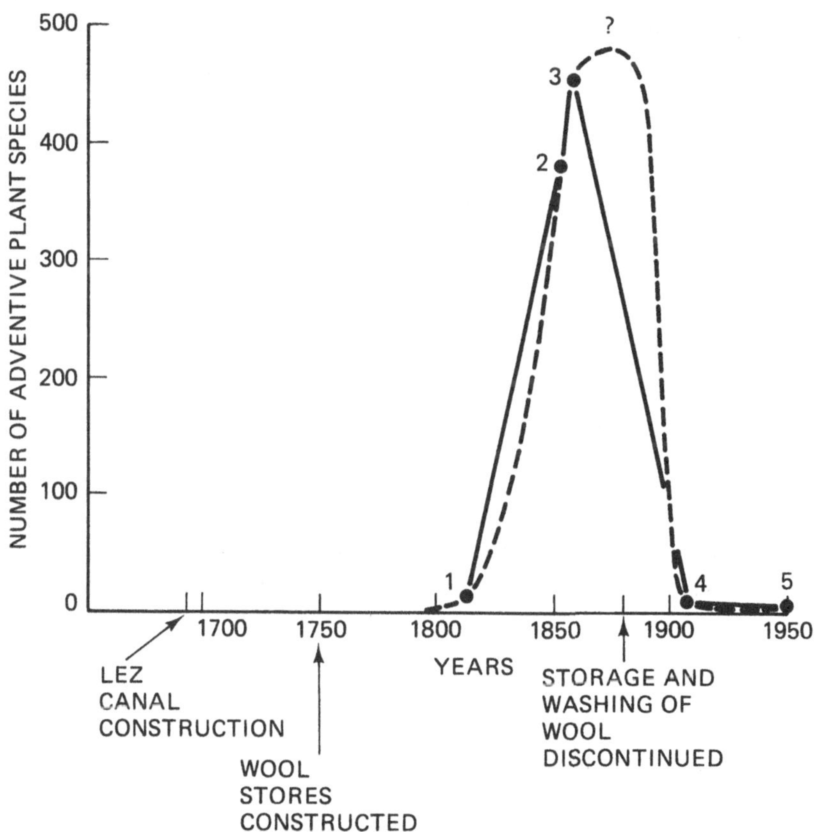

Figure 30.1. Increase and decrease in number of adventive plant species at Port Juvenal, near Montpellier from the period 1686–1950 (see Le Floc'h, this volume). 1, 1813 de Candolle in Rioux & Quézel, 1950); 2, 1853 (Godron, 1853); 3, 1859 (Cosson, 1860); 4, 1905–10 (Thellung, 1911–23); 5, 1950 (Rioux & Quézel, 1950). The broken line represents the hypothetical increase until 1880 when storage and washing of wool was discontinued at Port Juvenal.

1859). The example is also interesting, however, because it shows even more clearly that once the change in technology ceases to operate (in this case, cessation of the washing and drying of wool in the open) the number of species may drop rapidly – for the *Flora Juvenalis* from 458 to 10 within 50 years. Only a few of these plant species continue at the site currently. Only one, *Ludwigia uruguayensis*, a wetland species from North America, is still spreading and becoming a nuisance in the Camargue, a region some distance from its site of introduction to Montpellier in 1854 (Godron, 1854).

The introduced floras of most regions of mediterranean climate are increasing at the rate of about four to six species a year (Frenkel, 1970; Specht, 1981; Kloot, this volume). Rejmanek *et al.* (this volume) suggest that the rate of increase in the introduced flora of California may be decreasing, although it is still positive, unlike that of the *Flora Juvenalis* over the last 100 years which has been strongly negative, as is also the case for the introduced flora of Israel (see Figure 1, Dafni & Heller, 1982). The *Flora Juvenalis*, on the other hand, decreased drastically over the last 100 years (Figure 30.1). Presumably, this difference means that basic patterns of agricultural and pastoral land use have not changed drastically in California since plant collection and storage of specimens in herbaria commenced. In the case of the *Flora Juvenalis* it was, admittedly, a rather specialised change in land use that led initially to the plant invasion and, subsequently, to its demise.

The *Flora Juvenalis* appears to have had four main phases. The first was a latent or lag period after introduction to a site when numbers of propagules of the different species probably increased but usually only imperceptibly; this phase may vary in time depending on interactions between species characteristics and land-use patterns. The latent phase was followed by a period of rapid and detectable increase in the numbers of different species (often called the exponential phase). The third phase, that of a decrease in number of plant species, occurred as a result of changed land use or technology. In other situations it more often happens as a result of a deliberate decision to institute control measures against one or a group of species. The fourth and final phase of an invasion is usually a continuing low population level of a few invasive plant species. The final numerical level attained will vary, depending mainly on the economic costs associated with the different plant species comprising the introduced flora. Only rarely is eradication achieved, even of an individual species (but see Rejmanek *et al.*, this volume, and subsequently for some Californian and other examples).

Some biogeographical features of introduced invasive floras

The origins of the flora introduced to the different regions of mediterranean climate vary between the regions, depending mainly on the history of discovery, human settlement patterns and trade in agricultural products. One major feature common to all regions of mediterranean climate is the predominance of European plants in the invasive flora (Figure 30.2). This feature is most pronounced for Chile (Figure 30.2*b*) where invasive plants are mostly annual European herbs (see Figure 9.1 of Montenegro *et al.*, this volume). What is surprising, however, is that the predominance of European plants applies equally to the invasive flora of the western Mediterranean Basin (Guillerm *et al.*, 1990). Fifty-five per cent of the total flora introduced to the western Mediterranean either comes from northern Europe (20 per cent of the total) or from other areas of the Mediterranean Basin (35 per cent of the total). As agriculture moved from the eastern Mediterranean, so too presumably did certain plants, especially those winter-growing annual species associated with cropping (Le Floc'h, this volume). This feature of the western Mediterranean invasive flora, which reflects not only agriculture but also trade and human migrations over millennia, will not decrease as regional barriers to trade are broken down still further with the advent of the European Economic Community.

The data presented in Figure 30.2 are aggregates for Europe as a whole and for the Americas. Within Europe, several regional climates are represented and the divisions between the native floras of these regions are sometimes unclear. Human settlement by Europeans of the 'new world' regions differs in that predominantly northern Europeans (British, Dutch and Germans) settled South Africa and South Australia, whereas predominantly Mediterraneans (Spanish) settled both Chile and, initially, California. Are these differences in human settlement reflected in the invasive floras of these two sets of regions?

The answer to this question partly depends on the time since settlement. Initially, there was a predominance of northern European plants represented in the invasive flora of South Australia, for instance (Kloot, this volume). With time, however, the proportion of invasive plants in South Australia originating from similar climatic areas, e.g. the Mediterranean Basin, has increased. Such a change is also occurring in the invasive flora of Queensland which has a subtropical/tropical climate. Because Queensland was settled initially by northern Europeans, so was there once a high proportion of European plants in its invasive flora (Everist, 1960). Of late, however, the proportion of invasive plants from other tropical and subtropical regions has increased in Queensland.

A major biogeographical feature of the flora introduced to all regions of mediterranean climate is thus its European origin. But the distinctions between its origins in northern temperate Europe or in the Mediterranean region of Europe are not always clear, at least at the taxonomic level of species as used in the above analysis.

A further major biogeographical feature of the invasive flora of mediterranean regions, though less well documented, is the predominance of species introduced intentionally. For instance, Kloot (Table 11.4, this volume) gives a figure of 57 per cent intentional introductions. Plants introduced for culinary or medicinal (herbal) purposes, as fodder plants, or as a source of wood or as ornamentals have always been well represented in the invasive flora of all regions, including those with mediterranean climates. These intentionally introduced plants often include the dominant invasive plants of agriculture or horticulture, which we term 'weeds' (see Guillerm, this volume; Le Houérou, this volume). Deacon (this volume) reminds us that European colonisation of regions such as the western Cape of South Africa, but also of eastern Australia, was a horticultural event. Just as large numbers of novel South African and Australian plants were transported to Europe as horticultural curiosities, so too were the plants familiar to Europeans introduced to the parks and gardens of South Africa and Australia. Whilst few of the former seem to have become invasive in Europe (except for plants such as *Oxalis pes-caprae*), many of the latter have come to dominate disturbed areas in South Africa and Australia. The exchange has not always been equal.

A third major feature of the invasive flora of mediterranean-climate regions has been the numerical predominance of certain plant families (Figure 30.3). Species belonging to the large families Gramineae, Compositae and Cruciferae are consistently well represented in the invasive floras of all regions. In the invasive flora of the western Mediterranean Basin, Polygonaceae and Umbelliferae are also well represented, but not Leguminosae (Table 1 of Guillerm *et al.*, 1990). The proportions differ in each region, but only in a minor way. The biggest difference seems to be the relative predominance of grasses and legumes in the introduced flora of southern Australia (Figure 30.3*d*, 30.3*e*) – a result that reflects the history of introductions of potential pasture species to that region over a period of 150 years.

Some generalisations are thus possible about the biogeography of plant invasions in mediterranean-climate regions. They include:

1. Most introduced species have come from Europe (but not necessarily from the Mediterranean Basin).
2. The majority of species have been introduced intentionally, except possibly in California (M. Rejmanek, personal communication).
3. Most introduced species are likely to belong to the families Gramineae, Compositae, Cruciferae or Leguminosae, if only because of their numerical abundance in the world's flora (see Table 8.1, this volume). These families are not necessarily the ones having the highest percentages of weeds in them, however.

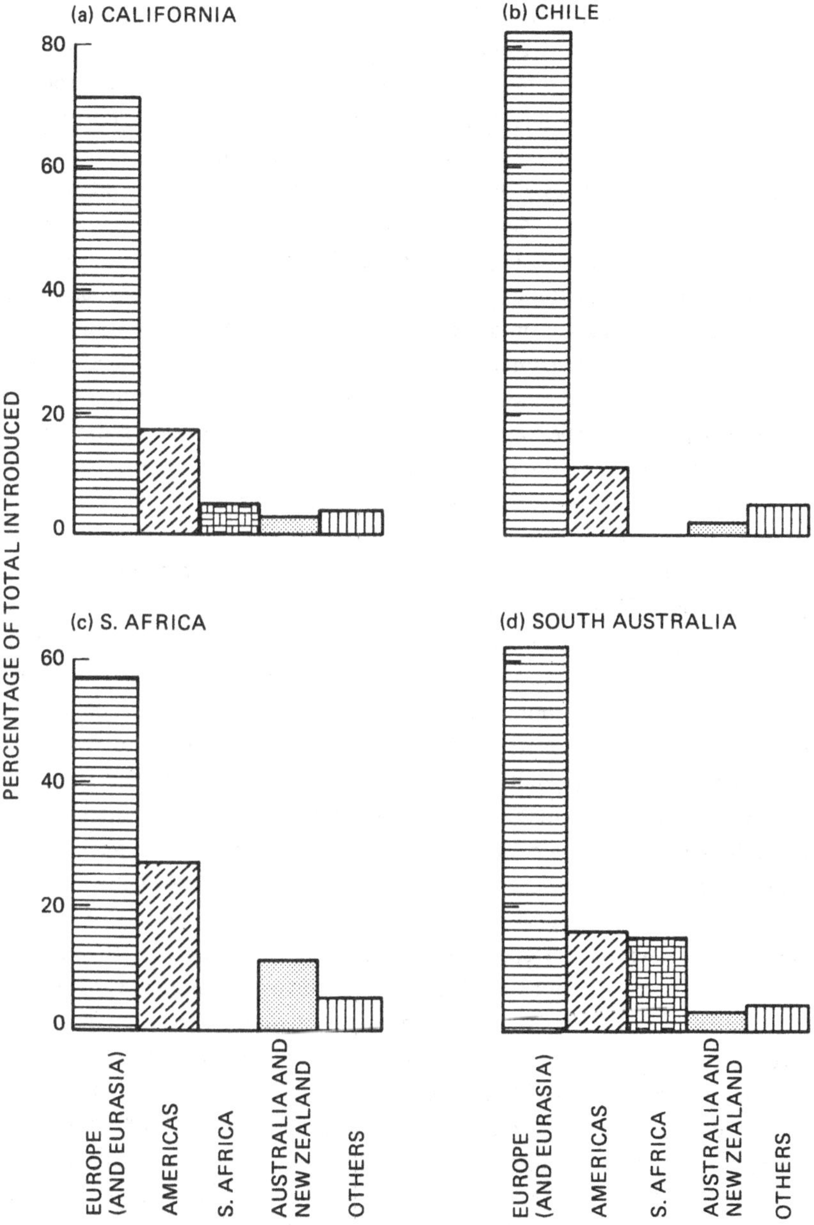

Figure 30.2. The origin of the introduced floras of (a) California (after Table 8.1, this volume), (b) Chile (after Appendixes 9.1 and 9.2, this volume), (c) South Africa (after Table 10.1, this volume) and (d) South Australia (after Table 11.1, this volume), expressed as a percentage of the total introduced flora of each region.

Management of invasive plants

Costs of controlling invasive plants are high. For instance Combellack (1987) has estimated the costs of invasive plants to Australian agriculture as totalling A$2000 million per year. Though high, Combellack's estimates do not cover the costs of maintaining plant quarantine services to reduce the introduction of new invasive plants, nor does it take into account the indirect costs of so-called 'environmental weeds'. Plants such as *Chrysanthemoides monilifera* and *Mimosa pigra* are invading large national parks and nature reserves in different climatic regions of Australia. It is difficult to estimate the indirect costs associated with such plants in non-agricultural areas, but certainly they add significantly to Combellack's estimate for plant invasions. It is also difficult to assign costs to only the mediterranean-climate region of southern Australia. The costs of invasive plants may be even higher in other regions of mediterranean climate, such as California. Clearly, plant invasions are costly. The aim of management of such plants is to reduce these costs as much as possible.

I have reviewed elsewhere (Groves, 1989) aspects of the management of invasive terrestrial plants, including some plants invasive in regions of mediterranean climate. From that review several generalisations were advanced, of which the most significant to this overview chapter is that rarely is only one method of control effective in limiting populations of undesirable plants. Control is more effective when more than one method is used – whether it be physical, chemical, biological or exclusion from a region by quarantine (Groves, 1989).

A second and related point relevant to the management of mediterranean plant invasions has also been made earlier and elsewhere (Groves & Cullen, 1981). It is that for *Chondrilla juncea*, a plant of Mediterranean origin and a weed of cereal crops in southern Australia and California, control was greater when it combined the effects of a competing pasture plant *Trifolium subterraneum*, also of Mediterranean origin, with the effects of a specific fungus *Puccinia chondrillina* introduced deliberately from the Mediterranean Basin as a biological control agent. Together the two agents from the region of origin of the weed were more effective at limiting growth of *C. juncea* than either alone (Groves & Williams, 1975). In other words, the more 'Mediterranean-like' the agro-ecosystem became, the more weed growth was limited. From among the 1000 to 1500 other plants invasive in regions of mediterranean climate it would be interesting to find some further examples of deliberate ecosystem management leading to reduced costs similar to that obtained for *C. juncea* in southern Australia.

This volume is concerned with the biogeography of mediterranean invasions. Hence few chapters mention management, the applied aspects of invasions. Rejmanek *et al.* (this volume), however, cite cases of some plants invasive to California that presently have a limited distribution – those rated as 'A' or 'B' weeds – and for which control and even eradication programs are feasible. They mention at least 18 candidates for eradication from California, including *Alhagi pseudalhagi* and *Crupina vulgaris*. The latter species, along with 9 other species, were eradicated from California between 1950 and 1986. These 10 species all had very limited distributions when eradication efforts began (M. Rejmanek, personal communication). The Greek thistle *Onopordum tauricum* has been introduced to the south of France where it occurs sporadically and where it may hybridise locally with the indigene *O. illyricum* (A. Sheppherd, personal communication). *Onopordum tauricum* has also been introduced to two different regions of southern Australia from which it has now been eradicated (W. T. Parsons, personal communication). Eradication is a feasible management option for invasive plants having only a limited distribution at present. Usually, however, widely distributed plants that interfere in some way with human well-being

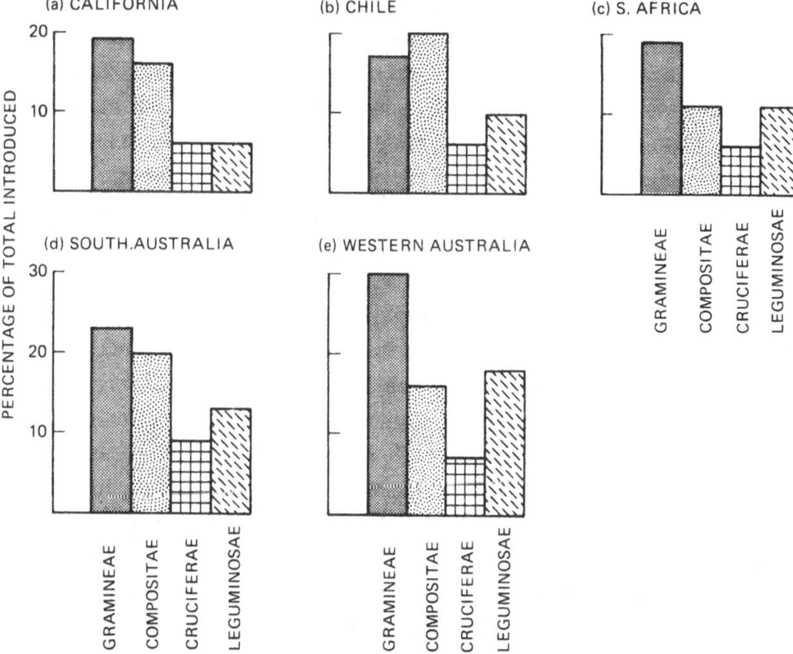

Figure 30.3. The four major plant families represented in the invasive floras of (a) California (after Table 8.1, this volume), (b) Chile (after Appendixes 9.1 and 9.2, this volume), (c) South Africa (Wells, this volume, p. 122), (d) South Australia (after Table 11.2, this volume) and (e) Western Australia (after Table 11.2, this volume), expressed as a percentage of the total introduced flora of each region.

will be subject to a variety of control methods in an attempt to limit their cost to humans.

If different control methods can be applied at critical stages in the growth and development of the plant, then invasive plant populations should reduce. Further, if such management itself can be integrated with the ecological characteristics of the ecosystem being invaded then plant numbers should be even more reduced in the long term (Groves, 1989). This volume provides many examples where control of invasive plants needs to be much more effective if mediterranean-climate ecosystems are not to be modified even further by the effects of invasive plants.

Are mediterranean ecosystems especially invasible?

Plants have invaded mediterranean ecosystems at the rate of about 4 to 6 species per year (Kloot, this volume; Rejmanek *et al.*, this volume). Specht (1981) showed, however, that the rate of invasion for Queensland, which has both tropical and subtropical climates, was the same as the rate for the southern Australian states with a predominantly mediterranean climate. Esler & Astridge (1987) showed a rate of invasion of 4.12 species per year to the Auckland area in New Zealand, an island region with a subtropical climate, for the period 1870–1970. I conclude that the rate of plant invasion is no higher in regions of mediterranean climate than in regions characterised by other climatic regimes (although the island nature of New Zealand may invalidate this conclusion somewhat).

Mediterranean-type ecosystems are characterised not only by summer-dry, winter-wet climate, but they are razed regularly by fire (Trabaud, this volume) and they occur on soils low in nutrients, especially phosphorus and nitrogen (Groves, 1981). Such ecosystems are subject to considerable disturbance by other factors, especially grazing by domesticated herbivores. Some plants originating in the eastern Mediterranean Basin are pre-adapted to various combinations of such ecological factors; these plants have become particularly invasive when introduced to other regions of mediterranean climate. I suggest that the success of these invasive plants is as much because of their pre-adaptation to the same factors – fire, low nutrients, grazing – operating in their 'new' environments as to their response to a particular climatic regime of rainfall and temperature.

Montenegro *et al.* (this volume) make the point that the matorral in Chile is more subject to invasion than the other vegetation types of mediterranean Chile (see Figure 9.3). On the other hand, Gullan (1988) showed that Victorian heathland (the structural analogue of matorral in southern Australia) had the lowest percentage of invasive species of all vegetation types in Victoria. Natural grasslands in Victoria are the most invaded vegetation type. Such grasslands occur on soils of higher nutrient status and therefore have always been more desirable for agriculture than the heathland soils with their low nutrient status and micronutrient deficiencies in copper and zinc. Grasslands and heathlands in Victoria have had very different histories of grazing over the less than 200 years of European settlement. I agree with Wells (this volume) that it remains a moot point whether fynbos and its ecological analogues in the other regions of mediterranean climate are more invasible than any other vegetation type. Factors other than climate will always interact to influence the degree of invasibility of an ecosystem, especially those containing species that have co-evolved with humans and their agriculture.

Acknowledgements. I wish to thank E. Le Floc'h, J. L. Guillerm, F. D. Panetta and M. Rejmanek for their helpful comments on an earlier draft of this chapter.

References

Combellack, J. H. (1987). Weed control pursuits in Australia. *Chemistry & Industry, 20 April 1987*, 273–80.

Cosson, E. (1860). Appendix Flora Juvenalis ou liste des plantes étrangères recémment observées au Port Juvenal près Montpellier. *Bulletin de la Societé Botanique de France*, **6**, 605–15.

Dafni, A. & Heller, D. (1982). Adventive flora of Israel – phytogeographical, ecological and agricultural aspects. *Plant Systematics & Evolution*, **140**, 1–18.

Esler, A. E. & Astridge, S. J. (1987). The naturalisation of plants in urban Auckland, New Zealand. 2. Records of introduction and naturalisation. *New Zealand Journal of Botany*, **25**, 523–37.

Everist, S. L. (1960). Strangers within the gates. *Queensland Naturalist*, **16**, 49–60.

Frenkel, R. E. (1970). Ruderal vegetation along some California roadsides. *University of California Publications in Geography*, **20**, 1–163.

Godron, D. A. (1853). *Flora Juvenalis seu enumeratis et descriptis plantarum e seminibus exoticis interlaus altatis, enatarum in campestribus Portus Juvenalis, prope Monspelium*, 1ère edn. *Memoirs de l'Academies des Sciences et Lettres – Montpellier*, 1–48.

Godron, D. A. (1854). *Flora Juvenalis, ou Enumeration des Plantes Etrangères qui Croissent Naturellement au Port Juvenal, près de Montpellier. Precédés de Considerations sur les Migrations des Végétaux*, 2ème edn. Nancy: Grimblot & Verve Raybois.

Groves, R. H. (1981). Heathland soils and their fertility status. In *Ecosystems of the World*, vol. 9B, *Heathlands and Related Shrublands*, ed. R. L. Specht, pp. 143–50. Amsterdam: Elsevier.

Groves, R. H. (1989). Ecological control of invasive terrestrial plants. In *Ecology of Biological Invasions: A Global Perspective*, ed. J. A. Drake, H. A. Mooney,

F. di Castri, R. H. Groves, F. Kruger, M. Rejmanek & M. Williamson, pp. 437–61. Chichester: Wiley.

Groves, R. H. & Cullen, J. M. (1981). *Chondrilla juncea*: the ecological control of a weed. In *The Ecology of Pests*, ed. R. L. Kitching & R. E. Jones, pp. 6–17. Melbourne: CSIRO Australia.

Groves, R. H. & Williams, J. D. (1975). Growth of skeleton weed (*Chondrilla juncea* L.) as affected by growth of subterranean clover (*Trifolium subterraneum* L.) and infection by *Puccinia chondrillina* Bubak and Syd. *Australian Journal of Agricultural Research*, **26**, 975–83.

Guillerm, J. L., Le Floc'h, E., Maillet, J. & Boulet, C. (1990). The invading weeds within the western Mediterranean Basin. In *Biological Invasions in Europe and the Mediterranean Basin*, ed. F. di Castri, A. J. Hansen & M. Debussche, pp. 61–84. Dordrecht: Kluwer.

Gullan, P. (1988). Weeds in Victoria: where are we? In *Weeds on Public Land – An Action Plan for Today*, ed. R. G. Richardson, pp. 7–14. Melbourne: Weed Science Society of Victoria & School of Environmental Science, Monash University.

Mirov, N. T. (1967). *The Genus* Pinus. New York: Ronald Press.

Rioux, J. & Quézel, P. (1950). La 'Flora Juvenalis' en 1950. *Le Monde des Plantes*, **272**, 73–4.

Specht, R. L. (1981). Major vegetation formations in Australia. In *Ecological Biogeography of Australia*, vol. 1, ed. A. Keast, pp. 165–297. The Hague: Junk.

Thellung, A. (1911–12). La flore adventice de Montpellier. *Bulletin de la Societé des Sciences Naturelles du Cherbourg*, **38**, 57–72.

Thirgood, J. V. (1981). *Man and the Mediterranean Forest: A History of Resource Depletion*. London: Academic Press.

31

The biogeography of mediterranean animal invasions

F. DI CASTRI

The preceding chapters of this book that deal with animals are restricted to mammals and birds, with only one chapter on beetles and brief mentions of the Argentine ant (*Iridomyrmex humilis*) and *Drosophila suboscura* in Chile. Despite the rich collection of facts and the interpretations given by the authors of the preceding chapters, any attempt at predictive generalisation is very difficult. The same conclusion holds after reviewing the many chapters on animal invasions that have been produced during the overall SCOPE project on biological invasions (see, for instance, Brooke *et al.*, 1986; Bruton & van As, 1986; Ehrlich, 1986; Macdonald *et al.*, 1986; Mooney *et al.*, 1986; Moran *et al.*, 1986; Moyle, 1986; Myers, 1986; Pimentel, 1986; Simberloff, 1986; Zimmermann *et al.*, 1986; Lawton & Brown, 1987; Ross & Tittensor, 1987; Brown, 1989; Usher, 1989). Almost exclusively, the invasion of oceanic islands seems to follow consistent patterns (see Cooper & Brooke, 1986; Moulton & Pimm, 1986; Coope, 1987; Holdgate, 1987; Loope & Mueller-Dombois, 1989), a result that could be predicted and in accordance with Williamson (1981). But for larger land masses, it is difficult to discern patterns as clearly as for islands. In his concluding remarks on vertebrate invasions, Ehrlich (1989) draws attention to our inability to make some kinds of predictions, but optimistically (and pragmatically) he contends that 'considering the enormous complexity of the problem, what can already be predicted is far from trivial' (p. 326).

Deacon (this volume) points out that invasion is a process, rather than an event (as it may appear at a first level of approximation); given this, the main challenge is to single out common mechanisms governing that process. In other words, we need to link the various events within a phenomenological framework; this should be done despite the large amount of environmental, demographic and historic stochasticity (di Castri, 1989). In this chapter, I shall firstly make some preliminary observations concerning animal invasions in general, and then I will refer to some case studies of mediterranean invasions, as outlined in this volume.

Plant and animal invasions

My first observation derives from a comparison of animal invasions with plant invasions. In the five mediterranean regions plant invasions seem to be more frequent, more often successful and to conform to more predictable patterns than do animal invasions. Two almost contradictory statements can be made. Firstly, the number of animal species is higher than that of plants by at least an order of magnitude or even greater (according to rough estimates of the total number of species on the planet that range from 4 to 40 millions). In animal life, there is greater taxonomic complexity and, accordingly, a much greater difference between the behaviour of a soil nematode and that of an eagle or an elephant than between, say, a moss and a giant tree. Such a difference is particularly true as regards patterns of reproduction. Secondly, if there are many more animal species, most of them are badly known or totally unknown and – as regards small invertebrates – records of invasions are conspicuously missing from most countries, with the exception of the most common plant pests. Furthermore, animals move actively and their current area of biogeographical distribution (and even more so, their geological area of distribution, when known) normally transcends the narrow boundaries of a mediterranean-climate region *sensu stricto*. Finally, mammals and birds, the two taxa treated in this volume, are endowed with various degrees of psychic and social attributes, and even of evolving psychic conditions; interactions between the behaviour of animals and humans may possibly give rise to a certain 'determinism' that is very difficult to comprehend at our present stage of knowledge of ethology and genetics.

Conversely, the history of introductions of given vertebrates, mostly of those introduced for hunting and fishing purposes or those released as pets, is in general better known than for most plants, so that the timing and rate of spread of some animals (e.g. rabbits in Australia and Chile) provide valuable information to better understand the invasion process.

The debatable rules of animal invasions

Among the hypotheses more often evoked to explain invasions, some are related to the animal itself, such as body size, lack of food specialisation and commensalism, whereas others deal with environmental patterns, such as homoclimes or disturbance.

The hypothesis on body size, that is the fact that large animals have a greater possibility of establishment than smaller ones, is based on the postulate of Leigh (1981) that the coefficient of variation of population size is a key factor in the persistence of a founding population. Small animals vary much more than large animals in population density through time (Hassell *et al.*, 1976). In fact, Lawton & Brown (1987) found a significant positive correlation between mean body size and the probability of establishment of taxa introduced to Britain. There are surprising and disturbing features in this correlation and Lawton & Brown presented it very cautiously. In support of the idea, it can be said that large organisms such as mammals and birds are warm-blooded and, in most cases, less susceptible to differences in climate than are invertebrates. They are also in general less specialised in food preferences than, say, phytophagous insects and they tend to have less predators than small invertebrates. On the other hand, the correlation of Lawton & Brown is totally inconsistent with the theory of *r* and *K* selection strategies and, above all, is based exclusively on observations of the most visible taxa. For instance, it is odd to see in their Figure 1, the same probability of establishment for insects as for amphibians and reptiles, the latter groups being among the least invasive of all animals, at least in mediterranean regions.

At this point, I would like to comment briefly on biocenoses that have practically been ignored during the SCOPE project on invasions, despite the fact that they constitute some of the most species-rich of communities – I refer to soil organisms. Admittedly, it is an almost intractable problem from the viewpoint of invasions, since most of the numerous soil taxa are very poorly known. Furthermore, it would be almost impossible to differentiate between 'old' invaders that have colonised soils through the continuous input of aerial plankton, as cysts, spores and small living animals, and those accidentally introduced by humans through the expansion of a European and a Mediterranean-like agriculture and forestry. Chilean soils, particularly the irrigated ones, after a few years of cultivation show a largely cosmopolitan fauna very similar to that of analogous fields in Europe (di Castri, 1973). di Castri & Vitali-di Castri (1981), in discussing the biogeographical roots of adaptation of soil animals, illustrated the many mechanisms of 'escape' (dormancy, diapause, anhydrobiosis, anabiosis, resistant eggs, cysts, spores) that allow soil organisms to survive long periods of transportation over several months before regaining activity. From the point of view of invasions such soil animals could be considered to behave more like plants than like large vertebrates.

The situation is far more complex as regards the microscopic Tardigrada, which are often considered to be the most ubiquitous and cosmopolitan of animal groups (Ehrlich, 1986) because of the ease with which they are transported passively. The terms 'cosmopolitan' and 'ubiquitous' are not synonymous. Cosmopolitan is a biogeographical term referring to species that exist in all parts of the world; its opposite is 'endemic'. Ubiquitous is a more ecological term and refers to a species that pervades all types of biotopes, even if its area of distribution is restricted; its opposite is 'stenotype'. Obviously, a cosmopolitan species can also be a ubiquitous one, and vice versa, but not necessarily so. Ramazzotti (1962*a*, 1962*b*) described hundreds of species of Tardigrades; all categories were present from cosmopolitan to endemic species, from ubiquitous species to those confined to only one habitat, including a number of possibly invasive species. This situation provides yet another warning against excessive generalisation and indiscriminate mixing of disparate groups of living organisms.

Commensalism with humans, particularly the evolution of commensalism as it relates to genetic and ethological factors, is one of the most challenging topics on which to build a theoretical background. With the anthropisation of the biosphere and a strong increase in the human population and of urbanised environments, commensalism is likely to become one of the main evolutionary trends in the future. For instance, a particularly important body of knowledge is being assembled on genetic and ethological patterns of wild and commensal populations of *Mus musculus* and of the closely related *M. spretus* and *M. spirilegus*, two strictly Mediterranean and feral species (see Britton & Thaler, 1978; Bonhomme *et al.*, 1983; Cassaing & Croset, 1985; Tchernov, 1985). The effects of social organization in determining genetic structure of populations have also been studied for the wild rabbit *Oryctolagus cuniculus* (Daly, 1981).

Among the environmental factors facilitating invasion, a common hypothesis is that homoclimatic conditions between the native territory and the region where a given species is introduced constitute a predictive tool by which to evaluate probability of naturalisation. This is strongly supported by some results for animal invasions (see Nix & Wapshere, 1986; Coope, 1987). The concept of homoclimes is widely taken into account in biological control programs and has also been considered for the deliberate introduction of domestic animals from advanced countries to many developing countries in an attempt to improve the quality of their animal husbandry (di Castri *et al.*, 1962; Nazar *et al.*, 1966). The homoclime approach has been rightly challenged (see Roy *et al.*, this volume). Too often the homoclime approach consists only of superficial comparisons of a few climatic diagrams or, on more substantive grounds, of arguing that a climatic analysis has low predictive power for evaluating success or failure of the introduction of a species. In the present volume we have compared the phenomenon of invasion in five 'homoclimatic' regions. Following Crawley (1987, Table 5), climate accounts for 44 per cent of the variation as a factor limiting (or facilitating) species introduction (of insects); it ranks first, followed by incompatibility (33 per cent), predators (22 per cent) and competition (12 per cent). One generalisation, in stating that climate has no satisfactory predictive power, could be that it nevertheless is the best of all factors considered.

Conversely, disturbance is almost unquestioned as being the main prerequisite for invasibility (see, for instance, the disturbance hypothesis of Fox & Fox, 1986). Here again I would like to introduce a word of caution. Disturbance depends on scales of time and space (Delcourt *et al.*, 1983; di Castri & Hadley, 1988); what is considered to be disturbance by one researcher (or for a given population of animals) may well not be applicable to another researcher or to another taxonomic group. A given disturbance that affects a few square centimetres of soil may have as much or more impact on some specific soil organisms as a killing frost affecting large areas. In other words, there is an important 'perceptional' component linked to the overall concept of disturbance. Orians (1986) vividly illustrated the strengths and limitations of this approach when he titled two subsections of his paper 'Ecosystems are in the mind of the beholder' and 'Disturbance is also in the mind of the beholder' (pp. 135 & 136).

In conclusion, several rules may predict or explain the process of invasion by animals. Nevertheless, no one of them is valid in isolation. Such rules should be treated as much as possible as 'group-specific' and 'site-specific'; in addition, they should be embedded in a background of historic conditions.

Mediterranean invasions by vertebrates

In planning this volume, it was the editors' intention to concentrate on vascular plants and on terrestrial vertebrates, the most 'visible' of organisms and the taxonomically better known groups. We decided intentionally not to include the complex and sometimes controversial aspects of biological control of pests and weeds because these aspects have received adequate coverage elsewhere (see, for instance, DeBach, 1964).

It is unfortunate for our purposes that invasions by amphibians and reptiles followed by successful naturalisations are so rare that they could not have justified an independent treatment. In the Mediterranean Basin a large amount of information exists on the latter aspect but it deals rather with migrations of 'old' invaders within the region, and the biogeography of small islands. In California, of 129 species of reptiles and amphibians, 124 are native (Jennings, 1983), and there have been 46 known but unsuccessful introductions. Results are very limited on invasions by these taxa in the other regions of mediterranean climate.

Before concentrating discussion on mammals and birds, attention should be drawn to biological invasions by inland fishes, most of which were introduced intentionally to improve sport and commercial fishing. There is an impressive record of successful invasions that have contributed to loss of native fish species. Mooney *et al.* (1986, Table 15.1) indicated that the percentage of successful invasions and naturalisations by fishes (19 per cent of the total species) is the highest compared to all other animal groups and vascular plants. di Castri (1989) also stressed that the most invasible ecosystems in Eurasia seem to be inland waters. In Chile, based on my own observations, I have hardly seen any native fish in rivers, ponds and lakes, where introduced species, such as Argentinian pejerrey (*Odonthestes bonariensis*), rainbow trout (*Salmo gairdneri*) and carp (*Cyprinus carpio*), predominate. In 1958 I even found in the northernmost Chilean river, Rio Loa, which runs through the Atacama Desert, very dense populations of rainbow trout, with some individuals only a few centimetres long. Fish invasion is by no means only a mediterranean phenomenon, however; depauperate native faunas with great biogeographical isolation and artificial reservoirs seem to be the most vulnerable environments.

My first observation in reviewing the taxa treated in this volume is that the absolute number of successful invasions by animals is far lower than for plants. This was to be expected since species of vascular plants outnumber by about an order of magnitude the number of species of mammals and birds. However, even the percentage of successful invasions is in general higher for plants, with the exception of fish. This conclusion does not imply that the damage caused by animal invasions has been lower; for instance, the presence of introduced commensal rodents and the expansion of rabbits in Australia and Chile are certainly of high biological and economic relevance.

My second general observation is that most authors evoke omnivory, commensalism, very high fecundity, climatic analogies and colonisation of suburban environments as the main reasons for a successful animal invasion. Body size does not seem to play a great role; in fact, there is often an inverse relation. Disturbance is not always a key factor. Chance, because of human historical and cultural factors, is a persistent theme.

In my comparative analysis, I shall consider firstly commensal rodents, then feral domestic animals, commensal birds and, finally, two species that are more closely 'mediterranean', viz. the European rabbit (*Oryctolagus cuniculus*) and the California quail (*Callipepla californica*).

The house mouse (*Mus musculus domesticus*), the brown rat (*Rattus rattus*) and the Norway rat (*R. norvegicus*) are largely implanted and occur generally in the five mediterranean-climate regions; they are not Mediterranean animals, however, considering both their origin (Pakistan to China) and their present distribution. Commensalism of *Mus musculus* with humans started some 15000 to 10000 years BP in western Asia and the eastern Mediterranean region. There are also feral populations of *Mus musculus* in all mediterranean-climate regions, with the apparent exception of South Africa. It is noteworthy that only in the cereal-growing regions of eastern and southern Australia are there reports of mouse plagues (in addition to China where the climate is not mediterranean). Feral populations of *Rattus rattus* exist in all five regions, but with different degrees of interaction with native rodents. There are only minor differences in the regional behaviour of these three species, the most recently invasive – *R. norvegicus* – being the most uniform.

Feral populations of domestic animals exist mostly in California and Australia, probably because these are both large regions with relatively low human population pressures. Wild pig (*Sus scrofa*), sometimes mixed with European wild boars, produces major negative effects in California and Australia. Human population pressure is also low in South Africa, but introduction of feral pigs and wild boars has been unsuccessful, probably because of inadequate food resources in the fynbos.

Commensal populations of birds refer mostly to the house sparrow (*Passer domesticus*), European starling (*Sturnus vulgaris*) and pigeon (*Columba livia*), the first two being the most relevant for their competitive interactions with native species. It is very interesting to note that *S. vulgaris*, so common and damaging in most regions, has never been reported for Chile.

The best known invasion by a truly Mediterranean mammal is that of the European rabbit *Oryctolagus cuniculus*, native to Spain. Invasion of Australia has been comprehensively described by Myers (1986) and by Redhead *et al.* (this volume). Invasion of Chile has been studied mainly by Jaksic & Fuentes (this volume); aspects of the population biology of rabbits (fecundity, sexual maturity, density, size, life expectancy) have changed very markedly in Chile compared with the native populations in Spain. Grazing behaviour and lower predatory pressure are among the hypotheses proposed by Jaksic & Fuentes to explain rabbit invasion in Chile. Herbivory by rabbits, together with livestock impact and invasion by Mediterranean grasses, also may have helped to change the overall landscape of the Chilean Intermediate Depression region to the north of Santiago from a woodland of *Prosopis chilensis* to a savanna of *Acacia caven* (Fuentes *et al.*, 1989).

I wish to stress that the very invasive European rabbit has never become naturalised in mainland California and South Africa, being restricted in both these regions to some offshore islands. In some countries, the rabbit invades regions of non-mediterranean climate. In Australia, areas with a non-mediterranean climate have been invaded; in Chile, the rabbit has crossed the Cordillera and invaded territories of Argentina with a non-mediterranean climate. The European rabbit has been introduced successfully from France to the steppes of Tierra del Fuego. The latter introduction was followed by the introduction of the fox to biologically control rabbits (Jaksic & Fuentes, this volume).

Another intriguing introduction is that of the California quail (*Callipepla californica*) to Chile (Vuilleumier, this volume). *Callipepla* is extremely successful in central Chile as an invasive species; it is also found in natural ecosystems where it coexists with the native *Nothoprocta perdicaria*. It should rightly now be considered as part of the Chilean bird fauna and valued for hunting. Nevertheless, attempts at introducing *Callipepla californica* to Mediterranean Europe and Australia have always failed.

The above are examples that illustrate the first part of this chapter, namely the debatable rules of invasion by animals. From a strictly scientific point of view, however, all these populations of invasive species provide an invaluable chance to study population dynamics and genetics because their area of origin, timing of introduction, chronology and rate of spread are well known and able to be compared intercontinentally between the five climatically similar regions. To a certain extent, the process of continuing evolution will be able to be followed and monitored.

In looking again at the five regions of mediterranean climate with the new insights coming from the study of vertebrate invasions, one is again impressed by the 'co-existence' of some predictable rules with apparently inexplicable events. Chance and human cultural differences are at least as important as definable biological patterns.

Phenomena in the Mediterranean Basin cannot be understood unless a long time scale is adopted, thereby allowing the overlap over time of successful invasions to be followed. Human presence must be considered. The eastern Mediterranean Basin is the key area in which to visualise patterns of invasion by mammals, birds and grasses. As a biogeographic crossroads, the Mediterranean has 'old' invasives present that range biogeographically from Euro-Siberian elements to Ethiopian elements.

A specific case is that of islands within the Mediterranean Sea which had an almost saturated fauna of invasive species by the time of the Romans. The 'Mediterranean Atlantic' or Fortunate Isles (Crosby, 1986), i.e. the Canary, Madeira and Azores islands (Macaronesia), all possess a mediterranean climate and have been settled by Europeans from the Mediterranean Basin. These islands show the usual suite of invasive animals and plants and particularly the impenetrable stands of *Rubus ulmifolius* on the Canary Islands as in Chile.

California and Chile have some common patterns and linkages, with Chile showing a lower percentage of introduced birds (3 per cent) than California (7 per cent). Nevertheless, the comparative analysis made by Vuilleumier (this volume) confirms that the Chilean avifauna has more 'insular' characteristics than the analogous Californian one. Major differences between the two regions are the spread of the European starling *Sturnus vulgaris* in California and, for mammals, the spread of the European rabbit (*Oryctolagus cuniculus*) in Chile.

Of particular importance in Chile is the massive invasion of one of the most outstanding biogeographical sites of the world, namely the Juan Fernandez (Robinson Crusoe) islands (Skottsberg, 1956). Feral goats and European rabbits, in particular, are practically destroying all natural ecosystems on these islands. The islands are characterised by extremely steep slopes; erosion is the most dramatic I have ever seen. *Eucalyptus globulus* and *Rubus ulmifolius* are changing the face of the landscape, particularly in the lower parts of these islands. Among other invasive species, the presence and abundance of *Callipepla californica* is noteworthy.

As regards South Africa, compared with its vulnerability to the introduction of plants, it seems surprisingly to have the most robust ecosystems able to resist the introduction of vertebrates. There are no invasive bird species in natural fynbos vegetation, the house mouse is not feral, the European rabbit is not on the mainland and wild pigs and fallow deer (*Dama dama*) are not successful, probably because in the case of the latter of the poor quality forage in fynbos (Bigalke & Pepler, this volume). Only the Himalayan tahr (*Hemitragus jemlahicus*) is vigorously invasive, although in restricted areas. An interesting case of obligatory interaction between two invasive species is that of the grey squirrel (*Sciurus carolinensis*) from North America; this squirrel is successful in South Africa as long as there are plantations of introduced *Pinus pinaster*, *P. pinea* and *Quercus robur* as well as orchards and vineyards; it does not survive in exotic plantations of *Eucalyptus* from Australia (the introduction of squirrels to Australia has failed!). It seems that appropriate niches for squirrels are missing from fynbos (and, by analogy, from comparable Australian sclerophyll communities). In conclusion, for South Africa there have been relatively few introductions of vertebrates, and most of them have been unsuccessful. It can be postulated that the low level of nutrients in fynbos, and the greater pressure of predators and pathogens in a mostly tropical country, are the main causes leading to the resistance of South African ecosystems to invasion by vertebrates.

Conversely, the activities of the Acclimatization Societies in the colonial Australian states accounts in part for the unusually high numbers of attempts at introductions of mammals and birds. According to Redhead *et al.* (this volume), 60 species of mammal were introduced between 1860 and 1880. As regards birds, Long & Mawson (this volume) claim that from 70 to 100 species were introduced, 22 of which are now naturalised and at least half of which originate from the Mediterranean Basin. These figures imply that Australia has been the most invaded of the mediterranean-climate regions and that the percentage of successes is greater there than in any of the other regions. After the introduction of the dingo (*Canis familiaris dingo*) some 3500 years ago, the most conspicuous introductions have been those of the commensal rodents and of the European rabbit, followed by the predatory species to control the latter (viz. the red fox,

Vulpes vulpes, and the feral cat, *Felis catus*). The red fox is threatening native species and, in particular, the rock wallabies. The several introduced bird species, including among them the European tree sparrow (*Passer montanus*), are rarely found in areas with natural vegetation; they tend to use European-type plants as food and habitat. The higher rate of successful introductions is attributable to the evolutionary isolation of Australia, although it is by no means comparable to the massive displacement of the native fauna by invasive species that happened in New Zealand (Wodzicki, 1950). Australia is an isolated continent but it can hardly be equated with an island.

I would like to end with two more general observations. First of all, many introductions should be considered now as an irreversible and harmonious part of local ecosystems and landscapes. I take as my example the most common and resilient ecosystem of central Chile – the *Acacia caven* savanna. The overstorey of *A. caven* is itself invasive, having originated in the neotropical region where its centre of speciation is considered by J. Aronson (personal communication) to be the Chaco region (at present, southern Paraguay and northern Argentina). Aronson also postulates that its expansion is relatively recent and that domestic ungulates have accelerated this process. The understorey is a continuous layer of Mediterranean grasses, introduced by the Spanish, and having evolved mostly in the eastern Mediterranean in the Pliocene–Pleistocene. These grasses are primarily grazed by domestic European cattle, sheep and goats, as well as by the European rabbit and by phytophagous insects of neotropical origin. The soil organisms have largely a very old Gondwanan and Palaeantarctic origin. This 'superpositioning' of biogeographic realms in one ecosystem has nevertheless produced a system of high productivity and stability.

My second and final observation is that the exceedingly high human population of the near future and the very rapid rate of urbanisation will result in most of the food webs originating from urban and suburban environments. Scavengers may often be the initial elements in those webs. Such conditions are ideal for commensal and invasive species that are likely to rise progressively to the most widespread and dominant proportion of terrestrial biota.

References

Bonhomme, F., Catalan, J., Gerasimov, S., Orsini, P. & Thaler, L. (1983). Le complexe d'espèces du genre *Mus* en Europe centrale et orientale. I. Génétique. *Zeitschrift für Saügetierkunde*, **48**, 78–85.

Britton, J. & Thaler, L. (1978). Evidence for the presence of two sympatric species of mice (genus *Mus* L.) in southern France based on the biochemical genetics. *Biochemical Genetics*, **16**, 213.

Brooke, R. K., Lloyd, P. H. & de Villiers, A. L. (1986). Alien and translocated terrestrial vertebrates in South Africa. In *The Ecology and Management of Biological Invasions in Southern Africa*, ed. I. A. W. Macdonald, F. J. Kruger & A. A. Ferrar, pp. 63–74. Cape Town: Oxford University Press.

Brown, J. H. (1989). Patterns, modes and extents of invasions by vertebrates. In *Biological Invasions. A Global Perspective*, ed. J. A. Drake, H. A. Mooney, F. di Castri, R. H. Groves, F. J. Kruger, M. Rejmanek & M. Williamson, pp. 85–109. Chichester: Wiley.

Bruton, M. N. & van As, J. G. (1986). Faunal invasions of aquatic ecosystems in southern Africa, with suggestions for their management. In *The Ecology and Management of Biological Invasions in Southern Africa*, ed. I. A. W. Macdonald, F. J. Kruger & A. A. Ferrar, pp. 47–61. Cape Town: Oxford University Press.

Cassaing, J. & Croset, H. (1985). Organisation spatiale, compétition et dynamique des populations sauvages de souris (*Mus spretus* Lataste et *Mus musculus domesticus* Rutty) du Midi de la France. *Zeitschrift für Saügetierkunde*, **50**, 271–84.

Coope, G. R. (1987). The invasion and colonization of the North Atlantic islands: a palaeoecological solution to a biogeographic problem. In *Quantitative Aspects of the Ecology of Biological Invasions*, ed. H. Kornberg & M. H. Williamson, pp. 117–33. London: The Royal Society.

Cooper, J. & Brooke, R. K. (1986). Alien plants and animals on South African continental and oceanic islands: species richness, ecological impacts and management. In *The Ecology and Management of Biological Invasions in Southern Africa*, ed. I. A. W. Macdonald, F. J. Kruger & A. A. Ferrar, pp. 133–42. Cape Town: Oxford University Press.

Crawley, M. J. (1987). The population biology of invaders. In *Quantitative Aspects of the Ecology of Biological Invasions*, ed. H. Kornberg & M. H. Williamson, pp. 209–29. London: The Royal Society.

Crosby, A. W. (1986). *Ecological Imperialism. The Biological Expansion of Europe, 900–1900*. Cambridge: Cambridge University Press.

Daly, J. C. (1981). Effects of social organization and environmental diversity on determining the genetic structure of a population of wild rabbit, *Oryctolagus cuniculus*. *Evolution*, **35**, 689–706.

DeBach, P. H. (ed.) (1964). *Biological Control of Insect Pests and Weeds*. New York: Reinhold.

Delcourt, H. R., Delcourt, P. A. & Webb, T. (1983). Dynamic plant ecology: the spectrum of vegetational change in space and time. *Quaternary Science Reviews*, **1**, 153–75.

di Castri, F. (1973). Soil animals in latitudinal and topographical gradients of mediterranean ecosystems. In *Mediterranean-type Ecosystems: Origin and Structure*, ed. F. di Castri & H. A. Mooney, pp. 171–90. Berlin: Springer-Verlag.

di Castri, F. (1989). History of biological invasions with special emphasis on the Old World. In *Biological Invasions. A Global Perspective*, ed. J. A. Drake, H. A. Mooney, F. di Castri, R. H. Groves, F. J. Kruger, M. Rejmanek & M. Williamson, pp. 1–30. Chichester: Wiley.

di Castri, F. & Hadley, M. (1988). Enhancing the credibility of ecology: interacting along and across hierarchical scales. *Geojournal*, **17**, 5–35.

di Castri, F. & Vitali-di Castri, V. (1981). Soil fauna of mediterranean-climate regions. In

Mediterranean-type Shrublands, ed. F. di Castri, D. W. Goodall & R. L. Specht, pp. 445–78. Amsterdam: Elsevier.

di Castri, F., Hajek, E. R. & Astudillo, V. (1962). Importancia pecuaria de los ambientes desfavorables chilenos. *Boletín de Producción Animal (Chile)*, **1**, 7–20.

Ehrlich, P. R. (1986). Which animal will invade? In *Ecology of Biological Invasions of North America and Hawaii*, ed. H. A. Mooney & J. A. Drake, pp. 79–95. New York: Springer-Verlag.

Ehrlich, P. R. (1989). Attributes of invaders and the invading processes: vertebrates. In *Biological Invasions. A Global Perspective*, ed. J. A. Drake, H. A. Mooney, F. di Castri, R. H. Groves, F. J. Kruger, M. Rejmanek & M. Williamson, pp. 315–28. Chichester: Wiley.

Fox, M. D. & Fox, B. J. (1986). The susceptibility of natural communities to invasion. In *Ecology of Biological Invasions: An Australian Perspective*, ed. R. H. Groves & J. J. Burdon, pp. 57–66. Canberra: Australian Academy of Science.

Fuentes, E. R., Avilés, R. & Segura, A. (1989). Landscape change under indirect effects of human use: the savanna of central Chile. *Landscape Ecology*, **2**, 73–80.

Hassell, M. P., Lawton, J. H. & May, R. M. (1976). Patterns of dynamical behaviour in single-species populations. *Journal of Animal Ecology*, **45**, 471–86.

Holdgate, M. W. (1987). Summary and conclusions: characteristics and consequences of biological invasions. In *Quantitative Aspects of the Ecology of Biological Invasions*, ed. H. Kornberg & M. H. Williamson, pp. 231–40. London: The Royal Society.

Jennings, M. R. (1983). An annotated check list of the amphibians and reptiles of California. *California Fish & Game*, **69**, 151–71.

Lawton, J. H. & Brown, K. C. (1987). The population and community ecology of invading insects. In *Quantitative Aspects of the Ecology of Biological Invasions*, ed. H. Kornberg & M. H. Williamson, pp. 105–15. London: The Royal Society.

Leigh, E. G. Jr (1981). The average lifetime of a population in a varying environment. *Journal of Theoretical Biology*, **90**, 213–39.

Loope, L. L. & Mueller-Dombois, D. (1989). Characteristics of invaded islands, with special reference to Hawaii. In *Biological Invasions. A Global Perspective*, ed. J. A. Drake, H. A. Mooney, F. di Castri, R. H. Groves, F. J. Kruger, M. Rejmanek & M. Williamson, pp. 257–80. Chichester: Wiley.

Macdonald, I. A. W., Powrie, F. J. & Siegfried, W. R. (1986). The differential invasion of southern Africa's biomes and ecosystems by alien plants and animals. In *The Ecology and Management of Biological Invasions in Southern Africa*, ed. I. A. W. Macdonald, F. J. Kruger & A. A. Ferrar, pp. 209–25. Cape Town: Oxford University Press.

Mooney, H. A., Hamburg, S. P. & Drake, J. A. (1986). The invasions of plants and animals into California. In *Ecology of Biological Invasions of North America and Hawaii*, ed. H. A. Mooney & J. A. Drake, pp. 250–72. New York: Springer-Verlag.

Moran, V. C., Neser, S. & Hoffmann, J. H. (1986). The potential of insect herbivores for the biological control of invasive plants in South Africa. In *The Ecology and Management of Biological Invasions in Southern Africa*, ed. I. A. W. Macdonald, F. J. Kruger & A. A. Ferrar, pp. 261–8. Cape Town: Oxford University Press.

Moulton, M. P. & Pimm, S. L. (1986). Species introductions to Hawaii. In *Ecology of Biological Invasions of North America and Hawaii*, ed. H. A. Mooney & J. A. Drake, pp. 231–49. New York: Springer-Verlag.

Moyle, P. B. (1986). Fish introductions into North America: patterns and ecological impact. In *Ecology of Biological Invasions of North America and Hawaii*, ed. H. A. Mooney & J. A. Drake, pp. 27–43. New York: Springer-Verlag.

Myers, K. (1986). Introduced vertebrates in Australia, with emphasis on the mammals. In *Ecology of Biological Invasions: An Australian Perspective*, ed. R. H. Groves & J. J. Burdon, pp. 120–36. Canberra: Australian Academy of Science.

Nazar, J., Hajek, E. R. & di Castri, F. (1966). Determinación para Chile de algunas analogías bioclimáticas mundiales. *Boletín de Producción Animal (Chile)*, **4**, 103–73.

Nix, H. A. & Wapshere, A. J. (1986). Biogeographic origins of invading species. In *Ecology of Biological Invasions: An Australian Perspective*, ed. R. H. Groves & J. J. Burdon, p. 155. Canberra: Australian Academy of Science.

Orians, G. H. (1986). Site characteristics favoring invasions. In *Ecology of Biological Invasions of North America and Hawaii*, ed. H. A. Mooney & J. A. Drake, pp. 133–48. New York: Springer-Verlag.

Pimentel, D. (1986). Biological invasions of plants and animals in agriculture and forestry. In *Ecology of Biological Invasions of North America and Hawaii*, ed. H. A. Mooney & J. A. Drake, pp. 149–62. New York: Springer-Verlag.

Ramazzotti, G. (1962*a*). Il phylum Tardigrada. *Memorie dell'Istituto Italiano di Idrobiologia*, **14**, 11–59.

Ramazzotti, G. (1962*b*). Tardigradi del Cile, con descrizione di quattro nuove specie e di una nuova varietà. *Atti Società Italiana Scienze Naturali e Museo Storia Naturale Milano*, **101**, 275–87.

Ross, J. & Tittensor, A. M. (1987). The establishment and spread of myxomatosis and its effect on rabbit populations. In *Quantitative Aspects of the Ecology of Biological Invasions*, ed. H. Kornberg & M. H. Williamson, pp. 97–104. London: The Royal Society.

Skottsberg, C. (1956). Derivation of the flora and fauna of Juan Fernandez and Easter Island. In *The Natural History of Juan Fernandez and Easter Island*, ed. C. Skottsberg, pp. 193–438. Uppsala: Almquist & Wiksell.

Simberloff, D. (1986). Introduced insects: a biogeographic and systematic perspective. In *Ecology of Biological Invasions of North America and Hawaii*, ed. H. A. Mooney & J. A. Drake, pp. 3–26. New York: Springer-Verlag.

Tchernov, E. (1985). Rodent faunas and environmental changes in the Pleistocene of Israel. In *Rodents in Desert Environments*, ed. I. Prakash & P. K. Ghosh, pp. 331–62. The Hague: Junk.

Usher, M. B. (1989). Ecological effects of controlling invasive terrestrial vertebrates. In *Biological Invasions. A Global Perspective*, ed. J. A. Drake, H. A. Mooney, F. di Castri, R. H. Groves, F. J. Kruger, M. Rejmanek & M. Williamson, pp. 463–89. Chichester: Wiley.

Williamson, M. (1981). *Island Populations*. Oxford: Oxford University Press.

Wodzicki, K. A. (1950). Introduced mammals of New Zealand: an ecological and economic survey. *Bulletin of the Department of Scientific & Industrial Research, New Zealand*, **98**, 1–255.

Zimmermann, H. G., Moran, V. C. & Hoffmann, J. H. (1986). Insect herbivores as determinants of the present distribution and abundance of invasive cacti in South Africa. In *The Ecology and Management of Biological Invasions in Southern Africa*, ed. I. A. W. Macdonald, F. J. Kruger & A. A. Ferrar, pp. 269–74. Cape Town: Oxford University Press.

Index of scientific names

Subject index